NUCLEASES

NUCLEASES

Edited by

Stuart M. Linn
University of California, Berkeley

Richard J. Roberts
Cold Spring Harbor Laboratory

Cold Spring Harbor Laboratory
1982

COLD SPRING HARBOR MONOGRAPH SERIES

The Lactose Operon
The Bacteriophage Lambda
The Molecular Biology of Tumour Viruses
Ribosomes
RNA Phages
RNA Polymerase
The Operon
The Single-Stranded DNA Phages
Transfer RNA:
  Structure, Properties, and Recognition
  Biological Aspects
Molecular Biology of Tumor Viruses, Second Edition:
  DNA Tumor Viruses
  RNA Tumor Viruses
The Molecular Biology of the Yeast Saccharomyces:
  Life Cycle and Inheritance
  Metabolism and Gene Expression
Mitochondrial Genes
Nucleases

NUCLEASES

Printed in the United States of America
Book design by Emily Harste

Library of Congress Cataloging in Publication Data

Main entry under title:

Nucleases.

(Cold Spring Harbor monograph series ; 14)
Includes bibliographical references and index.
1. Nucleases—Congresses. I. Linn, Stuart M.
II. Roberts, Richard J. III. Cold Spring Harbor
Laboratory. IV. Series. [DNLM: 1. Deoxyribonuclease—Congresses.
2. Ribonuclease—Congresses.
3. Phosphatases—Congresses. W1 CO133C v.14 / QU 136
N964 1981]
QP609.N78N8 1982   591.19′25   82-71651
ISBN 0-87969-155-7

All Cold Spring Harbor Laboratory publications are available through booksellers or may be ordered directly from Cold Spring Harbor Laboratory, Box 100, Cold Spring Harbor, New York 11724.

SAN 203-6185

Contents

Appendixes

Preface

In August 1981, at the suggestion of Jim Watson, a meeting was held at Cold Spring Harbor on the topic of nucleases. The idea was to bring together people who study these enzymes, but who do not usually meet as a group because their individual interests in nucleases are primarily in the context of some other field. Historically, nucleases have not usually been considered a field in their own right, rather they have been studied because of their role in some particular biological phenomena such as the four R's — Replication, Recombination, Repair, and Restriction.

Within the last few years, considerable progress has been made in this area. Several nucleases, whose existence was not even suspected a few years ago, such as the resolvases involved in transposition and the splicing enzymes needed to process RNA in eukaryotes, are now the subject of intense study. Some nucleases show exquisite specificity, and nowhere has this been more apparent than in the elegant writhings of DNA induced by the topoisomerases. Since meddling with DNA in vitro is a popular pursuit, some nucleases such as the type-II restriction enzymes have been widely used as reagents. The demand for more such reagents is unlikely to abate.

The idea for this book was also Jim Watson's and the meeting provided a timely mechanism to bring the authors together and intimidate them into writing chapters. The objective of the book is to provide comprehensive coverage of the many facets of the nucleases that would prove both interesting and useful to those already immersed in this field. We hope it will also stimulate those who view these enzymes as "stoichiometric reagents living in their refrigerators" to explore new technical avenues and so to exploit their catalytic potential fully.

We thank the National Institutes of Health and the National Science Foundation, as well as Bethesda Research Laboratories, New England Biolabs, PL-

Biochemicals, Enzo, and New England Nuclear for generously providing financial support for the meeting. We also thank the authors for their contributions and for their forbearance in accepting our editorial idiosyncrasies. Finally, we wish to acknowledge the help of Nancy Ford, Director of Publications at Cold Spring Harbor Laboratory, for providing valuable guidance in the production of this book; Judy Cuddihy, for her conscientious editing; and Nancy D'Anna, for assistance in preparing the manuscripts for the printer.

Michael Laskowski Sr., one of the great pioneers in this field, had been scheduled to deliver the keynote address at the meeting. His untimely death prevented this. Fortunately, Dr. Laskowski belonged to the old school, and unlike most of us, had prepared his manuscript for the book in good time. It is presented as Chapter 1. We are pleased to dedicate this book to his memory.

Stuart M. Linn
Richard J. Roberts

M. Laskowski, Sr.

1905–1981

NUCLEASES

Nucleases: Historical Perspectives

M. Laskowski, Sr.
Laboratory of Enzymology
Roswell Park Memorial Institute
Buffalo,New York 14263

PERSPECTIVES AND THE "PERIOD OF DISCOVERY"

There are fairly rigid boundaries on which I should not trespass, and they are given by the program of this meeting. Repair, recombination, replication, and particularly restriction are the essence of this meeting, and so to avoid repetitions they are off limits for my introduction. Only passing references will be made to these topics to show their roots. The exception to this rule is single-strand-specific nucleases. In the past 10 years, they have been our major occupation and cannot be totally eliminated from the Introduction. I apologize to Dr. Ando for encroachment upon his territory.

To avoid sounding like a telephone book, a decision was made arbitrarily to abandon accurate chronology and, instead, to discuss periods of development through which the reviewer has lived. These are:

1. period of discovery,
2. period of crystal chasing that evolved into a search for an ever-increasing purity (no end),
3. period of classification (fades away),
4. period of structural studies (primary and three dimensional) aimed at understanding the mechanism of enzyme action,
5. period of specificity (restriction),
6. period of fitting a nuclease into a general physiological function.

The last two periods are now in full bloom and constitute the essence of the present meeting.

Another arbitrary decision was made to refer to the reviews, whenever possible, starting with a charming little booklet by the late Prof. J.N. Davidson (Davidson 1950) that had a chapter on nucleases. This was followed by several reviews (Laskowski 1951, 1971; Schmidt 1955; Lehman 1963, 1971) and one full-size book devoted only to nucleases (Privat de Garilhe 1964).

For the sake of humility, it may just be the right place to remark that nucleases did not significantly contribute either to the annihilation of the "tetranucleotide structure" (Chargaff 1955) or to the establishment of the 3′-5′ linkage in DNA by Lord Todd and co-workers (Brown and Todd 1955). Both of these important issues of DNA structure were solved using chemical methods. Similarly, the Watson and Crick (1953) model was conceived with-

The editors are indebted to Dr. D. Kowalski for his help in preparing this manuscript.

out a significant contribution of nucleases. By contrast, probably the greatest (and certainly the least rewarded) discovery by Avery and co-workers (1944) of the transforming substance relied on crude pancreatic DNase to establish the chemical nature of this substance.

With respect to old-fashioned nucleases, the period of discovery is now coming to a close. As everywhere, exceptions exist. On occasion a new nuclease is still being described, and presumably new nucleases or even new classes of nucleases will be found as we dare to investigate increasingly more complex biological substrates like chromatin and chromosomes.

over. New restriction nucleases are being discovered with increasing rapidity.

PERIOD OF "CRYSTAL CHASING"

During this period, under the impact of the discovery by the late Prof. J.B. Sumner that an enzyme (urease) is a protein (Sumner 1926), it was generally accepted that the crystallization of an enzyme is also evidence of its purity.

And now, the first digression with a few words about Prof. Sumner. He offered me hospitality in his laboratory although he knew I was a poor risk. After 2 years of service with the Polish Army, I was out of touch with the laboratory and the library. Contrary to the general opinion that he was cool and impersonal to his students, I found him friendly and helpful. In general, he was a personality full of contrasts: a scion of a Mayflower family, a millionaire from birth, an amputee from early adolescence (lost an arm at age 15), a most ardent polemicist starting with his first paper, and a Nobel prize winner. His lectures were rather dull and overloaded with detail. His mind kept an enormous amount of information dealing with laboratory facts from many branches of sciences related to biochemistry and resembled a biochemical encyclopedia rather than a textbook. This quality is probably responsible for the fact that he left a lasting imprint not only in enzymology (urease) but also in immunology (concanavalin). He had a set of rules, some of which I still remember.

1. Never purify an enzyme until you have located a source that contains at least 1%.
2. In the process of purification, never allow the concentration of protein to fall below 1%.
3. Do not use adsorbants.
4. Use as many denaturing solvents as your enzyme tolerates—the more drastic the solvent the better.

The rules were short-lived, but they produced numerous successes in his and his students' hands.

Let us now return to the initial attempts at crystallizing nucleases. In 1940, Kunitz crystallized pancreatic RNase (Kunitz 1940). The enzyme withstood several criteria of purity available at that time and became the first enzymatic tool to distinguish between RNA and DNA. The discovery by Schmidt and co-workers (1951) that RNase cleaves after pyrimidine residues extended the use of RNase to more detailed studies of the structure of RNA.

Shortly after the crystallization of RNase, the hunt for crystalline pancreatic DNase started. Two methods for following DNase activity were available: viscosimetric (Laskowski and Seidel 1945) and the liberation of acid-soluble phosphorus (Laskowski 1946), which was determined after ashing (Martland and Robison 1926). A third method, based on hyperchromicity, was published 5 years later (Kunitz 1950). By following the first two methods, a crystalline protein was obtained (Laskowski 1946a), as shown in Figure 1. It had high specific activity, but there were three things wrong with it: mother liquor and crystals had similar specific activities, crystals were losing activity on recrystallization, and a small proteolytic activity was detectable. A few weeks after this paper appeared, I received a long, friendly letter from Dr. Kunitz stating that he too had obtained similar crystals (using a surprisingly similar procedure) and had decided that they were a pancreatic globulin that strongly adsorbed DNase. Kunitz described in detail his experiments that removed DNase from the crystals. By the time the letter arrived, we already knew that recrystallization at a little lower pH leads to the accumulation of DNase in the mother liquor (Laskowski and Kazenko 1947).

Since I was privileged to know Kunitz, I will make a second digression. He had an idiosyncrasy for tobacco. As a result, he never went to a meeting, refused all invitations for seminars, and limited his students to one girl (Margaret McDonald). From the gossip that reached me, I found out that Leon Heppel (a nonsmoker) holds the record for time spent by a male in Kunitz's laboratory; he was invited for a whole week. I would not be surprised to find that I hold second place—I was invited for a whole afternoon.

Back to crystals. It took me almost 2 years to establish that the predominant component of my crystalline material was chymotrypsinogen B, a novel proteolytic zymogen (Keith et al. 1947; Brown et al. 1948). Kunitz (1950)

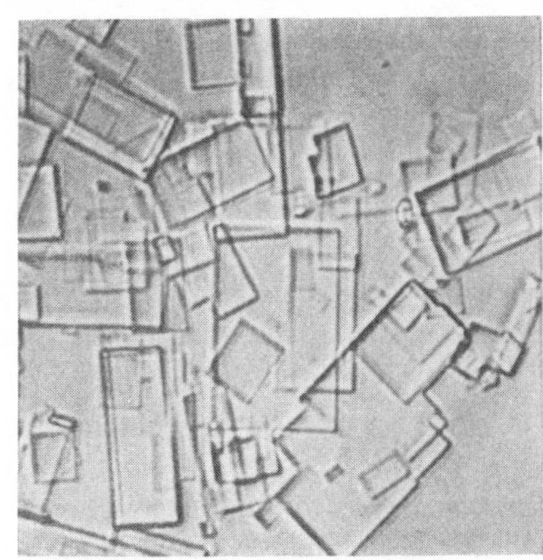

Figure 1 Freshly prepared crystals from extracts of beef pancreas, Crystals contain up to 20% DNase (by activity), 2% chymotrypsin B, and at least 78% chymotrypsinogen B. Magnified 600×. (Reprinted. with permission, from Laskowski 1936a.)

then published his method of crystallization of DNase. His crystals had more than twice the specific activity that mine had. I felt lower than a worm. That Kunitz's crystals were also impure was shown many years later in the laboratory of Stanford Moore (Price et al. 1969).

At this point, the bankruptcy of "crystal chasers" became obvious. The limited number of survivors was forced to use more selective methods of purification. The "purifiers" had to use more sophisticated methods to detect and remove impurities. A new branch of "crystallographers" *sensu strictu* developed. Starting with a highly purified enzyme, they made good-quality crystals useful for the study of the three-dimensional structure. Both branches are doing well, and there is no end in sight for either of them.

The purifiers deserve credit for introducing nucleases in a form that may be safely used as a specific reagent. I am quoting a few examples with which I deal daily. Thus, the DNase-free RNase was prepared rather a long time ago (Kleczkowski 1948), whereas the RNase-free DNase is a relatively late newcomer (Wang and Moore 1978; see also an excellent review on DNase by Moore [1981]). Micrococcal nuclease of the highest purity and free from monophosphatase has been known for several years (Mikulski et al. 1969). Preparations of mung bean nuclease and venom phosphodiesterase that are homogeneous by several criteria and free of interfering activities are available (Kowalski et al. 1976; Pritchard et al. 1977; Laskowski 1980a,b). The availability of pure restriction enzymes produced a revolution and created a major reason for organizing this meeting.

CLASSIFICATION

Most sciences go through a period of classification. An example of successful and lasting classification is Mendeleev's Table. The division of nucleases according to Kunitz into DNases and RNases (the nomenclature survived 40 years, and is likely to survive another 40) fulfilled the requirements, so long as the specificity toward sugar was the only important property of a nuclease; however, this stage was short-lived.

Still in the 1940s a second DNase (II) was discovered by Maver and Greco (1949a,b) in thymus and in spleen. It differed from DNase I in optimal pH and did not require addition of Mg^{2+} for optimal activity. At about the same time, an inhibitor of pancreatic DNase I was discovered in the hypertrophic pigeon crop gland, which was the fastest growing tissue known to us; it doubles its weight in 24 hours (Dabrowska et al. 1949). The inhibitor was also found in other fast-growing tissues, including tumors (Cooper et al. 1950), but the degree of malignancy and the content of inhibitor did not correlate. The analogy between intracellular and digestive pancreatic nucleases and proteinases led Allfrey and Mirsky (1952) to postulate that DNase-I-type enzymes perform only a digestive function, whereas all other tissues contain only DNase II. A simple experiment (Cunningham and Laskowski 1953) shown in

Figure 2 revealed that this was not the case. The homogenate of veal kidney showed both activities of DNase I (curve *C*) and II (curve *B*). The activity of DNase I increased significantly after the homogenate was exposed for 24 hours to pH 3 (curve *A*), conditions known to destroy the inhibitor (Dabrowska et al. 1949). The claim of Mazia (1949) that intracellular distribution of an enzyme and its substrate should follow the same pattern was not confirmed (Brown et al. 1952). A systematic study of the distribution of intracellular hydrolases in the cell organelles by de Duve and coworkers (Wattiaux and de Duve 1955) ruled out this hypothesis, and earned a Nobel prize.

As a result of the DNase-II discovery, the work on purification was stimulated (Shimomura and Laskowski 1957). In addition, even with a partly purified preparation it was obvious that products formed by DNase I and DNase II differ. DNase I produced predominantly dinucleotides and only a few mononucleotides. We originally thought that two chromatographic spots (on paper at that time), which had a clear spectrum of cytidylic acid, might be dpC and either dpCpC or dCp (Potter et al. 1952). When our manuscript was in press, Sinsheimer and Koerner (1952), using the Dowex column (Cohn 1950), showed that dpCpC was the second product. Thus, the alternative that the same enzyme can form 3′- and 5′-phosphomononucleotides was ruled out.

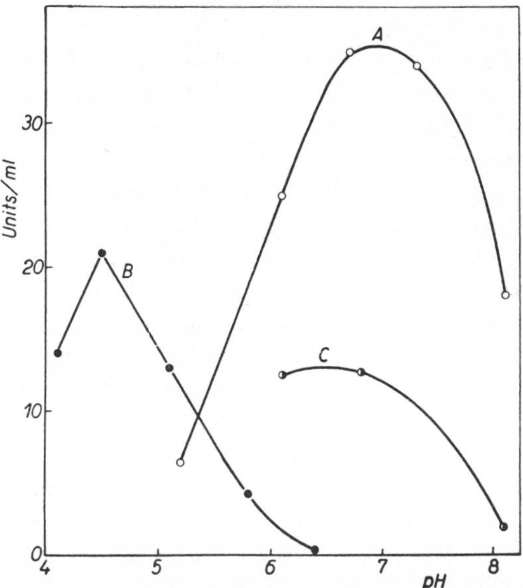

Figure 2 Relationship between pH and DNase activity of kidney homogenates expressed in viscosimetric units. Curve *A*, homogenate exposed to pH 3 for 24 hours, analyzed in the presence of 20 mM $MgSO_4$ (DNase I); curve *B*, homogenate exposed to pH 3 for 24 hours, analyzed in the presence of 10 mM citrate (DNase II); curve *C*, homogenate not exposed to pH 3, analyzed in the presence of 20 mM $MgSO_4$ (DNase I + inhibitor). (Reprinted, with permission, from Cunningham and Laskowski 1953.)

To extend this observation to other products, a number of mono-, di-, and trinucleotides were analyzed in Sinsheimer's laboratory (Koerner and Sinsheimer 1951, 1957; Sinsheimer 1954, 1955) and in our laboratory (Privat de Garilhe and Laskowski 1954, 1955; Laurila and Laskowski 1957; Privat de Garilhe et al. 1957; Potter et al. 1958; Vanecko and Laskowski, 1961a,b, 1962). The answer was always the same: the position of the phosphoryl group was decided by the nature of the nuclease. The results from these two laboratories were published almost simultaneously. They are best summarized by stating that dinucleotides are the predominant products of the DNase I action and that all of the 16 possible combinations were found, but the combinations pY-pR predominated. DNase II produced a larger proportion of mononucleotides of the type of dNp, and among dinucleotides the combination Rp-Yp predominated.

In our laboratory the turning point occurred around 1955. Cunningham discovered micrococcal nuclease and proved that it forms 3′-phosphonucleotides (Cunningham et al. 1956). The discovery of this enzyme came a bit too late for the establishment of the internucleotide 3′-5′ linkage by Lord Todd (Brown and Todd 1955), but the enzyme was used by him as supporting evidence of the already established linkage. The enzyme was not too late for Kornberg and co-workers to devise the nearest-neighbor method (Josse et al. 1961). The enzyme was also used by Holley and co-workers (1965) to sequence tRNA and by Van Holde (Rill and Van Holde 1973), Noll (1974), Felsenfeld (1978), and many others to prepare mononucleosomes. We are presently inclined to think (T.K. Dziegielewski and M. Laskowski, Sr., unpubl.) that the reason for micrococcal nuclease being a rather specific reagent for preparing mononucleosomes from chromatin is its inability to hydrolyze the 5′P-5′P linkage present in poly(ADP-Rib). I will return to this observation in the discussion of specificity. Returning to 1955, we had been looking for 3′-monoester-forming nucleases since 1951, always in the wrong places. After Cunningham's discovery, it was a routine check of DNase II that confirmed the existence of 3′-phosphate-forming nucleases and extended it to longer fragments (Laurila and Laskowski 1957; Privat de Garilhe et al. 1957). It came simultaneously with the conclusion arrived at in Sinsheimer's laboratory (Koerner and Sinsheimer 1951, 1957). However, we overlooked the second very important characteristic of micrococcal nuclease—that it was capable of hydrolyzing both DNA and RNA. This observation was made by J.D. Smith of Cambridge and was communicated to Cunningham just before the symposium organized by the New York Academy. It did not come as a surprise; the fact that venom phosphodiesterase hydrolyzes ribo- and deoxyribooligonucleotides had been known for many years (Schmidt 1955).

A second event in our laboratory that occurred in 1955 was the observation of Privat de Garilhe (Privat de Garilhe and Laskowski 1956) that venom phosphodiesterase acts as an exonuclease by consecutively liberating 5′-mononucleotides from the same terminus. In this case, we made an error of assigning the direction 5′-3′ for its action. The error was promptly corrected by Singer et al. (1958) and by Razzell and Khorana (1959). The reason for our error was explained, and optimal conditions for hydrolyzing 3′-phosphate-terminated chains were described several years later (Richards and Laskowski 1969).

In the mid-1950s, it became obvious that Kunitz's division of nucleases into DNases and RNases was an oversimplification and that additional criteria were required to characterize a nuclease. During the symposium organized by the New York Academy, I proposed four criteria to characterize a nuclease (Laskowski 1959). Of the four criteria that are listed in Table 1, only one (the third) survived the proof of time. No enzyme is yet known that can form both 3′- and 5′-phosphate-terminated products.

For historical reasons, criterion 1 involving sugar specificity is still used. Names like RNase and DNase became everyday terms and will not be changed. However, it is now quite clear that many enzymes exist that attack both substrates. Furthermore, gradation of the "blindness to sugar" is known. Mung bean nuclease and venom phosphodiesterase hydrolyze DNA, RNA, and derivatives of arabinose, while micrococcal nuclease attacks DNA and RNA but not the oligomers containing arabinose (Wechter et al. 1968). To complicate the definition of a DNase, indirect evidence suggests that pancreatic DNase may be capable of reluctantly hydrolyzing rN-dN linkage (Pruch and Laskowski 1980). If confirmed, the specificity toward sugar of this enzyme is limited to the nucleotide following the sensitive linkage. However, providing solid evidence that the last trace of RNase activity has been removed from the DNase I preparation is a very difficult task. Differences were also observed in cleaving 2′-5′, and 5′-5′ bonds; both are susceptible to venom phosphodiesterase (Richards et al. 1967), but not to micrococcal nuclease.

This table is also misleading with respect to the concept of exonucleases. At the time of presentation of the four criteria in Table 1, an exonuclease was defined as an enzyme that consecutively liberated a mononucleotide from the same terminus. This definition became inadequate when viral circular DNA replaced calf thymus DNA as a popular substrate. Frenkel and Richardson (1971) corrected the definition of exonucleases to describe enzymes that require a free terminus. With purer venom phosphodiesterase (Laskowski 1980b) and viral supercoiled PM2 DNA, Pritchard et al. (1977) found that

Table 1 Classification of nucleolytic enzymes

1. Susceptible substrates	RNA, DNA
2. Type of attack	exonuclease, endonuclease
3. Products	3′-terminated, 5′-terminated
4. Preferential linkage	Pu ↓ Py, Py ↓ Pu

Reprinted from Laskowski (1959)

this classical exonuclease is capable of rapidly opening supercoiled DNA. It opens the relaxed circular DNA 10,000 times slower. Thus, the term ''exonuclease'' has only a relative significance, whether the earlier definition or that of Frenkel and Richardson (1971) is used. Probably a better term would be ''exophilic nuclease'' to indicate a preference for the open termini. A lack of sharp boundaries between endo- and exonucleases started to appear when we compared the action of different enzymes using two different methods (Williams et al. 1961). A relatively large number of phosphodiester bond cleavages as determined by pH-stat had little effect on the viscosity (see Fig. 3, Δ) when native thymus DNA was digested with venom phosphodiesterase (typical exonuclease). In contrast, when the digestion was performed with DNase I, the drop in viscosity was very rapid (Fig. 3, ○), suggesting that centrally located linkages were preferentially attacked. Micrococcal nuclease, then believed to be a typical endonuclease, did not give an equally steep curve (Fig. 3, □); the endonucleolytic linkages must have been followed by some terminally located ones.

No provision for a difference in double-strand cleavage and single-strand nicking was made in the original table, yet it became a very important property. The first to realize its importance were Young and Sinsheimer (1965) who used as a substrate λ phage DNA specifically prepared not to contain single-strand breaks. This DNA was digested with either DNase I or DNase II and the reaction was followed by ultracentrifugation in neutral and in alkaline media. The results showed that DNase II produced single-strand and double-strand breaks at or near the same level, while DNase I produced approximately four times more single-strand than double-strand cleavages. A simplified

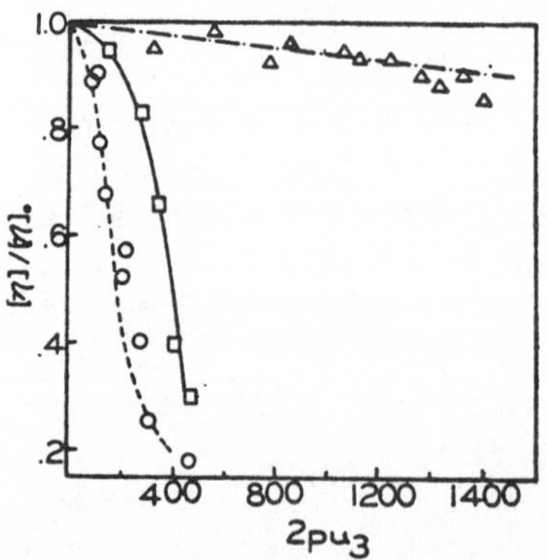

Figure 3 Correlation of the change in the approximated intrinsic viscosity ($[\eta]/[\eta]o$) and the number of diester bond cleavages per parent Z-average-molecular-weight DNA molecule ($2pu_3$). (Δ) Snake venom phosphodiesterase; (○) DNase I; (□) micrococcal nuclease. The theoretical molecular weight decay curves for DNase I (---) and phosphodiesterase (-•-) are shown. (Reprinted, with permission, from Williams et al. 1961.)

procedure using phosphatase and polynucleotide kinase recently led to very similar numbers for DNase I (Kroeker and Laskowski 1977).

An additional complication was encountered when it was found that the rate of hydrolysis depends on the stage of the reaction; some nucleases show autoacceleration while the others show autoretardation (Laskowski 1967). In conclusion, all these results show that the concept of endo- and exonucleases is an oversimplification, and neither the original nor the corrected definitions have sufficiently sharp boundaries to justify the retention of either class.

The inclusion of category 4 was an overemphasis. It arose because of the findings of Schmidt and co-workers (1951) on specific cleavage by RNase after a pyrimidine. Since, in the early work on DNases, it looked as if both the nature of the preceding base and the nature of the following one affected the susceptibility of the linkage, it was formulated to indicate both partners.

A factor that undermined my interest (and probably of many other workers) in a rational classification was the decision of International Union of Biochemistry (IUB) to include ribonucleases that form as a first product an intermediate 2′,3′-cyclic phosphate into transferases while leaving other nucleases as hydrolases. This decision found some defenders (Barnard 1969) but did not survive. In the early 1970s, IUB returned all RNases to hydrolases.

Already in 1958, one omission from the table should have been evident to me. No provision was made for the enzymes reacting with the pigeon crop gland type of DNase-I inhibitor. Only a few years had passed before Lindberg (1967a,b,c) obtained a pure inhibitor. A few years later, Lazarides and Lindberg (1974) identified it as actin G. The next step was the finding that 5′-nucleotidase also forms a complex with actin (Mannherz and Rohr, 1978; Rohr and Mannherz 1979), and that 5′-nucleotidase slowly displaces DNase I from the complex with actin. In 1958, I underestimated the significance of inhibition of DNase I by an inhibitor with a then unknown chemical identity. When the nature of the inhibitor was still unknown, I attempted to correct for it and emphasized this property (Laskowski 1971). Further developments justified the latter decision, and it appears now to be highly significant in the understanding of physiological functions such as passage through a membrane or stopping by a barrier (placenta). A recent finding of Lacks (1981) that rabbit's actin does not inhibit DNase-I-like enzymes of a rat opens a new aspect of this problem.

The originally presented table had one virtue—it was transparent. It also had numerous faults, among them many omissions caused by ignorance. An attempt to correct for the information acquired at a later date (Laskowski 1967) by introducing five additional criteria brought the table up to date, but could not succeed as a system of classification. It was too clumsy and did not provide for the rapid growth of the restriction enzymes, even though a group to accommodate them was provided. It also contained an additional category of single-strand-specific nucleases. The credit for their discovery goes to Lehman (1960), but with due embarassment; because I also was among them, I must state that many people have worked with this class of enzymes without realizing their single-strand specificity, e.g., venom phosphodiesterase. In our laboratory, the realization that we are working with single-strand-specific enzymes came with the work of Johnson and Laskowski (1970) who showed that mung bean nuclease had a marked preference of the order 30,000:1 for the single-stranded substrates. We started working on this enzyme some 8 years earlier with the hope of purifying a nuclease that was specific for A residues (Sung and Laskowski 1962).

My review of 1967 was the last attempt of the saving usefulness of classification. The issue expired with little regret from anyone. The progress just outgrew all boundaries.

STRUCTURE

The work on structure started surprisingly early considering the general knowledge of nucleases. It glitters with Nobel prizes more than any other branch of nuclease research. It is now way past its infancy, but I am unable to detect any signs of aging, and I cannot conceive that in the foreseeable future it will be losing its position of relative dominance. The reason for not assigning a special section to this topic in our program is explainable by the impossibility of avoiding trespassing. Even though I have lived through the period in which structural studies of nuclease started to emerge, I never joined the players, and my role was limited to that of an eager spectator and not a member of the team.

Pancreatic RNase was the second protein and the first enzyme to be sequenced. Strange as it may sound, the work on the primary structure of RNase started before the structure of insulin was completed by Sanger and his colleagues (Sanger and Tuppy 1951; Sanger and Thompson 1953; Ryle et al. 1955). The rivalry was between Anfinsen's team and the team of Moore and Stein. The results of this rivalry were spectacular. Completely new techniques of amino acid analysis were developed in the latter laboratory and these techniques, with some modernization, are still being widely used. Fortunately, an excellent review by Richards and Wyckoff (1971) covers not only the essential facts of primary structure but also an historical introduction and the detailed description of the work on the three-dimensional structure.

With regard to the three-dimensional structure of RNase, the rivalry was between the Roswell Park group (Kartha et al. 1967) who worked on RNase A and the Yale group (Richards and Wyckoff 1971) who worked on RNase S. The fact that of all available enzymes pancreatic RNase was selected as the first enzyme for studies of the three-dimensional structure was due primarily to its small molecular weight. As soon as the methods for sequencing were developed, comparative studies started. The conservative character of the structure throughout evolution became evident (Barnard 1969).

The second nuclease whose primary structure was elucidated was micrococcal nuclease. Our original attempts at purification of this enzyme were hampered by the unusual toxicity of the strain (it was penicillin and streptomycin resistant). In spite of that, a reasonable purification was achieved even though I landed in the hospital for a few days with acute toxemia. Anfinsen and his colleagues joined the hunt by using a "Foggi" strain which was less toxic, penicillin sensitive, and, in addition, 10-fold better in producing nuclease. A disadvantage of the Foggi strain was that it also was a 10-fold better producer of nonspecific phosphatase. Purification in both laboratories ran a rather close race.

Now I beg your indulgence for one more digression. During the 1966 Federation Meeting, Dr. Eugene Sulkowski met Dr. Anfinsen and at the first opportunity informed him that "we just crystallized a phosphatase-free micrococcal nuclease and sent the manuscript to JBC." A day or so later at the same Federation Meeting Dr. Anfinsen told me that he had asked the *Journal of Biological Chemistry* to hold his already-accepted paper for a couple of months so that both papers could appear side by side. It seems that chivalry is not dead among scientists!

Efforts in our laboratory had been directed toward a nuclease as free from phosphatase as possible, and our preparation was cleaner in this respect. However, our method was clumsier, and commercial preparation is now done by Anfinsen's method. A small but detectable content of phosphatase is not a serious handicap in most experiments, and no handicap at all for the purpose of amino acid sequence (Anfinsen et al. 1971) and three-dimensional structure (Cotton and Hazen 1971); in both respects Anfinsen's preparation was superior and was used.

The molecular weight of micrococcal nuclease is only slightly higher than that of pancreatic RNase. The sequence was established (Anfinsen et al. 1971), and elucidation of the three-dimensional structure followed (Cotton and Hazen 1971). The three-dimensional structure readily explained the observation made independently in Anfinsen's laboratory (Cuatrecases et al. 1967) and in ours (Sulkowski and Laskowski 1968) that dpTp in the presence of Ca^{2+} is a strong inhibitor of micrococcal nuclease. The shape of the active center accommodated dpTp but not dpCp. Last year we faced the dilemma of distinguishing between small amounts of adsorbed micrococcal nuclease and an endogenous nuclease in oligonucleosomes prepared from chicken erythrocyte nuclei (Kalinski et al. 1980). To our surprise, dpTp inhibited both enzymes. As yet we have not started purifying the endogenous nuclease, but we expect that the purified enzyme will show a spatial arrangement of the active center similar to that of micrococcal nuclease.

DNase I, the starting point of all research on nucleases, was sequenced in the laboratory of Stanford Moore (Liao et al. 1973; Salnikow et al. 1973; Moore 1981). Besides the major component DNase A, three additional isoenzymes were also sequenced. The position of the glycosidic chain is known. A number of active residues, as well as functionally expendable ones, have been identified (Moore 1981). As yet the three-dimensional structure is not known.

SPECIFICITY

Strict sequence specificity is what this meeting is all about and as a consequence every aspect that deals with specificity toward bases is taboo for me. I will only touch the periphery of the subject and only those aspects that are not connected with bases.

Historically, studies on specificity started very early, with sugar moieties attracting most of the attention. Even before Kunitz's studies, it was known that some nucleases are sugar-specific, e.g., boiling of pancreatic juice (Davidson 1950) destroyed DNase-type but not RNase-type activity. It had also been known for a long time that phosphodiesterase hydrolyzes derivatives of both RNA and DNA (Davidson 1950), but in the pre-Kunitz era phosphodiesterase was not considered a *sensu strictu* nuclease. The concept of nuclease as opposed to oligonucleotidase died a natural death early. The only survivors that remain are the observed changes in rates of hydrolysis, be it autoretardation or autoacceleration (Laskowski 1971). After the discovery of micrococcal nuclease (Cunningham et al. 1956), it became clear that at least three types of nucleases exist: RNA specific, DNA specific, and nonspecific to sugar. The finding that some nonspecific nucleases also hydrolyze derivatives of arabinose (Wechter et al. 1968) added an additional category. Furthermore, Richards et al. (1967) found that the rate of hydrolysis by phosphodiesterase changes when arabinose is linked either in a 3′- or a 5′-position; the 2′→5′ internucleotide linkage was the most resistant of all. The specificity toward sugar has a potentially useful function in enzymatic analysis for nucleotides that occur in small quantities either as a contaminant or a built-in component (Pruch and Laskowski 1980). So far, this characteristic has not been widely utilized because of the difficulties involved in isolation of an enzyme free of all interfering contaminants.

The paper of Richards et al. (1967) also touches another aspect of specificity—the effect of the location of the monophosphoryl group. The old observations of the different susceptibilities of 3′-phosphate and 5′-phosphate-terminated oligonucleotides to venom phosphodiesterase were mentioned earlier (Laskowski 1971). The elegant method of determination of the nearest neighbor frequencies (Kornberg 1961) utilized DNA polymerase and individual α-[^{32}P]-triphosphates, one at a time, the other three nucleotides being unlabeled. The newly synthesized polymer was degraded with 3′-phosphate-forming enzymes (micrococcal nuclease + spleen phosphodiesterase) and each of the liberated 3′-mononulceotides was analyzed for the content of ^{32}P.

Specificity toward the position of the monophosphoryl group is usually expressed in terms of the products as 3′-phosphate-formers or 5′-phosphate-formers. This type of nomenclature assumes that the normal 3′→5′ internucleotide linkage is the only linkage. It also assumes that nucleases are exclusively phosphodiesterases as opposed to phosphomonoesterases. The distinction of diesterases versus monoesterases has been generally accepted since the 1930s. An unexpected exception to this distinction is mung bean nuclease (Laskowski 1980a), which is able to hydrolyze the terminal 3′-phosphate (ω-monophosphatase activity). Since the two activities increase in parallel throughout the whole course of purification (by the two different methods), it seems probable that both activities are located at the same active center. Additional confirmation comes from the finding that both activities have the same ionic requirements and optimal pH.

The history of mung bean nuclease started in 1962, when we attempted to purify a nuclease specific for the ApN linkage. The purified enzyme indeed preferentially attacked this linkage but was not totally specific (Sung and Laskowski 1962). A few years later, Johnson and Laskowski (1970) showed that the enzyme was highly specific for single-stranded substrates. Ardelt and Laskowski (1971) obtained it as a homogeneous preparation. The method was simplified and Zn^{2+} was recognized by Kowalski et al. (1976) as a required component. This latter form is now commercially available.

More surprising was the realization that venom phosphodiesterase, our standard enzyme that led us to a concept of exonucleases (Laskowski 1971), shows many properties common to single-strand-specific enzymes (Pritchard et al. 1977). Among them is the ability to hydrolyze supercoiled PM2 DNA rapidly, while the relaxed circular form is hydrolyzed at a 10,000-fold slower rate. We have now purified this enzyme to a state of homogeneity as judged by gel electrophoresis with and without SDS. The interfering enzymes were excluded to less than one part in 1000, the limit of our testing (Laskowski 1980b). The finding that phosphodiesterase rapidly opens the supercoiled DNA buried the concept of strict definitions of endo- and exonucleases, regardless of whether the earlier definition (Laskowski 1959) or that of Frenkel and Richardson (1971) was used.

Since I have already committed the crime of trespassing into single-strand-specific nucleases, I will proceed with a premeditated violation of my assignment to stick to history, and show some unpublished material.

This story starts with the paper of Fujimoto et al. (1979) who found that three single-strand-specific enzymes (venom phosphodiesterase, P_1 nuclease, and mung bean nuclease) attack chromatin (chicken erythrocyte nuclei) by producing fragments ranging from 10–80 nucleosomes with an average size of 30. The conclusion was drawn that only a limited number of sites are susceptible to these single-strand-specific enzymes.

We were not wholly satisfied with this conclusion and decided to look a bit closer. Takamatsu (1981) prepared oligonucleosome fractions with micrococcal nuclease, then exposed them to venom phosphodiesterase. The enzyme attacked linkers preferentially, and in spite of that did not form a sharply defined series of oligonucleosomes. If, then, linkers are preferentially susceptible, even in dinucleosomes, the reason for the inability to form and maintain mononucleosomes cannot be due to the linker versus core preference. T.K. Dziegielewski (unpubl.) performed a series of experiments with micrococcal nuclease. Figure 4 shows that mononucleosomes produced by micrococcal nuclease represent an intermediate form between polynucleosomes and acid-soluble products. We analyzed the composition of each fraction for the content of ribonucleic acid derivatives by the method of Duch and Laskowski (1971). When working with chromatin, deoxy-derivatives predominate so heavily that the sample must be overloaded and deoxy-derivatives are off-scale before the ribo-derivatives are seen. We have analyzed all three fractions (oligo- and mononucleosomes, as well as the acid-soluble fraction) and the results appeared to be interesting. Ribo-derivatives were found in all fractions. Furthermore, only rG and rA were detected. I have verified Murphy's Law many times in the past, and we verified it again. The authentic sample of 2′(1″ ribosyl)adenosine, r_2A, obtained from Dr. Miwa, appeared in our chromatogram exactly in the position of rG. The same result was observed when we digested a polymer of ADP-ribose obtained from Dr. Kidwell with our best phosphodiesterase and dephosphorylated the product with alkaline phospha-

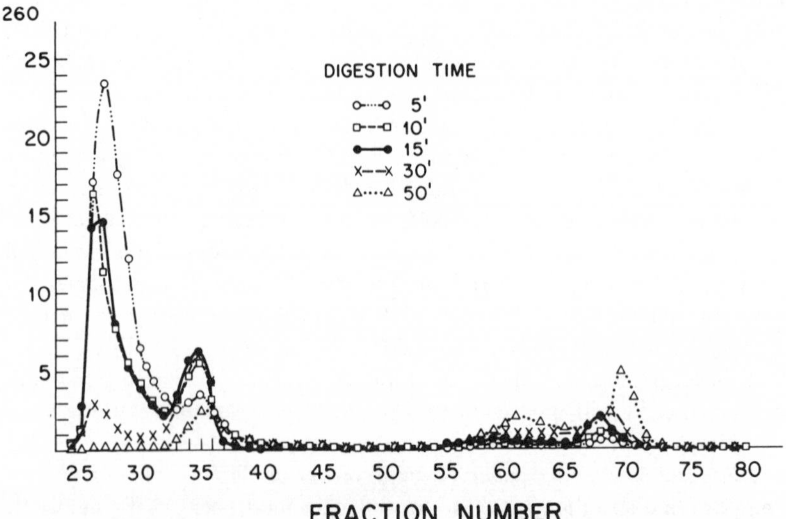

Figure 4 Gel-filtration patterns of nuclei digested with micrococcal nuclease. Chicken erythrocyte nuclei were incubated at 37°C with micrococcal nuclease (7.5 units/ml) for various times (see figure for digestion times) in a solution containing 0.3 M sucrose, 10 mM Tris-HCl (pH 7.2), 0.75 mM $CaCl_2$, and 0.1 mM phenylmethanesulfonyl fluoride. The reaction was terminated by cooling the solution to 4°C and adding EDTA. After centrifugation (15,000 rpm, 10 min) the supernatant was applied to a Bio-Gel A-5m column (1.4 × 80 cm) and eluted in 1.8-ml fractions (T.K. Dziegielewski and M. Laskowski, pers. comm.).

tase. A simple way to decide between rG and r_2A was to pool the peak, treat it with adenosine deaminase, and rechromatograph; rG did not change the position of elution, and r_2A shifted to the left because of deamination. From then on it was easy to prove that only rA and r_2A were present in our digests. Comparison of the results shown in Figure 5 with those of Figure 4 shows that a minimal amount of poly(ADP-Rib) is required to preserve the structure of mononucleosomes. All single-strand-specific enzymes easily cleave the 5′-phosphate-phosphate-5′ linkage, but micrococcal nuclease does not. This is the reason why micrococcal nuclease preserved nucleosomal structure. Another nuclease known to preserve nucleosomal structure is the endogenous nuclease described by Burgoyne and co-workers (Burgoyne et al. 1970, 1974; Hewish and Burgoyne 1973). The enzyme has not been purified as yet, but when it is purified, I predict it will be unable to cleave 5′-phosphate-phosphate-5′ linkage.

To complete the discussion on the side issues of specificity, the functional as opposed to structural specificity must be mentioned. A classic example is the

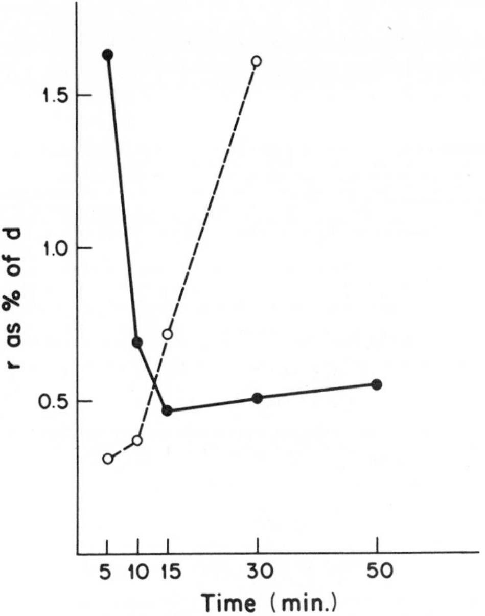

Figure 5 Ribonucleoside distribution in oligo- (○) and mononucleosomes (●). Oligonucleosomes (fractions 24–32) and mononucleosomes (fractions 33–40) were isolated as described in Fig. 4. After deproteinization, the ethanol-insoluble material was digested to nucleosides and analyzed by high-pressure liquid chromatography (Duch and Laskowski 1971). The ribonucleosides present as a function of time of micrococcal nuclease digestion are expressed as % of deoxyribonucleosides in oligo- and mononucleosomes. See Fig. 4 for the distribution of oligo- and mononucleosomes as a function of digestion time (T.K. Dziegielewski and M. Laskowski, pers. comm.).

processive mechanism discovered by Maxine Singer (1966). Only after Young and Sinsheimer showed (1965) that nicked DNA could be prepared with pancreatic DNase was it possible to start looking for the "gapping" nucleases (Masamune et al. 1971).

PHYSIOLOGICAL FUNCTION

Attempts to fit nucleases into physiologically important functions started rather early—right after their discovery. Obviously, the digestive function of nucleases was first stressed because the major known representatives were located and secreted by the pancreas. After the discovery of nucleases in tissue other than the pancreas, the next short-lived theory of dual roles was proposed: digestive and metabolic (whatever that meant) functions. The only part of this theory that is still useful is that different nucleases are indeed involved in different functions. The tissue of origin and the intracellular distribution are not the most important aspects of fitting a particular nuclease to a particular function. During the rest of this meeting, we will see how many special functions depend upon different nucleases. The plurality of vital processes in which nucleases participate is now taken for granted to the extent that it seems harder to think of a significant biological function in which nucleases do not participate than one in which they do.

ENDING

Before I close I would like to make a last trespass. The restriction enzymes opened doors not only to a new scientific domain and a new industry but also to new fears of creating a biological monster. I do not share those fears and I do not believe there is a need for any imposed legal limitations. Historically, introduction of limitations based on the ethical standards of a particular society at a particular time led to domination by Egyptian priests, the Inquisition, and "Lysenkoism." Accidents in the laboratory do and will happen, but only a deliberate action directed toward massive destruction could lead to a world catastrophe. Such action invariably finds a retaliation, and in all probability a biological monster, if ever created, would be a single-shot weapon.

The summary of 40 years of observation and participation in research on nucleases leads to a conclusion that several great discoveries were made at a cost of producing a mass of trivialities and many serious errors, the correction of which was time consuming. I believe, however, that such a proportion is a chronic disease of scientific progress and that my successor reviewing the next 40 years will come to a similar conclusion. The reason for my belief is that this generation of my colleagues and rivals has had its normal share of very gifted people. Specifically, I would like to pay tribute to a friend and archrival, the late Dr. Moses Kunitz.

REFERENCES

Allfrey, V. and A.E. Mirsky. 1952. Some aspects of the desoxyribonuclease activities of animal tissues. *J. Gen. Physiol.* **36:**227.

Anfinsen, C.B., P. Cuatrecases, and H. Taniuchi. 1971. Staphylococcal nuclease, chemical properties and catalysis. In *The enzymes,* vol. 4, p. 177. Academic Press, New York.

Ardelt, W. and M. Laskowski, Sr.. 1971. Mung bean nuclease I. IV. An improved method of preparation. *Biochem. Biophys. Res. Comm.* **44:**1205.

Avery, O.T., C.M. MacLeod, and M. McCarty. 1944. Studies on the chemical nature of the substance inducing transformation of pneumococcal types. Induction of transformation by a desoxyribonucleic acid fraction isolated from Pneumococcus type III. *J. Exp. Med.* **79:**137.

Barnard, E.A.. 1969. Ribonucleases. *Annu. Rev. Biochem.* **38:**677.

Brown, D.M. and A.R. Todd. 1955. Evidence on the nature of the chemical bonds in nucleic acids. In *The nucleic acids* (ed. E. Chargaff and J.N. Davidson), vol. 1, p. 409. Academic Press, New York.

Brown, K.D., G. Jacobs, and M. Laskowski, Sr. 1952. The distribution of nucleodepolymerases in calf thymus fractions. *J. Biol. Chem.* **194:**445.

Brown, K.D., R.E. Shupe, and M. Laskowski. 1948. Crystalline activated protein B (chymotrypsin B). *J. Biol. Chem.* **173:**99.

Burgoyne, L.A., D.R. Hewish, and J. Mobbs. 1974. Mammalian chromatin substructure studies with the calcium-magnesium endonuclease and two-dimensional polyacrylamide-gel electrophoresis. *Biochem. J.* **143:**67.

Burgoyne, L.A., M.A. Waqar, and M.R. Atkinson. 1970. Initiation of DNA synthesis in rat thymus: Correlation of calcium-dependent initiation in thymocytes and in isolated thymus nuclei. *Biochem. Biophys. Res. Comm.* **39:**918.

Chargaff, E. 1955. Isolation and composition of the deoxypentose nucleic acids and of the corresponding nucleoproteins. In *The nucleic acids* (ed. E. Chargaff and J.N. Davidson), vol. 1, p. 307. Academic Press, New York.

Cohn, W.E. 1950. Heterogeneity in pyrimidine nucleotides. *J. Am. Chem. Soc.* **72:**2811.

Cooper, E.J., M.L. Trautmann, and M. Laskowski, Sr. 1950. Occurrence and distribution of an inhibitor for desoxyribonuclease in animal tissues. *Proc. Soc. Expt. Biol. Med.* **73:**219.

Cotton, F.A. and E.E. Hazen, Jr. 1971. Staphylococcal nuclease X-ray structure. In *The Enzymes* (ed. P.D. Boyer), vol. 4, p. 153. Academic Press, New York.

Cuatrecases, P., S. Fuchs, and C.B. Anfinsen. 1967. Catalytic properties and specificity of the extracellular nuclease of *Staphylococcus aureus. J. Biol. Chem.* **242:**1541.

Cunningham, L. and M. Laskowski, Sr. 1953. Presence of two different desoxyribonucleodepolymerases in veal kidney. *Biochim. Biophys. Acta* **11:**590.

Cunningham, L., B.W. Catlin, and M. Privat de Garilhe. 1956. A deoxyribonuclease of *Micrococcus pyogenes. Am. Chem. Soc.* **78:**4642.

Dabrowska, W., E.J. Cooper, and M. Laskowski, Sr. 1949. A specific inhibitor for desoxyribonuclease. *J. Biol. Chem.* **177:**991.

Davidson, J.N. 1950. *The biochemistry of the nucleic acid.* Methuen, London.

Duch, D.S. and M. Laskowski, Sr. 1971. A sensitive method for the determination of RNA in DNA and vice versa. *Anal. Biochem.* **44:**42.

Felsenfeld, G. 1978. Chromatin. *Nature* **271:**115.

Frenkel, G.D. and C.C. Richardson. 1971. The deoxyribonuclease induced after infection of *Escherichia coli* by bacteriophage T5. I. Characterization of the enzyme as a 5′-exonuclease. *J. Biol. Chem.* **246:**4839.

Fujimoto, M., A. Kalinski, A.E. Pritchard, D. Kowalski, and M. Laskowski, Sr. 1979. Accessibility of some regions of DNA in chromatin (chicken erythrocytes) to single-strand specific nucleases. *J. Biol. Chem.* **254:**7405.

Hewish, D.R. and L.A. Burgoyne. 1973. The calcium dependent endonuclease activity of isolated nuclear preparations. Relationships between its occurrence and the occurrence of other classes of enzymes found in nuclear preparations. *Biochem. Biophys. Res. Comm.* **52:**475.

Holley, R.W., J. Apgar, G.A. Everett, J.T. Maddison, M. Marquisee, S.H. Merrill, J.R. Penswick, and A. Zamir. 1965. Structure of a ribonucleic acid. *Science* **147:**1462.

Johnson, P.H. and M. Laskowski, Sr. 1970. Mung bean nuclease I. II. Resistance of double-stranded deoxyribonucleic acid and susceptibility of regions rich in adenosine and thymidine to enmzymatic hydrolysis. *J. Biol. Chem.* **245:**891.

Josse, J., A.D. Kauser, and A. Kornberg. 1961. Enzymatic synthesis of deoxyribonucleic acid VIII. Frequencies of nearest neighbor base sequences in deoxyribonucleic acid. *J. Biol. Chem.* **236:**864.

Kalinski, A., H. Takamatsu, and M. Laskowski, Sr. 1980. An endonuclease activity of chicken erythrocyte nuclei and mononucleosomes. *J. Biol. Chem.* **255:**10542.

Kartha, G., J. Bello, and D. Harker. 1967. Structure of ribonuclease. *Nature* **213:**557.

Keith, C.K., A. Kazenko, and M. Laskowski. 1947. Studies on proteolytic activity of crystalline protein B from pancrease. *J. Biol. Chem.* **170:**227.

Kleczkowski, A. 1948. Proteolytic activity of preparations of crystallized ribonuclease. *Biochem. J.* **42:**523.

Koerner, J.F. and R.L. Sinsheimer. 1951. A deoxyribonuclease from calf spleen. I. Purification and properties. *J. Biol. Chem.* **228:**1039.

_________. 1957. A deoxyribonuclease from calf spleen. *J. Biol Chem.* **228:**1049.

Kornberg, A. 1961. *Enzymatic synthesis of DNA.* Ciba Lectures, John Wiley and Sons, New York.

Kowalski, D., W.D. Kroeker, and M. Laskowski, Sr. 1976. Mung bean nuclease I. Physical, chemical and catalytic properties. *Biochemistry* **15:**4457.

Kroeker, W.D. and M. Laskowski, Sr. 1977. Polynucleotide kinase: Functional purification and use in the direct kinetic measurement of single- and double-strand cleavages of DNA by restriction and other endonucleases of limited action. *Anal. Biochem.* **79:**63.

Kunitz, M. 1940. Crystalline ribonuclease. *J. Gen. Physiol.* **24:**15.

_________. 1950. Crystalline desoxyribonuclease I. Isolation and general properties. Spectrophotometric method for the measurement of desoxyribonuclease activity. *J. Gen. Physiol.* **33:**349.

Lacks, S.A. 1981. Deoxyribonuclease I in mammalian tissues specificity of inhibition by actin. *J. Biol. Chem.* **256:**2644.

Laskowski, M. 1946b. Crystalline protein with thymonucleodepolymerase activity isolated from beef pancreas. *J. Biol. Chem.* **166:**555.

_________. 1946a. Studies on thymonucleodepolymerase. *Arch. Biochem.* **11:**41.

_________. 1951. Nucleolytic enzymes. In *The Enzymes* (eds. J.B. Sumner and K. Myrbàck), vol. 1, part 2, p. 956. Academic Press, New York.

Laskowski, M., Sr. 1959. Enzymes hydrolyzing DNA. *Ann. N.Y. Acad. Sci.* **81:**776.

_________. 1967. DNases and their use in the studies of primary structure of nucleic acids. *Adv. Enzymol.* **29:**165.

_________. 1971. Venom exonuclease. In *The enzymes*, 3rd edition (ed. P. Boyer), vol. 4, p. 311. Academic Press, New York.

_________. 1980a. Purification and properties of the mung bean nuclease. *Methods Enzymol.* **65:**263.

_________. 1980b. Purification and properties of venom phosphodiesterase. *Methods Enzymol.* **65:**276.

Laskowski, M. and A. Kazenko. 1947. Nucleolytic and proteolytic activities of the new crystalline protein from beef pancrease. *J. Biol. Chem.* **167:**617.

Laskowski, M. and M.D. Seidel. 1945. Viscosimetric determination of thymonucleodepolymerase. *Arch. Biochem.* **7:**465.

Laurila, U.-R. and M. Laskowski, Sr. 1957. Studies of specificity of deoxyribonuclease II from thymus. *J. Biol. Chem.* **228:**49.

Lazarides, E. and U. Lindberg. 1974. Actin is the naturally occurring inhibitor of deoxyribonuclease I. *Proc. Natl. Acad. Sci.* **71:**4742.

Lehman, I.R. 1960. The deoxyribonuclease of *Escherichia coli* I. Purification and properties of a phosphodiesterase. *J. Biol. Chem.* **235:**1479.

_________. 1963. The nucleases of *Escherichia coli*. *Prog. Nucleic Acid Res.* **2:**83.

_________. 1971. Bacterial DNases. In *The enzymes*, 3rd edition (ed. P. Boyer), vol. 4, p. 251. Academic Press, New York.

Liao, T.-H., J. Salnikow, S. Moore, and W.H. Stein. 1973. Bovine pancreatic deoxyribonuclease A. Isolation of cyanogen bromide peptides; complete covalent structure of the polypeptide chain. *J. Biol. Chem.* **248:**1489.

Lindberg, U. 1967a. Purification from calf spleen of two inhibitors of deoxyribonuclease I. Physical and chemical characterization of the inhibitor II. *Biochemistry* **6:**323.

_________. 1967b. Molecular weight and amino acid composition of deoxyribonuclease I. *Biochemistry* **6:**335.

_________. 1967c. Studies on the complex formation between deoxyribonuclease I and spleen inhibitor II. *Biochemistry* **6:**343.

Mannherz, G. and G. Rohr. 1978. 5′-Nucleotidase reverses the inhibitory action of actin on pancreatic deoxyribonuclease I. *FEBS Lett.* **95:**284.

Martland, M. and R. Robison. 1926. CVI. Possible significance of hexosephosphoric esters in ossification. Part VI. Phosphoric esters in blood-plasma. *Biochem. J.* **20:**847.

Masamune, Y., R.A. Fleischman, and C.C. Richardson. 1971. Enzymatic removal and replacement of nucleotides at single-strand breaks in deoxyribonucleic acid. *J. Biol. Chem.* **246:**2680.

Maver, M.E. and A. Greco. 1949a. The hydrolysis of nucleoproteins by cathepsins from calf thymus. *J. Biol. Chem.* **181:**853.

_________. 1949b. The nuclease activities of cathepsin preparations from calf spleen and thymus. *J. Biol. Chem.* **181:**861.

Mazia, D. 1949. The distribution of desoxyribonuclease in the developing embryo *(Arbacia punctulata)*. *J. Cell Comp. Physiol.* **34:**17.

Mikulski, A., E. Sulkowski, L. Stasiuk, and M. Laskowski, Sr. 1969. Susceptibility of dinucleotides bearing either 3′- or 5′-monophosphate to micrococcal nuclease. *J. Biol. Chem.* **244:**6559.

Moore, S. 1981. Pancreatic DNase. In *The enzymes*, vol. 14A, p. 281. Academic Press, New York.

Noll, M. 1974. Subunit structure of chromatin. *Nature* **251:**249.

Potter, J.L., K.D. Brown, and M. Laskowski, Sr. 1952. Enzymatic degradation of desoxyribonucleic acid by crystalline desoxyribonuclease. *Biochim. Biophys. Acta* **9:**150.

Potter, J.L., U.-R. Laurila, and M. Laskowski, Sr. 1958. Studies of the specificity of deoxyribonuclease I. I. Hydrolysis of a trinucleotide. *J. Biol. Chem.* **233:**915.

Price, P.A., T.-Y. Lin, W.H. Stein, and S. Moore. 1969. Properties of chromatographically purified bovine pancreatic deoxyribonuclease. *J. Biol. Chem.* **244:**917.

Pritchard, A.E., D. Kowalski, and M. Laskowski, Sr. 1977. An endonuclease activity of venom phosphodiesterase specific for single-stranded and superhelical DNA. *J. Biol. Chem.* **252:**8652.

Privat de Garilhe, M. 1964. *Les nucleases*. Hermann, Paris.

Privat de Garilhe, M. and M. Laskowski, Sr. 1954. Evidence for different specificities of DNase I and DNase II. *Biochim. Biophys. Acta.* **14:**147.

_________. 1955. Study of the enzymatic degradation of deoxyribonucleic acid by two different deoxyribonucleodepolymerases. *J. Biol. Chem.* **215:**269.

_________. 1956. Optical changes occurring during the action of phosphodiesterase on oligonucleotides derived from deoxyribonucleic acid. *J. Biol. Chem.* **223:**661.

Privat de Garilhe, M., L. Cunningham, U.-R. Laurila, and M. Laskowski, Sr. 1957. Studies on isomeric dinucleotides derived from deoxyribonucleic acid. *J. Biol. Chem.* **224:**751.

Pruch, J.M. and M. Laskowski, Sr. 1980. Covalently bound ribonucleotides in crab d(A-T) polymer. *J. Biol. Chem.* **255:**9409.

Razzell, W.E. and H.G. Khorana. 1959. Studies on polynucleotides IV. Enzymic degradation. The stepwise action of venom phosphodiesterase on deoxyribo-oligonucleotides. *J. Biol. Chem.* **234:**2114.

Richards, F.M. and H.W. Wyckoff. 1971. Bovine pancreatic ribonuclease. In *The enzymes*, vol. 4, p. 647. Academic Press, New York.

Richards, G.M. and M. Laskowski, Sr. 1969. Negative charge at the 3′-terminus of oligonucleotides and resistance to venom exonuclease. *Biochemistry* **8:**1786.

Richards, G.M., D.J. Tutas, W.J. Wechter, and M. Laskowski, Sr. 1967. Hydrolysis of dinucleoside monophosphates containing arabinose in various internucleotide linkages by exonuclease from the venom of *Crotalus adamanteus*. *Biochemistry* **6:**2908.

Rill, R. and K.E. Van Holde. 1973. Properties of nuclease-resistant fragments of calf thymus chromatin. *J. Biol. Chem.* **248:**1080.

Rohr, G. and G. Mannherz. 1979. The activation of actin: DNase I complex with rat liver plasma membranes. The possible role of 5′ nucleotidase. *FEBS Lett.* **99:**351.

Ryle, A.P., F. Sanger, L.F. Smith, and R. Kitai. 1955. The disulphide bonds of insulin. *Biochem. J.* **60:**541.

Salnikow, J., T.-H. Liao, S. Moore, and W.H. Stein. 1973. Bovine pancreatic deoxyribonuclease A. Isolation, composition, and amino acid sequences of the tryptic and chymotryptic peptides. *J. Biol. Chem.* **248:**1480.

Sanger, F. and E.O.P. Thompson. 1953. The amino-acid sequence in the glycyl chain of insulin. 2. The investigation of peptides from enzymic hydrolysates. *Biochem. J.* **53:**366.

Sanger, F. and H. Tuppy. 1951. The amino acid sequence in the phenylalanyl chain of insulin. 1. The identification of lower peptides from partial hydrolysates. *Biochem. J.* **49:**463.

Schmidt, G. 1955. Nucleases and enzymes attacking nucleic acid components. In *The nucleic acids* (ed. E. Chargaff and J.N. Davidson), vol. 1, p. 555. Academic Press, New York.

Schmidt, G., R. Cubilis, N. Zollner, L. Hecht, N. Strickler, K. Seraidarian, M. Seraidarian, and S.J. Tannhauser. 1951. On the action of ribonuclease. *J. Biol. Chem.* **192:**715.

Shimomura, M. and M. Laskowski, Sr. 1957. Purification of deoxyribonuclease II from spleen. *Biochim. Biophys. Acta* **26:**198.

Singer, M.F. 1966. Potassium-activated phosphodiesterase (ribonuclease II) from *Escherichia coli*. *Prog. Nucleic Acid Res.* (ed. L. Cantoni and Davis), p. 192. Harper and Row, New York.

Singer, M.F., R.J. Hilmoe, and L.A. Heppel. 1958. Oligonucleotides as primers for polynucleotide phosphorylase. *Fed. Proc.* **17:**312 (Abstr.).

Sinsheimer, R.L. 1954. The action of pancreatic desoxyribonuclease I. Isolation of mono- and dinucleotides. *J. Biol. Chem.* **208:**445.

———. 1955. The action of pancreatic deoxyribonuclease II. Isomeric dinucleotides. *J. Biol. Chem.* **215:**579.

Sinsheimer, R.L. and J.F. Koerner. 1952. Di-desoxyribonucleotides. *J. Am. Chem. Soc.* **72:**283.

Sulkowski, E. and M. Laskowski, Sr. 1968. Protection of micrococcal nuclease against thermal inactivation. *J. Biol. Chem.* **243:**651.

Sumner, J.B. 1926. The isolation and crystallization of the enzyme urease. *J. Biol. Chem.* **69:**435.

Sung, S.-C. and M. Laskowski, Sr. 1962. A nuclease from mung bean sprouts. *J. Biol. Chem.* **237:**506.

Takamatsu, H. 1981. Action of venom phosphodiesterase on chromatin (chicken erythrocyte nuclei) fragmented with micrococcal nuclease. *Fed. Proc.* **40:**1567 (Abstr.).

Vanecko, S. and M. Laskowski, Sr. 1961. Studies of the specificity of deoxyribonuclease I. II. Hydrolysis of oligonucleotides carrying a monoesterified phosphate on carbon 3′. *J. Biol. Chem.* **236:**1135.

__________ . 1961. Studies of the specificity of deoxyribonuclease I. III. Hydrolysis of chains carrying a monoesterified phosphate on carbon 5′ *J. Biol. Chem.* **236:**3312.

__________ . 1962. Terminal nucleosides in the fragments of deoxyribonculeic acid produced by the action of splenic deoxyribonuclease II. *Biochim. Biophys. Acta* **61:**547.

Wang, D. and S. Moore. 1978. Preparation of protease-free and ribonuclease-free pancreatic deoxyribonuclease. *J. Biol. Chem.* **253:**7216.

Watson, J.D. and F.H.C. Crick. 1953. Molecular structure of nucleic acids. A structure for deoxyribose nucleic acid. *Nature* **171:**737.

Wattiaux, R. and C. de Duve. 1955. A hexose-1-phosphatase in silkworm blood. *Biochem. J.* **60:**590.

Wechter, W.J., A.J. Mikulski, and M. Laskowski, Sr. 1968. Gradation of specificity with regard to sugar among nucleases. *Biochem. Biophys. Res. Commun.* **30:**318.

Williams, E.J., S.-C. Sung, and M. Laskowski, Sr. 1961. Action of venom phosphodiesterase on deoxyribonculeic acid. *J. Biol. Chem.* **236:**1130.

Young, E.T., III and R.L. Sinsheimer. 1965. A comparison of the initial actions of spleen deoxyribonuclease and pancreatic deoxyribonuclease. *J. Biol. Chem.* **240:**1274.

Role of Nucleases in Genetic Recombination

Paul D. Sadowski
Department of Medical Genetics
University of Toronto
Toronto, Ontario, Canada M5S 1A8

INTRODUCTION

Genetic recombination has been defined by Clark as the interaction of nucleic acids so as to produce a change in the linkage relationships between genes or parts of genes (Clark 1971). While the phenomenon has been recognized in higher organisms for nearly three-quarters of a century, the discovery of genetic recombination in bacteria and bacteriophage some 35 years ago led to considerable impetus towards the determination of the mechanisms by which it occurred.

Initial mechanisms were of the copy-choice variety (Levinthal 1954), which invoked the alternate copying of one DNA molecule and then another to produce a new, recombinant DNA structure. However, copy-choice models fell from favor with the discovery that DNA replicated by a semiconservative mechanism (Meselson and Stahl 1957) and with the density-shift experiments of Meselson and Weigle (1961) which showed that phage λ recombination involved breakage and reunion of DNA molecules. The pendulum has swung back somewhat with the realization that there is a definite correlation between recombination events and DNA replication in bacteriophages λ (Stahl 1979) and T7 (Burck and Miller 1978; Smith and Miller 1981). Furthermore, models for the movement of transposons and bacteriophage Mu (Arthur and Sherratt 1979; Shapiro 1979) also place a heavy emphasis on a role for DNA replication.

Genetic recombination can be broadly classified into two types: (1) general recombination and (2) site-specific recombination. General recombination implies that the breakage and rejoining of DNA takes place between homologous DNA molecules and that the likelihood of recombination is comparable in all regions of homologous DNA molecules. The validity of the latter assumption will be discussed later. The prototype for this kind of recombination is that promoted by the *recA*, *B*, and *C* genes of *Escherichia coli*. Site-specific recombination occurs between DNA molecules bearing only small regions of homology (e.g., 15 bp for the λ-integrase reaction) or even between molecules with no apparent homology (as typified by transposable elements). The λ integration-excision reaction is further classified as a "conservative" subtype of site-specific recombination (Campbell 1981), since the recombining chromosomes do not undergo replication during the recombination act. This is distinguished from replicative recombination exhibited by insertion elements, transposons, and phage Mu where the transposition of these elements is closely related to their duplication (Calos and Miller 1980). While the sites to which these elements transpose are not specific, this kind of recombination is classified as site-specific because resolution of the duplicated intermediates is thought to come about by a site-specific recombination event between the two elements (Arthur and Sherratt 1979; Shapiro 1979). Direct evidence for such site-specific resolution of transposition intermediates is now available from in vivo and in vitro studies (Reed 1981a,b; Reed and Grindley 1981; Sherratt et al. 1981).

GENERAL RECOMBINATION

Recombination Models

Workers in the field of general genetic recombination have long been faced with somewhat of a dilemma. A considerable amount of data about recombination mechanisms has become available over the past two decades from the study of meiotic tetrads of fungi. Tetrad analysis enables one to examine the direct products of meiosis and thereby to formulate very elegant hypotheses about the mechanisms of recombination (e.g., Holliday 1964). At the same time, prokaryotic molecular biologists and biochemists have amassed an impressive body of data concerning prokaryotic recombination but of necessity most of it concerns events in a large population of recombining molecules and very little of it deals with single recombination events. Thus, one of the important functions of a recombination model should be to marry the genetic data of the fungal geneticist with the biochemical studies of the prokaryotic molecular biologist.

One of the most useful models of general recombination is that proposed by Meselson and Radding (1975). The features of this model serve as a framework for a discussion of specific nucleases that could possibly be involved in steps postulated by the model.[1] In the latter part of the chapter, brief reference will be made to in vitro systems promoting site-specific recombination, which promise to throw light on the nucleolytic events involved in this type of recombination.

The Meselson-Radding (Aviemore) Model

This model was born in 1973 at a meeting on genetic recombination near a village in Scotland called Aviemore. It is a modification of the Holliday model for gene conversion in fungi (Holliday 1964) and was developed to accomodate data in yeast suggesting that formation of heteroduplex DNA occurs asymmetrically. The model has been adequately explained and reviewed (Meselson and Radding 1975; Radding 1978) so that only a summary of its features, with particular attention to those steps at which nucleases are postulated to act, will be presented here (Fig. 1). Subsequent discussion will focus on some of the nucleases that have properties which might enable them to promote some of the steps of the Aviemore model.

Nicking. One of the pair of homologous duplex DNA molecules receives a single-strand break. Although there seem to be numerous nicking enzymes in cell extracts, very few of them have actually been implicated in genetic recombination. One exception is the gene-*3* endonuclease of phage T7 (Center et al. 1970; Center and Richardson 1970a,b; Sadowski 1971), whose properties will be detailed later.

Polymerization and Strand Displacement. The 3′-OH terminus created by the nicking enzyme serves as a site for initiation of DNA polymerization which in turn displaces a 5′-phosphorylated single strand ahead of it (Fig. 1a). Enzyme systems promoting polymerization and strand displacement, as part of in vitro DNA replication reactions, are well characterized for *E. coli*, T4, and

[1]Other models, such as that of Broker and Lehman (1971) also invoke the action of endo- and exonucleases, pairing of homologous regions, and branch migration. That model is perhaps more aptly applied to recombination of phage DNAs such as T4 (Broker 1973) and T7 (Tsujimoto and Ogawa 1977).

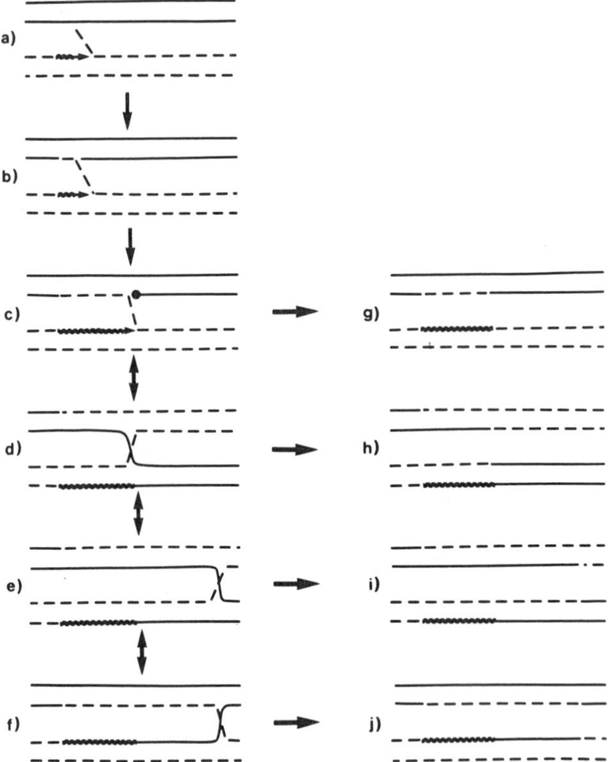

Figure 1 The Meselson-Radding (Meselson and Radding 1975) model for genetic recombination. (*a*) Strand displacement. (*b*) Single-strand aggression. (*c*) Strand assimilation. (*d*) Isomerization. (*e*) Branch migration. (*f*) Isomerization. (*g–j*) Endonucleolytic scission of the cross-connection. In structures *g* and *j*, the flanking markers are in the parental configuration. In structures *h* and *i*, the flanking markers are in the recombinant configuration. Structures *g* and *h* have undergone only asymmetric strand transfer. Structures *i* and *j* have undergone both asymmetric and symmetric strand transfer.

T7 and will not be dealt with further (see Alberts 1980). The key point is that a 5′ single strand is displaced which "invades" a homologous region of the opposite duplex (Fig. 1b).

Strand Uptake. The invasion of supercoiled DNA by a homologous single strand was shown by Holloman et al. (1975) to occur spontaneously, albeit rather slowly. The plausibility of this step increased with the discovery that the *recA* protein of *E. coli* promotes single-strand uptake with the formation of structures called D loops (McEntee et al. 1979; Shibata et al. 1979). The function of this remarkable protein has come under intense scrutiny during the past 3 years and its properties have been summarized by Radding (1981).

Suffice it to say that the *recA* protein is capable of promoting the kind of strand invasion demanded by the model as well as a host of other pairing reactions. The common feature of the pairing reactions seems to be a requirement that single-stranded DNA and a free end exist somewhere in one of the two partners.

D-loop Cleavage. The single-stranded loop resulting from the strand invasion is cleaved by a single-strand-specific endonuclease (Fig. 1c). This reaction has been shown to occur when D loops created by strand invasion are cleaved by the *recBC* nuclease of *E. coli* or S1 endonuclease (Wiegand et al. 1977). In principle, T7 endonuclease I might carry out such a reaction, although this has yet to be tested.

Strand Assimilation. A 5′-specific exonuclease removes nucleotides ahead of the invading strand and thus promotes strand uptake (Fig. 1, d and e). Such a reaction has been demonstrated by Cassuto and Radding (1971) for phage λ exonuclease and in principle might also be catalyzed by the gene-*6* exonuclease of phage T7 and *E. coli* exonuclease VIII (*recE* product).

Isomerization and Branch Migration. Isomerization (Sigal and Alberts 1972; Sobell 1974) converts an asymmetric structure to a symmetric one (Fig. 1g) and branch migration (Fig. 1, g and h) may shift the position of the cross-strand exchange in either direction. One important result of branch migration is that regions of heteroduplex can be varied in length.

The repair of mismatched bases (Fig. 1, h and i) in regions of heteroduplex DNA is thought to play a role in gene conversion in fungi (Holliday 1964; Radding 1978; Stahl 1979) and in recombination of closely linked markers of phage λ (Russo 1973; White and Fox 1974), at least when replication of DNA is restricted. However, the involvement of mismatch repair in recombination of closely linked markers may be negligible if DNA replication is normal (Gussin et al. 1980). While the repair of mismatched bases may involve nucleases of the S1 endonuclease type (Ahmad et al. 1975), virtually nothing about the enzymology of mismatch repair is known. (For a detailed discussion of DNA repair, see Chapter 4.)

Resolution. The cross-connections holding the two duplexes together must be cut to resolve the Holliday structure prior to ligation of the DNAs (Fig. 1, j and m). Little is known about the enzymology of this step, although the gene-*49* product of phage T4 (endonuclease VII) has recently been shown to resolve Holliday junctions accurately in vitro (B. Kemper and R. Weissberg, pers. comm.). The availability of recombination intermediates isolated in vivo or made in vitro (Potter and Dressler 1980; Radding 1981) should make this problem amenable to solution.

Nucleases Involved in General Recombination

Endonucleases

T7 Endonuclease I. This enzyme is the product of gene *3* of phage T7 and has been implicated in such intracellular processes as destruction of host DNA (Sadowski and Kerr 1970), DNA maturation (Paetkau et al. 1977), and genetic recombination (Powling and Knippers 1974; Kerr and Sadowski 1975; Lee et al. 1976; Lee and Sadowski 1981).

The enzyme degrades single-stranded DNA to acid-soluble oligonucleotides, but does not produce acid-soluble material from duplex DNA (Center and Richardson 1970b; Sadowski 1971). However, it does introduce both nicks and double-strand breaks into duplex DNA. Panayotatos and Wells (1981) recently showed that the enzyme would cut the loops of cruciform structures in plasmid DNA (S1 nuclease did the same).

In view of the above properties, it is possible to envisage involvement of this enzyme at one or more of the stages of the recombination process. It could initiate recombination by the introduction of single-strand breaks. It could be involved in the cleavage of a D loop after strand transfer. Finally, it could be involved in the resolution of Holliday structures by cutting the strands involved in the crossed-strand exchange. The development of a simple physical assay for the measurement of in vitro recombination of T7 DNA and the finding that the system requires T7 endonuclease I (D. Lee and P.D. Sadowski, in prep.) should enable us to define better the role of this enzyme in recombination.

Other Single-strand-specific Nucleases. The DNase I of the smut fungus *Ustilago maydis* (Holloman and Holliday 1973) degrades single-stranded DNA to oligonucleotides and mononucleotides and nicks duplex DNA (Holloman 1973). Mutants deficient in DNase I are also recombination-deficient (Holliday et al. 1974). The number of nicks is enhanced by prior UV irradiation of the DNA (Holliday et al. 1974). Such an enzyme may be involved in mismatch repair (Ahmad et al. 1975) and, at high concentrations, S1 nuclease appears able to recognize single-base mismatches in DNA (Shenk et al. 1975). Thus, a role for single-strand-specific nucleases in the repair of mismatched bases is possible. Further information about this class of nucleases is contained in Chapter 7.

ATP-independent Exonucleases

The recombination exonucleases can be broadly grouped into two classes: the ATP-independent type such as λ exonuclease and the ATP-dependent variety typified by the *recBC* nuclease of *E. coli*. Members of the former class are all 5′-specific exonucleases that degrade their duplex DNA substrate to 50% of completion. The latter group has numerous unique properties in addition to their exonucleolytic activities. These are described in Section III.C.

λ *Exonuclease.* One of the first nucleases to be implicated in recombination, this enzyme was first purified and characterized by Radding (1966), Little et al. (1967), and Little (1967). It is the product of the λ *red* alpha gene, one of a group of λ gene products synthesized from the leftward λ promoter and involved in recombination. The enzyme has been crystallized (Little et al. 1967), has a subunit molecular weight of 24,000, and acts from 5′-phosphorylated termini to liberate mononucleotides from duplex DNA (Little 1967).

The enzyme has a distinct preference for duplex DNA (Little 1967) and acts in a processive manner (Carter and Radding 1970), degrading one DNA strand extensively before starting to degrade another. Although the enzyme degrades the protruding 5′ termini of phage λ DNA, it acts poorly on 5′ termini recessed for more than 100 nucleotides (Sriprakash et al. 1975) and also on 5′-hydroxyl termini (Little 1967). The enzyme has no requirement for ATP. The enzyme will not act at nicks, but it will bind to them (Radding and Carter 1971). The enzyme is often purified as a complex with the product of the *red β* gene, the 28K beta protein (Radding and Shreffler 1966). The function of this protein is an enigma, although it appears to increase the affinity of the λ exonuclease for DNA. Perhaps the recent discovery of DNA pairing reactions promoted by *recA* protein, DNA synaptases, and topoisomerases will revive interest in determining the function of beta protein.

It is not at all obvious how the enzymatic properties of λ exonuclease are related to its functions in vivo. Under replication-blocked conditions, λ recombination takes place due to mismatch repair in extremely long regions of heteroduplex DNA (Russo 1973; White and Fox 1974; Fox et al. 1979). The exonuclease could presumably create very long single-stranded tails by a processive attack on one of the 5′ termini. The resulting 3′ tail would then have to invade another duplex molecule to form a heteroduplex. It is uncertain whether long heteroduplex regions exist during normal λ DNA replication. In fact, Gussin et al. (1980) have shown that under normal conditions of DNA replication, mismatch correction contributes little to the recombination of closely linked markers.

As already mentioned, λ exonuclease is capable of promoting strand assimilation, although it is not clear whether this reaction occurs during λ recombination. A self-regulating feature of the strand assimilation reaction is attributable to the fact that λ exonuclease cannot digest DNA at a nick. Thus, when a single strand has been completely assimilated, digestion by the exonuclease ceases, assuring that a gap will not be created in front of the incoming strand. On the other hand, the enzyme's inability to act at nicks would make it unsuitable for creating gapped substrates that are postulated intermediates in other models of genetic recombination (Broker and Lehman 1971; Broker 1973).

Exonuclease VIII. Exonuclease VIII was discovered as a result of the selection of suppressors of mutations of the *recB* and *recC* genes of *E. coli* (Barbour et al. 1970; Templin et al. 1972; Kushner et al. 1974). The nuclease is thought to be the product of the *recE* gene (Gillen et al. 1981), which is in turn contained in a cryptic lambdoid prophage (the "*rac*" prophage; Low 1973; Kaiser and Murray 1979, 1980) harbored by several strains of *E. coli* K12. The *sbcA* mutation results in constitutive expression of exonuclease VIII, presumably due to the deletion of the repressor of the *rac* prophage. Pseudorevertants of recombination-deficient λ phage can be selected ("λ reverse") that have acquired new recombination functions, including the ability to synthesize a new exonuclease, which is related to λ exonuclease and is probably exonuclease VIII (Gottesman et al. 1974; Gillen et al. 1977). *sbcA*$^-$ bacteria support very high levels of plasmid recombination that are independent of the *recA* and *recF* functions (Fishel et al. 1981).

Exonuclease VIII has been extensively purified and studied in the laboratory of R. Kolodner (pers. comm.). The enzyme is enormous, being apparently composed of two to four identical subunits of molecular weight 160,000. It is a 5′-specific, double-strand-specific exonuclease, which, like λ exonuclease, is very highly processive and releases 5′-nucleotides. It has absolutely no activity at nicks or gaps. Unlike λ exonuclease, exonuclease VIII will degrade DNA containing a 5′-hydroxyl end and will efficiently degrade DNA containing recessed 5′-termini. It is ATP-independent.

Biologically, the enzyme appears able to substitute for both the *recBC* nuclease of *E. coli* and λ exonuclease, although in the latter case the λ reverse phage appears to have acquired additional recombination functions (Gottesman et al. 1974). It will be interesting to see whether additional properties of exonuclease VIII emerge and whether the enzyme plays any role in recombination in *sbcA*$^+$ *E. coli*.

T7 Exonuclease. This enzyme is the product of T7 gene *6* (Kerr and Sadowski 1972a) and is an ATP-independent, 5′-specific, double-strand-specific exonuclease (Kerr and Sadowski 1972b). The enzyme has more recently been shown to have an RNaseH activity (Shinozaki and Okazaki 1978) and thereby has been implicated in the removal of RNA primers during DNA replication. This exonuclease is essential for T7 development and is required for the destruction of host DNA to provide nucleotides for phage DNA replication (Sadowski and Kerr 1970). The enzyme is also required for genetic recombination in vivo (Powling and Knippers 1974; Kerr and Sadowski 1975; Miller et al. 1976; Lee and Sadowski 1981) and also for the maintenance of the integrity of concatemeric DNA formed during DNA replication (Miller et al. 1976). It is not clear which of the above functions is the most important to maintain the viability of the phage.

T7 exonuclease, like λ exonuclease and exonuclease VIII, is a 5′ exonuclease that will degrade double-stranded DNA to about 50% completion. Since these enzymes all have a strong preference for double-stranded DNA, hydrolysis ceases when the molecule becomes entirely single-stranded. However, the gene-*6* exonuclease is much less fastidious about the DNA substrate than the two preceding enzymes. It acts equally well whether the 5′ terminus is phosphorylated or not and it acts quite happily at nicks as well as at external termini. Finally, the gene-*6* exonuclease acts ''distributively'' rather than processively. This means that it dissociates from the substrate between hydrolytic events, rather than completely hydrolyzing one molecule before moving on to the next (Kerr and Sadowski 1972b; Thomas and Olivera 1978).

At physiological salt concentrations, T7 exonuclease is markedly stimulated by T7 DNA-binding protein (Reuben and Gefter 1973; Sadowski et al. 1980), the product of T7 gene *2.5* (Dunn and Studier 1981). This effect appears to be due to a combined effect on K_m and V_{max} and is specific both for the T7 exonuclease and the T7 DNA-binding protein. It appears not to be due to the formation of either 3′ single-stranded tails or recombination intermediates (L. Roberts and P.D. Sadowski, in prep.).

T7 exonuclease is required for recombination promoted by T7 extracts in vitro (Sadowski and Vetter 1976; Sadowski 1977; Vlachopoulou and Sadowski 1977; Roeder and Sadowski 1979). In the two-stage recombination/packaging assay developed by Roeder and Sadowski, the exonuclease was required for both the wild-type and exonuclease pathways of recombination. We are now attempting to reconstruct the events in these pathways using purified components.

Treatment of T7 DNA with the exonuclease alone yields a tailed substrate that can anneal through the terminally redundant ends (Langman et al. 1978), in the presence of the T7 DNA-binding protein, to form concatemeric structures. The T7 DNA-binding protein has been shown to promote reannealing of complementary single strands of DNA (Sadowski et al. 1980). If the T7 DNA substrate contains nicks, the same two proteins cause the formation of highly complex, branched, rapidly sedimenting intermediates that resemble similar structures described in vivo (Serwer 1974; Paetkau et al. 1977). Also seen are H-shaped intermediates, which are thought to arise from annealing of complementary single-stranded DNA located in gaps produced by the exonuclease. These structures were originally seen as recombination intermediates in T4-infected cells (Broker and Lehman 1971) and have subsequently been found after T7 infection in vivo (Tsujimoto and Ogawa 1977, 1978) and after treatment of nicked circular DNAs with T7 extracts in vitro (Ogawa et al. 1979). Araki and Ogawa (1981) have recently used a filter binding assay to demonstrate that the T7 exonuclease and DNA-binding protein promote pairing of complementary duplex DNAs.

A similar role for the exonuclease and DNA-binding protein of phage T3 in the generation of concatemeric DNA, which is actively packaged in vitro, has been shown by Fujisawa et al. (1980) and Yamagishi et al. (1981).

A common feature of the above three ATP-independent exonucleases is that they all act from the 5′ terminus of DNA. The significance of this observation is unclear. However, it may mean that the resulting 3′ single-stranded terminus is somehow a preferred substrate in the strand invasion reaction. Recently Radding's group (Kahn et al. 1981) has found that *recA* promotes strand transfer from 3′ termini to duplex DNA in preference to the 5′ termini. Further studies, however, showed that this preferential transfer was due to vectorial transfer of the 3′ terminus promoted by *recA* protein (C.M. Radding, pers. comm.) rather than a bias in the strand invasion reaction in favor of 3′ termini (DasGupta et al. 1981). Such studies suggest that a similar kind of polarity might be worth seeking in λ- and T7-infected cells.

ATP-dependent Nucleases

The recBC *Nuclease of* E. coli. This enzyme is chosen as the prototype for a class of ATP-dependent nucleases which is also found in several other species of bacteria (see Clark 1973) and seems to have similar properties. The term "*recBC* nuclease" is perhaps more apt than exonuclease V, since this enzyme has both exonuclease and endonuclease activities.

The *recBC* nuclease consists of two subunits of 130,000 and 140,000 (Goldmark and Linn 1972). Preparations of the *recBC* nuclease also seem to contain another polypeptide of molecular weight 60,000 that appears important for the reassociation of the subunits to give active enzyme (Lieberman and Oishi 1974). The *recBC* enzyme has an endonucleolytic activity that is specific for single-stranded DNA and is stimulated by ATP (Goldmark and Linn 1972). An improved purification procedure yields an enzyme that is devoid of ATP-independent, single-strand-specific endonuclease activity (A. Karu and M. Urbauer, pers. comm.). In addition, the enzyme has a single- and double-strand exonuclease activity that requires ATP and hydrolyzes at least 20 ATP molecules per phosphodiester bond cleaved (Goldmark and Linn 1972). The enzyme acts exonucleolytically from both 3′ and 5′ termini and the products are acid-soluble oligonucleotides rather than 5′-mononucleotides produced by the ATP-independent class of exonucleases (Tanner et al. 1972). Mackay and Linn (1974) proposed that the enzyme hydrolyzed duplex DNA by a multistep mechanism which involves: (1) partial denaturation of the DNA with the consumption of ATP, (2) cutting off a single-strand by the single-strand-specific endonuclease activity of the enzyme, and (3) exonucleolytic hydrolysis of the single-stranded DNA.

Recently, another remarkable property of the *recBC* nuclease has come to light. Rosamond et al. (1979) and Taylor and Smith (1980) have shown that, in the presence of Ca^{2+} and the DNA-binding protein of *E. coli*, the enzyme has the capacity to track along DNA and produce "bubbles" of denatured DNA in its wake. These single-stranded loops are tantalizing candidates for the formation of recombination intermediates, as will be discussed below.

How do the in vitro properties of this enzyme relate to its role in vivo?

recBC mutants are sensitive to UV light and are recombination-deficient. One possible role for the enzyme is simply to degrade damaged DNA to provide the signal molecules (presumably oligonucleotides) which in turn lead to the induction of "SOS" functions (including the *recA* protein) involved in DNA repair (Witkin 1976). There is some in vitro evidence for this possibility from the work of Oishi et al. (1979), although in vivo the induction may involve other nucleases in addition to the *recBC* nuclease (Smith and Oishi 1978). The activity of the *recBC* nuclease is somehow regulated by the *recA* protein in vivo since UV irradiation of *recA*$^-$ bacteria leads to "reckless" destruction of cellular DNA by the *recBC* nuclease. The recent findings of Williams et al. (1981) that the *recA* protein protects single-stranded DNA from degradation by the *recBC* enzyme, supports this idea.

It would seem a pity if an enzyme with such a vast repertoire of functions had only a degradative role in recombination. The ability of the enzyme to generate looped single strands (Rosamond et al. 1979; Taylor and Smith 1980) suggests that these structures could be an ideal recipient for the initial stages of a strand-transfer reaction promoted by the *recA* protein. In fact, it is possible to imagine that two such complementary loops in different molecules could pair by the combined action of *recA* protein and topoisomerase I (Cunningham et al. 1981). The endonucleolytic activity of the *recBC* enzyme could be involved in the cleavage of displacement loops as demanded by the Aviemore model (Wiegand et al. 1977). The development of an in vitro system that promotes genetic recombination between plasmid molecules and is dependent upon the *recA*, *B*, *C*, and *F* gene products (Kolodner 1980) promises to shed light upon the role of these proteins in recombination in *E. coli*.

SITE-SPECIFIC RECOMBINATION

The λ Integration Reaction

The λ *int* system promotes conservative reciprocal recombination between two DNA molecules containing specialized crossover sites called attachment sites. Typically, the crossover occurs between a phage chromosome containing the phage attachment site (*att*P) and the bacterial chromosome containing the bacterial attachment site (*att*B).

Recently a great deal of information has become available regarding the structure of the attachment sites (Landy and Ross 1977) and the proteins that promote the reaction (Nash et al. 1981; Nash and Robertson 1981). The area has been well summarized in a recent review (Nash 1981) and therefore will only be briefly dealt with here.

The λ *int* system catalyzes recombination within the 15-bp core sequences that are present within both *att*B and *att*P. Mizuuchi et al. (1981) found that the recombination breaks were staggered such that a 5- or 7-bp overlap bearing 5′ extensions of parental strands was an intermediate step in the reaction.

While there is a 15-bp core common to both *att* sites, this does not mean that the cores are the only essential elements in *att*. In fact, *att*P is perhaps as long as 240 bp and *att*B at least 20–25 bp (Nash 1981).

The establishment of simple in vitro assays for the *int* reaction (e.g., Kikuchi and Nash 1978) has permitted the purification and characterization of the proteins involved. The entire reaction is carried out in vitro by two proteins: (1) the λ *int* gene product, a monomeric protein of molecular weight 40,000 (Kikuchi and Nash 1978; Kotewicz et al. 1980), and (2) an integration host factor (IHF), which is a heterodimer consisting of polypeptides of molecular weight 11,000 and 9500 (Nash and Robertson 1981). These two components are the likely products of the host *him*A and *hip* genes. The only other apparent requirements are for a supercoiled DNA substrate and a cation (Mg^{2} or spermidine).

While it is gratifying to have a reaction proceed to apparent completion in vitro, an understanding of the mechanism is aided by the detection of partial reactions or other activities of the proteins concerned. Nash and co-workers (Nash 1981) have searched for and found such activities. The *int* protein is a DNA-binding protein protecting sequences in both *att*P and *att*B in vitro. The binding of *int* to *att*P DNA is stimulated by IHF. A second property of the *int* protein is that it has an activity resembling topoisomerase I (Kikuchi and Nash 1979). This activity is not specific for *att*-containing DNA, but has led to a proposal for the involvement of the *int* protein in reciprocal recombination via a four-stranded DNA intermediate (see Nash 1981).

The λ *int* system at first glance does not seem like a typical nucleolytic enzyme because at no time are broken intermediates detectable during the reaction. However, it has recently been shown that the *int* protein demonstrates a weak cleavage reaction within the core sequence (N. Craig and H. Nash, pers. comm.). This reaction is probably a reflection of the topoisomerase-I activity, since the protein appears to remain bound to the DNA at the site of cleavage.

Other Site-specific Recombination Reactions

Recently Reed (1981b) has developed an in vitro assay for the detection of site-specific resolution of a substrate resembling a cointegrate structure. The reaction requires only the *tn*3-coded resolvase (the product of the *tnp*R gene), negatively supercoiled DNA, and magnesium ions. The product consists of catenated circular DNA molecules. The reaction proceeds via the formation of a covalent intermediate between the protein and the 5′ end of the DNA (Reed and Grindley 1981).

There are numerous other examples of site-specific recombination reactions that occur in both prokaryotes and eukaryotes. These include the inversion of DNA sequences such as the G loop of phage Mu by the *gin* gene product (Howe 1980), the phenomenon of phase variation in *Salmonella* which is promoted by the *hin* protein (Simon et al. 1980), the inversion of a segment of the 2-micron circle of yeast (Livingston and Klein 1977; Broach and Hicks 1980), and the joining of the *V* and *J* gene segments during immunodifferentiation (Sakano et al. 1980). Thus, the time seems propitious for the study of numerous important site-specific recombination reactions at the molecular level by the use of in vitro assays.

Site-specificity in General Recombination

In concluding a section on site-specific recombination, it might be appropriate to raise the possibility that even general recombination may exhibit some site-specificity. Specific sites for initiation or termination of recombination in fungi have long been implied by the finding of polarity of gene conversion frequencies (see Stahl 1979). Generally, for a series of intragenic two-factor crosses, there is a correlation between the frequency of conversion of a given marker (non-4:4 segregation) and the position of that marker relative to a fixed point. Thus, markers close to this point will give higher conversion frequencies than those further away in a given direction.

The discovery of recombination hot spots in bacteriophage λ (Stahl 1979) adds weight to the argument for site-specificity, even in general recombination. These sites are called "chi" sites and promote recombination catalyzed by the *recBC* system of *E. coli*. Chi sites have been found in *E. coli* DNA and contain a common 8-bp sequence (Smith et al. 1981). Chi-promoted recombination exhibits several remarkable properties (Stahl 1979), including the ability to act at a considerable distance and to act even opposite a long heterology. Chi sites may be involved in resolution or the formation of recombination intermediates (Stahl 1979). However, no chi-specific reactions involving the *recA*, *B*, or *C* gene products have yet been uncovered.

A new type of recombination hotspot that promotes high levels of recombination between plasmids has been discovered by Kolodner and co-workers (Kolodner et al. 1980). These elements were cloned from *E. coli* DNA. They were *cis*-acting and were distinct from chi sequences.

The ability of the gene-*3* endonuclease of phage T7 to cut the loops of hairpin structures in DNA (Panayotatos and Wells 1981) and the implication of this enzyme in genetic recombination (Powling and Knippers 1974; Kerr and Sadowski 1975; Miller et al. 1976) suggests that recombination in this phage may show site-specificity. It has recently been shown that the gene-*3* endonuclease apparently cuts T7 DNA in vivo at specific sites if the gene-*6* exonuclease is absent (D. Lee and P.D. Sadowski, in prep.).

ACKNOWLEDGMENTS

I thank Andy Becker, Richard Kolodner, Howard Nash, Charles Radding, and Frank Stahl for their critical reading of the manuscript and Josie Chapman-Smith for typing it. I am a Career Investigator of the Medical Research Council of Canada.

REFERENCES

Ahmad, A., W.K. Holloman, and R. Holliday. 1975. Nuclease that preferentially inactivates DNA containing mismatched bases. *Nature* **258:**54.

Alberts, B., ed. 1980. Mechanistic studies of DNA replication and genetic recombination. *ICN-UCLA Symp. Mol. Cell. Biol.* **19.**

Araki, H. and H. Ogawa. 1981. The participation of T7 DNA-binding protein in T7 genetic recombination. *Virology* **111:**509.

Arthur, A. and D. Sherratt. 1979. Dissection of the transposition process: A transposon-encoded site-specific recombination system. *Mol. Gen. Genet.* **175:**267.

Barbour, S.D., H. Nagaishi, A. Templin, and A.J. Clark. 1970. Biochemical and genetic studies of recombination proficiency in *Escherichia coli*. II. *Rec*$^+$ revertants caused by indirect suppression of *Rec*$^-$ mutations. *Proc. Natl. Acad. Sci.* **67:**128.

Broach, J.R. and J.B. Hicks. 1980. Replication and recombination functions associated with the yeast plasmid 2μ circle. *Cell* **21:**501.

Broker, T.R. 1973. An electron microscope analysis of pathways for bacteriophage T4 recombination. *J. Mol. Biol.* **81:**1.

Broker, T.R. and I.R. Lehman. 1971. Branched DNA molecules: Intermediates in T4 recombination. *J. Mol. Biol.* **60:**131.

Burck, K. and R.C. Miller, Jr. 1978. Marker rescue and partial replication of bacteriophage T7 DNA. *Proc. Natl. Acad. Sci.* **75:**6144.

Calos, M.P. and J.H. Miller. 1980. Transposable elements. *Cell* **20:**579.

Campbell, A. 1981. Some general questions about movable elements and their implications. *Cold Spring Harbor Symp. Quant. Biol.* **45:**1.

Carter, D.M. and C.M. Radding. 1970. The role of exonuclease and β protein of phage λ in genetic recombination. *J. Biol. Chem.* **246:**2502.

Cassuto, E. and C.M. Radding. 1971. Mechanism for the action of λ exonuclease in genetic recombination. *Nature New Biol.* **229:**13.

Center, M.S. and C.C. Richardson. 1970a. An endonuclease induced after infection of *Escherichia coli* with bacteriophage T7. I. Purification and properties of the enzyme. *J. Biol. Chem.* **245:**6285.

__________ . 1970b. An endonuclease induced after infection of *Escherichia coli* with bacteriophage T7. II. Specificity of the enzyme toward single- and double-stranded deoxyribonucleic acid. *J. Biol. Chem.* **245:**6292.

Center, M.S., F.W. Studier, and C.C. Richardson. 1970. The structural gene for a T7 endonuclease essential for phage DNA synthesis. *Proc. Natl. Acad. Sci.* **65:**242.

Clark, A.J. 1971. Towards a metabolic interpretation of genetic recombination of *E. coli* and its phages. *Ann. Rev. Microbiol.* **25:**437.

Clark, A.J. 1973. Recombination deficient mutants of *E. coli* and other bacteria. *Ann. Rev. Genet.* **7:**67.

Cunningham, R.P., A.M. Wu, T. Shibata, C. DasGupta, and C.M. Radding. 1981. Homologous pairing and topological linkage of DNA molecules by combined action of *E. coli recA* protein and topoisomerase I. *Cell* **24:**213.

DasGupta, C., A.M. Wu, R. Kahn, R.P. Cunningham, and C.M. Radding. 1981. Concerted strand exchange and formation of Holliday structures by *E. coli recA* protein. *Cell* **25:**507.

Dunn, J.J. and F.W. Studier. 1981. Nucleotide sequence from the genetic left end of bacteriophage T7 DNA to the beginning of gene 4. *J. Mol. Biol.* **148:**303.

Fishel, R.A., A.A. James, and R. Kolodner. 1981. *RecA*-independent general genetic recombination of plasmids. *Nature* **294:**184.

Fox, M.S., C.S. Dudney, and E.J. Sodergren. 1979. Heteroduplex regions in unduplicated bacteriophage λ recombinants. *Cold Spring Harbor Symp. Quant. Biol.* **43:**999.

Fujisawa, H., M. Yamagishi, H. Matsuo-kato, and T. Minagawa. 1980. Purification of DNA-binding proteins of bacteriophage T3 and their role in in vitro packaging of phage T3 DNA. *Virology* **105:**480.

Gillen, J.R., A.E. Kary, H. Nagaishi, and A.J. Clark. 1977. Characterization of the deoxyribonuclease determined by λ reverse as exonuclease VIII of *Escherichia coli*. *J. Mol. Biol.* **113:**27.

Gillen, J.R., D.K. Willis, and A.J. Clark. 1981. Genetic analysis of the *recE* pathway of genetic recombination in *Escherichia coli* K-12. *J. Bacteriol.* **145:**521.

Goldmark, P.J. and S. Linn. 1972. Purification and properties of the *recBC* DNase of *Escherichia coli* K12. *J. Biol. Chem.* **247:**1849.

Gottesman, M.M., M.E. Gottesman, S. Gottesman, and M. Gellert. 1974. Characterization of bacteriophage λ reverse as an *Escherichia coli* phage carrying a unique set of host-derived recombination functions. *J. Mol. Biol.* **88:**471.

Gussin, G.N., E.D. Rosen, and D.L. Wulff. 1980. Mapability of very close markers of bacteriophage λ. *Genetics* **96:**1.

Holliday, R. 1964. A mechanism for gene conversion in fungi. *Genet. Res.* **5:**282.

Holliday, R., W.K. Holloman, G.R. Banks, P. Unrau, and J.E. Pugh. 1974. Genetic and biochemical studies of recombination in *Ustilago maydis*. In *Mechanisms in recombination* (ed. R.F. Grell), p. 239. Plenum Press, New York.

Holloman, W.K. 1973. Studies on a nuclease from *Ustilago maydis*. II. Substrate specificity and mode of action of the enzyme. *J. Biol. Chem.* **248:**8114.

Holloman, W.K. and R. Holliday. 1973. Studies on a nuclease from *Ustilago maydis*. I. Purification, properties and implication in recombination of the enzyme. *J. Biol. Chem.* **248:**8107.

Holloman, W.K., R. Wiegand, C. Hoessli, and C.M. Radding. 1975. Uptake of homologous single-stranded fragments by superhelical DNA: A possible mechanism for initiation of genetic recombination. *Proc. Natl. Acad. Sci.* **72:**2394.

Kahn, R., R.P. Cunningham, C. DasGupta, and C.M. Radding. 1981. The polarity of heteroduplex formation promoted by *E. coli recA* protein. *Proc. Natl. Acad. Sci.* **78:** (in press).

Kaiser, K. and N.E. Murray. 1979. Physical characterization of the "Rac prophage" in *E. coli* K12. *Mol. Gen. Genet.* **175:**159.

Kaiser, K. and N.E. Murray. 1980. On the nature of *sbcA* mutations in *E. coli* K12. *Mol. Gen. Genet.* **179:**555.

Kerr, C. and P.D. Sadowski. 1972a. Gene 6 exonuclease of bacteriophage T7. I. Purification and properties of the enzyme. *J. Biol. Chem.* **247:**305.

———. 1972b. Gene 6 exonuclease of bacteriophage T7. II. Mechanism of the reaction. *J. Biol. Chem.* **247:**311.

———. 1975. The involvement of genes 3, 4, 5 and 6 in genetic recombination in bacteriophage T7. *Virology* **65:**281.

Kikuchi, Y. and H.A. Nash. 1978. The bacteriophage λ *int* gene product. A filter assay for genetic recombination, purification of *int* and specific binding to DNA. *J. Biol. Chem.* **253:**7149.

———. 1979. Nicking-closing activity associated with bacteriophage λ *int* gene product. *Proc. Natl. Acad. Sci.* **76:**3760.

Kolodner, R. 1980. Genetic recombination of bacterial plasmid DNA: Electron microscopic analysis of *in vitro* intramolecular recombination. *Proc. Natl. Acad. Sci.* **77:**4847.

Kolodner, R., J. Joseph, A. James, F. Dean, and M.J. Doherty. 1980. Genetic recombination of bacterial plasmid DNAs *in vivo* and *in vitro*. In *Mechanistic studies of DNA replication and genetic recombination* (ed. B. Alberts), p. 933. Academic Press, New York.

Kotewicz, M., E. Grzesiuk, W. Courschesne, R. Fischer, and H. Echols. 1980. Purification and characterization of the integration protein specified by bacteriophage λ. *J. Biol. Chem.* **255:**2433.

Kushner, S., H. Nagaishi, and A.J. Clark. 1974. Isolation of exonuclease VIII: The enzyme associated with the *sbcA* indirect suppressor. *Proc. Natl. Acad. Sci.* **71:**3593.

Landy, A. and W. Ross. 1977. Viral integration and excision: Structure of the lambda *att* sites. *Science* **197:**1147.

Langman, L., V. Paetkau, D. Scraba, R.C. Miller, Jr., G.S. Roeder, and P.D. Sadowski. 1978.

The structure and maturation of intermediates in bacteriophage T7 DNA replication. *Can. J. Biochem.* **56:**508.

Lee, D. and P.D. Sadowski. 1981. Genetic recombination of bacteriophage T7 *in vivo* studied by use of a simple physical assay. *J. Virol.* **40:**839.

Lee, M., R.C. Miller, Jr., D. Scraba, and V. Paetkau. 1976. The essential role of bacteriophage T7 endonuclease (gene 3) in molecular recombination. *J. Mol. Biol.* **104:**883.

Levinthal, C. 1954. Recombination in phage T2: Its relationship to heterozygosis and growth. *Genetics* **39:**169.

Lieberman, R.P. and M. Oishi. 1974. The *recBC* deoxyribonuclease of *Escherichia coli*: Isolation and characterization of the subunit proteins and reconstitution of the enzyme. *Proc. Natl. Acad. Sci.* **71:**4816.

Little, J.W. 1967. An exonuclease induced by bacteriophage λ. II. Nature of the enzyme reaction. *J. Biol. Chem.* **242:**679.

Little, J.W., I.R. Lehman, and A.D. Kaiser. 1967. An exonuclease induced by bacteriophage λ. I. Preparation of the crystalline enzyme. *J. Biol. Chem.* **242:**672.

Livingston, D.M. and H.L. Klein. 1977. Deoxyribonucleic acid sequence organization of a yeast plasmid. *J. Bacteriol.* **129:**472.

Low, B. 1973. Restoration by the *rac* locus of recombination forming ability in *recB*$^-$ and *recC*$^-$ merozygotes of *Escherichia coli* K-12. *Mol. Gen. Genet.* **122:**119.

MacKay, V. and S. Linn. 1974. The mechanism of degradation of duplex deoxyribonucleic acid by the *recBC* enzyme of *Escherichia coli* K-12. *J. Biol. Chem.* **249:**4286.

McEntee, K., G.H. Weinstock, and I.R. Lehman. 1979. Initiation of general recombination catalyzed *in vitro* by the *recA* protein of *Escherichia coli*. *Proc. Natl. Acad. Sci.* **76:**2615.

Meselson, M.S. and C.M. Radding. 1975. A general model for genetic recombination. *Proc. Natl. Acad. Sci.* **72:**358.

Meselson, M. and F.W. Stahl. 1958. The replication of DNA in *Escherichia coli*. *Proc. Natl. Acad. Sci.* **44:**671.

Meselson, M. and J.J. Weigle. 1961. Chromosome breakage accompanying genetic recombination in bacteriophage. *Proc. Natl. Acad. Sci.* **47:**857.

Miller, R.C. Jr., M. Lee, D.G. Scraba, and V. Paetkau. 1976. The role of bacteriophage T7 exonuclease (gene 6) in genetic recombination and production of concatemers. *J. Mol. Biol.* **101:**223.

Mizuuchi, K., R. Weisberg, L. Enquist, M. Mizuuchi, M. Buracynska, C. Foeller, P.-L. Hsu, W. Ross, and A. Landy. 1981. Structure and function of the phage λ *att* site: Size, *int* binding sites, and location of the crossover point. *Cold Spring Harbor Symp. Quant. Biol.* **45:**429.

Nash, H.A. 1981. Integration and excision of bacteriophage λ: The mechanism of conservative site-specific recombination. *Ann. Rev. Genet.* **15:**143.

Nash, H.A., K. Mizuuchi, L.W. Enquist, and R. Weisberg. 1981. Strand exchange in λ integrative recombination: Genetics, biochemistry, and models. *Cold Spring Harbor Symp. Quant. Biol.* **45:**417.

Nash, H.A. and C.A. Robertson. 1981. Purification and properties of the *E. coli* protein factor required for λ integrative recombination. *J. Biol. Chem.* **256:**9246.

Ogawa, H., H. Araki, and Y. Tsujimoto. 1979. Recombination intermediates formed in the extract from T7-infected cells. *Cold Spring Harbor Symp. Quant. Biol.* **43:**1033.

Oishi, M., C.L. Smith, and B. Friefeld. 1979. Molecular events and molecules that lead to induction of prophage and SOS functions. *Cold Spring Harbor Symp. Quant. Biol.* **43:**879.

Paetkau, V., L. Langman, R. Bradley, D. Scraba, and R.C. Miller, Jr. 1977. Folded catenated genomes as replication intermediates of bacteriophage T7 DNA. *J. Virol.* **22:**130.

Panayotatos, N. and R.D. Wells. 1981. Cruciform structures in supercoiled DNA. *Nature* **289:**466.

Potter, H. and D. Dressler. 1980. DNA synaptase: An enzyme that fuses DNA molecules at a region of homology. *Proc. Natl. Acad. Sci.* **77:**2390.

Powling, A. and R. Knippers. 1974. Some functions involved in bacteriophage T7 genetic recombination. *Mol. Gen. Genet.* **134:**173.

Radding, C.M. 1966. Regulation of λ exonuclease. I. Properties of λ exonuclease purified from lysogens of λ_{T11} and wild-type. *J. Mol. Biol.* **18:**235.

_________ . 1978. Genetic recombination: Strand transfer and mismatch repair. *Ann. Rev. Biochem.* **47:**847.

_________ . 1981. Recombination activities of *E. coli recA* protein. *Cell* **25:**3.

Radding, C.M. and D.M. Carter. 1971. The role of exonuclease and β protein of phage λ in genetic recombination. III. Binding to deoxyribonucleic acid. *J. Biol. Chem.* **246:**2513.

Radding, C.M. and D.C. Shreffler. 1966. Regulation of λ exonuclease. II. Joint regulation of exonuclease and a new λ antigen. *J. Mol. Biol.* **18:**251.

Reed, R.R. 1981a. Resolution of cointegrates between transposons γδ and Tn3 define the recombination site. *Proc. Natl. Acad. Sci.* **78:**3428.

_________ . 1981b. Transposon-mediated site-specific recombination: A defined *in vitro* system. *Cell* **25:**713.

Reed, R.R. and Gridley, N.D.F. 1981. Transposon mediated site-specific recombination *in vitro*: DNA cleavage and protein-DNA linkage at the recombination site. *Cell* **25:**721.

Reuben, R.C. and M.L. Gefter. 1973. A DNA-binding protein induced by bacteriophage T7. *Proc. Natl. Acad. Sci.* **70:**1846.

Roeder, G.S. and P.D. Sadowski. 1979. Pathways of recombination of bacteriophage T7 DNA *in vitro*. *Cold Spring Harbor Symp. Quant. Biol.* **43:**1023.

Rosamond, J., K.M. Telander, and S. Linn. 1979. Modulation of the action of the *recBC* enzyme of *Escherichia coli* K-12 by Ca^{2+}. *J. Biol. Chem.* **254:**8646.

Russo, V.E.A. 1973. On the physical structure of λ recombinant DNA. *Mol. Gen. Genet.* **122:**533.

Sadowski, P.D. 1971. Bacteriophage T7 endonuclease I. Properties of the enzyme purified from T7 phage-infected *Escherichia coli* B. *J. Biol. Chem.* **246:**209.

_________ . 1977. Genetic recombination of bacteriophage T7 DNA *in vitro*. II. Further properties of the *in vitro* recombination-packaging reaction. *Virology* **78:**192.

Sadowski, P.D., W. Bradley, D. Lee, and L. Roberts. 1980. Genetic recombination of bacteriophage T7 DNA *in vitro*. In *Mechanistic studies of DNA replication and genetic recombination* (ed. B. Alberts), p. 941. Academic Press, New York.

Sadowski, P.D. and C. Kerr. 1970. Degradation of *Escherichia coli* B deoxyribonucleic acid after infection with deoxyribonucleic acid-defective amber mutants of bacteriophage T7. *J. Virol.* **6:**149.

Sadowski, P.D. and D. Vetter. 1976. Genetic recombination of bacteriophage T7 DNA *in vitro*. *Proc. Natl. Acad. Sci.* **73:**692.

Sakano, H., R. Maki, Y. Kurosawa, R. Roeder, and S. Tonegawa. 1980. Two types of somatic recombination are necessary for the generation of complete immunoglobulin heavy-chain genes. *Nature* **286:**676.

Serwer, P. 1974. Fast sedimenting bacteriophage T7 DNA from T7-infected *Escherichia coli*. *Virology* **59:**70.

Shapiro, J.A. 1979. Molecular model for the transposition and replication of bacteriophage Mu and other transposable elements. *Proc. Natl. Acad. Sci.* **76:**1933.

Shenk, T.E., C. Rhodes, P.W.J. Rigby, and P. Berg. 1975. Biochemical method for mapping mutational alterations in DNA with S1 nuclease: The location of deletions and temperature-sensitive mutations in simian virus 40. *Proc. Natl. Acad. Sci.* **72:**989.

Sherratt, D.J., A. Arthur, and M. Burke. 1981. Transposon-specified, site-specific recombination systems. *Cold Spring Harbor Symp. Quant. Biol.* **45:**275.

Shibata, T., C. DasGupta, R.P. Cunningham, and C.M. Radding. 1979. Purified *Escherichia coli recA* protein catalyzes homologous pairing of superhelical DNA and single-stranded fragments. *Proc. Natl. Acad. Sci.* **76:**1638.

Shinozaki, K. and T. Okazaki. 1978. T7 gene 6 exonuclease has an RNase H activity. *Nucleic Acids Res.* **5:**4245.

Sigal, N. and B. Alberts. 1972. Genetic recombination: The nature of a crossed strand exchange between two homologous DNA molecules. *J. Mol. Biol.* **71:**789.

Simon, M., J. Zeig, M. Silverman, G. Mandel, and R. Doolittle. 1980. Phase variation: Evolution of a controlling element. *Science* **209:**1370.

Smith, R.D. and R.J. Miller. 1981. Replication and plasmid-bacteriophage recombination. I. Marker rescue analysis. *Virology* **115:**223–236.

Smith, C.L. and M. Oishi. 1978. Early events and mechanisms in the induction of bacterial SOS functions: Analysis of the phage repressor inactivation process *in vivo*. *Proc. Natl. Acad. Sci.* **75:**1657.

Smith, G.R., M. Comb, D.W. Schultz, D.L. Daniels, and F.R. Blattner. 1981. Nucleotide sequence of the Chi recombinational hot spot X^+D in bacteriophage λ. *J. Virol.* **37:**336.

Sobell, H.M. 1974. Concerning the stereochemistry of strand equivalence in genetic recombination. In *Mechanisms in recombination* (ed. R.F. Grell), p. 433. Plenum Press, New York.

Sriprakash, K.S., N. Lundh, M.M.-O. Huh, and C.M. Radding. 1975. The specificity of λ exonuclease interactions with single-stranded DNA. *J. Biol. Chem.* **250:**5438.

Stahl, F.W. 1979. Specialized sites in generalized recombination. *Ann. Rev. Genet.* **13:**7.

Tanner, D., F.G. Nobrega, and M. Oishi. 1972. 5′-oligonucleotides as the acid-soluble products of the ATP-dependent DNase from *Escherichia coli*. *J. Mol. Biol.* **67:**513.

Taylor, A. and G.R. Smith. 1980. Unwinding and rewinding of DNA by the *recBC* enzyme. *Cell* **22:**447.

Templin, A., S.R. Kushner, and A.J. Clark. 1972. Genetic analysis of mutation indirectly suppressing *recB* and *recC* mutations. *Genetics* **72:**205.

Thomas, K.R. and Olivera, B.M. 1978. The processivity of DNA exonucleases. *J. Biol. Chem.* **253:**424.

Tsujimoto, Y. and H. Ogawa. 1977. Intermediates in genetic recombination of bacteriophage T7. *J. Mol. Biol.* **109:**423.

———. 1978. Intermediates in genetic recombination of bacteriophage T7 DNA. Biological activity and roles of gene 3 and gene 5. *J. Mol. Biol.* **125:**255.

Vlachopoulou, P.J. and P.D. Sadowski. 1977. Genetic recombination of bacteriophage T7 DNA *in vitro*. III. A physical assay for recombinant DNA. *Virology* **78:**203.

White, R.L. and M.S. Fox. 1974. On the molecular basis of high negative interference. *Proc. Natl. Acad. Sci.* **71:**1544.

Wiegand, R.C., K.L. Beattie, W.K. Holloman, and C.M. Radding. 1977. Uptake of homologous single-stranded fragments by superhelical DNA. III. The product and its enzymatic conversion to a recombinant molecule. *J. Mol. Biol.* **116:**805.

Williams, J.G.K., T. Shibata, and C.M. Radding. 1981. *E. coli recA* protein protects single-stranded DNA or gapped duplex DNA from degradation by *recBC* DNase. *J. Biol. Chem.* **256:**7573.

Witkin, E.M. 1976. Ultraviolet mutagenesis and inducible DNA repair in *Escherichia coli*. *Bacteriol. Rev.* **40:**869.

Yamagishi, M., H. Fujisawa, and T. Minagawa. 1981. Bacteriophage T3 DNA binding protein interaction with DNA. *Virology* **109:**148.

DNA Topoisomerases

James C. Wang
Department of Biochemistry and Molecular Biology
Harvard University
Cambridge, Massachusetts 02138

In the past few years, much progress has been made in the study of DNA topoisomerases. The quickened pace of the field is reflected by the number of recent reviews on this subject (Bauer 1978; Champoux 1978; Wang and Liu 1979; Cozzarelli 1980; Gellert 1981a,b; Wang 1981; Wang and Kirkegaard 1981). The purpose of this chapter is not to provide an additional comprehensive review of the literature but to give a brief account of the present status of the field and to comment on a few selected topics.

REACTIONS THAT ARE CATALYZED BY DNA TOPOISOMERASES

Some of the reactions that are catalyzed by DNA topoisomerases are illustrated in Figure 1 and are listed in Table 1. All these topoisomerizations involve the transient breakage of DNA phosphodiester bonds and the reformation of the very same linkages; reactions that involve bond exchange will be discussed in a later section.

The forward reaction of Figure 1A, the relaxation of a supercoiled DNA, has been observed for all DNA topoisomerases. The various subclasses of the enzymes differ, however, in their dependence on the sense of DNA superhelicity. The major bacterial type-I enzyme, the *topA* gene product, is effective only when the DNA is negatively supercoiled (Wang 1971). The major type-I enzyme from eukaryotes, on the other hand, relaxes negatively and positively supercoiled DNA with equal efficiency (Champoux and Dulbecco 1972). Similarly, the known ATP-dependent type-II enzymes coded by phage T4 and

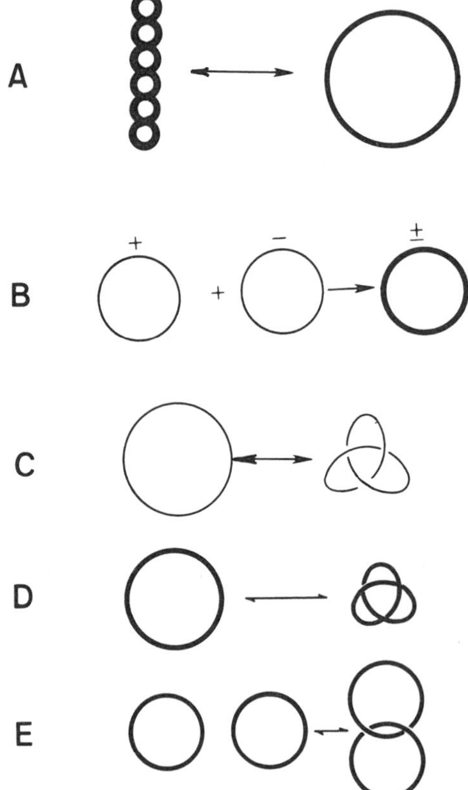

Figure 1 Topoisomerization reactions that are catalyzed by DNA topoisomerases. Single-stranded DNA is represented by a thin line and double-stranded DNA is represented by a thick line. (*A*) Relaxation and supercoiling. (*B*) Linking of single-stranded rings of complementary sequences into a duplex ring with the strands intertwined. (*C*) Knotting and unknotting of a single-stranded ring. (*D*) Knotting and unknotting of a double-stranded ring. (*E*) Catenation and decatenation of double-stranded rings. The knotted rings and catenane shown are of the simplest kinds for the sake of convenience. More complex structures are usually involved in the actual reactions. See text for details.

eukaryotes are all insensitive to the sense of DNA superhelicity (Liu et al. 1979; Stetler et al. 1979; Baldi et al. 1980; Hsieh and Brutlag 1980; Miller et al. 1981).

The reverse reaction of Figure 1A, the supercoiling of DNA, is known to be catalyzed only by bacterial gyrase (*gyrA* and *gyrB* gene products). This type-II enzyme catalyzes the negative supercoiling of DNA coupled to ATP hydrolysis (Gellert et al. 1976). In the absence of ATP, gyrase can relax negatively but not positively supercoiled DNA (Liu and Wang 1978a; Brown

Table 1 Examples of DNA topoisomerases and some of their characteristics

Type	Enzyme	Topoisomerization reaction[a]	Remarks
I	bacterial DNA topoisomerase I	relaxation of negatively supercoiled DNA; B; C; E only if one of the reacting pair has a nick; D not tested but expected to occur if nick is present	*topA* gene product; Mg^{2+} required; enzyme linked to 5′P when DNA is transiently cleaved
I	mouse DNA topoisomerase I or other known eukaryotic type-I enzymes	relaxation of negatively or positively supercoiled DNA; B–E same as above	Mg^{2+} stimulated; enzyme linked to 3′P when DNA is transiently cleaved
I	phage λ integrase	relaxation of negatively supercoiled DNA; B–E not tested	*int* gene product; normally catalyzes site-specific recombination that involves DNA strand exchanges; enzyme linked to 3′P when DNA is transiently cleaved
II	bacterial DNA gyrase (DNA topoisomerase II)	negative supercoiling; reverse of D; E; B, C, and forward of D not tested	*gyrA* and *gyrB* gene product; Mg^{2+} required; ATP-dependent; relaxes negatively supercoiled DNA slowly without ATP; enzyme linked to 5′P when DNA is transiently broken
II	phage T4 DNA topoisomerase	relaxation of negatively or positively supercoiled DNA; reverse of D; E; B and C not tested	product of genes *39, 52, 60;* Mg^{2+} required; ATP-dependent; forward of D observed at high enzyme level without ATP
II	*Drosophila* DNA topoisomerase II or other known eukaryotic type-II enzymes	same as above	Mg^{2+} required; ATP-dependent; forward of D not tested (forward of D observed at high enzyme level with ATP [T-S. Hsieh, pers. comm.])

[a]B, C, D, and E refer to the reactions shown in Fig. 1.

et al. 1979). The addition of a nonhydrolyzable β, γ-imido analog of ATP to gyrase, however, accelerates its relaxation of positively supercoiled DNA (Gellert et al. 1981). An activity that is probably a variant of the normal gyrase with part of the B subunit missing, DNA topoisomerase II′, has been found to relax both negatively and positively supercoiled DNA (Brown et al. 1979; Gellert et al. 1979). These findings are likely to be significant in the eventual elucidation of the mechanisms of DNA topoisomeration.

Reactions B and C (Fig. 1), the linking of single-stranded DNA rings of complementary sequences and the knotting/unknotting of single-stranded DNA rings, respectively, are known to be catalyzed by some of the type-I enzymes (Liu et al. 1976; Champoux 1977a; Kirkegaard and Wang 1978). Reaction B is a special case of reaction A, since the basic parameter that is changing in both reactions is the linking number. The driving force for this reaction is clearly base-pairing. It has been argued that the driving force for the formation of a single-stranded knot is also base-pairing (Liu et al. 1976). Knotting is envisioned as the intramolecular version of the intermolecular catenation: the single-stranded ring is assumed to be divided into two or more loops as a result of random strand crossover or intramolecular base-pairing.

The reverse of reaction D (Fig. 1), the unknotting of a knotted duplex DNA ring, has been observed for all type-II DNA topoisomerases (Liu et al. 1979, 1980; Mizuuchi et al. 1980) and is becoming the reaction of choice for the assay of this type of enzyme (Liu et al. 1981b). The forward reaction has been observed for the T4 enzyme under rather special conditions: supercoiled DNA, no ATP, and excess enzyme (Liu et al. 1979). These conditions are not unreasonable in view of the type of mechanisms first proposed for the knotting of a single-stranded ring. In the double-stranded case, supercoiling of the DNA presumably helps to divide the molecule into domains and the intramolecular linking of the domains, catalyzed by a very low level of topoisomerase activity in the absence of ATP, yields the knotted duplex ring. If ATP is present, rapid relaxation of the DNA occurs and the relaxed DNA gives knotted rings with a very low probability. Knot formation by DNA gyrase will be discussed in a later section.

Reaction E (Fig. 1), the catenation/decatenation of duplex DNA rings, has also been observed for all type-II DNA topoisomerases (Baldi et al. 1980; Hsieh and Brutlag 1980; Kreuzer and Cozzarelli 1980; Mizuuchi et al. 1980). At the DNA concentrations generally employed, catenation is thermodynamically unfavorable (Wang and Schwartz 1967). Therefore, the forward reaction in general requires the presence of an agent that promotes intermolecular association of DNA. Polyamines or certain proteins are usually required for the catenation reaction. The reverse reaction, the decatenation of the rings, is readily catalyzed by the type-II enzymes in the absence of the DNA "condensing agent."

It has been found that reaction E is also catalyzed by the major type-I topoisomerases from bacterial sources, provided that there is at least one preexisting nick in one of the pair of reacting rings (Tse and Wang 1980; Brown and Cozzarelli 1981). Brown and Cozzarelli (1981) also reported that type-I topoisomerases from eukaryotes can catalyze this reaction.

THE ASSAY OF DNA TOPOISOMERASES

Various assays for detecting DNA topoisomerases have been reviewed previously (Wang and Kirkegaard 1981). One frequently encountered problem is the detection of a type-II activity in the presence of excess type-I activity. From the discussions in the section above, it is clear that the two most useful reactions in this regard are the decatenation and unknotting of covalently closed, double-stranded DNA rings. A particularly convenient procedure that has been developed recently uses knotted DNA rings obtained from tailless phage P4 capsids (Liu et al. 1981b). It has been observed that the DNA extracted from tailless capsids of phage P2 or P4 is in the form of complex knots (Liu et al. 1981a,b). These knotted rings are not covalently closed; there are two staggered single-chain breaks in each molecule where the cohesive ends join. Because of the complexity of the knots, these molecules migrate much faster during electrophoresis in an agarose gel and are readily resolved from nicked rings without knots. All type-II DNA topoisomerases tested can unknot these complex knots; bacterial type-I enzymes show a very low level of unknotting activity with this substrate and eukaryotic type-I enzymes show virtually no activity. If these knotted rings are first incubated with DNA ligase to give the covalently closed form, then no type-I enzyme can untie the knots. In this laboratory, we routinely use a mixture of the nicked, knotted P4 DNA and a supercoiled plasmid DNA of a smaller size as the substrate; the simultaneous monitoring of relaxation and unknotting activities is advantageous during the isolation and purification of DNA topoisomerases.

SOME TOPOLOGICAL AND MECHANISTIC CONSIDERATIONS

Directionality of Strand Passage

One of the most significant recent developments in the study of DNA topoisomerases is the realization that catalysis of topoisomerization by the type-II enzymes involves transient double-stranded breakage and the passage of a double-stranded segment through the break (Brown and Cozzarelli 1979; Kreuzer and Cozzarelli 1980; Liu et al. 1980).

The directionality of the duplex passage event is of interest. Consider the supercoiling/relaxation reaction depicted in Figure 1. If the passage is nondirectional, a particular breakage-passage-reclosure event may either increase or decrease the linking number in steps of two (Fuller 1978), but eventually the DNA will relax into the thermodynamic valley. In the case of the negative supercoiling of DNA, catalyzed by gyrase and coupled to ATP hydrolysis, the passage must be directional. It is of interest to examine how this directionality might be achieved.

Figure 2a illustrates a crossover of two duplex segments: the complementary strands are shown as thick and thin lines and the antiparallel nature of the strands is indicated by the arrows. One can count the contribution of such a duplex crossing in a covalently closed DNA to the linking number. The duplex crossover contains two crossovers between complementary strands. The sign of the D′C′ and AB cross is positive according to the accepted convention (Crick

1976; Bauer et al. 1980), since a clockwise rotation is required to align the top piece, D′C′, with the lower piece AB (Bauer et al. 1980). Similarly, the CD and B′A′ cross is also positive. Thus, the duplex crossover contributes +2/2 or +1 to the linking number of the covalently closed circular molecule.

The above exercise gives the impression that a duplex crossover has an intrinsic sign and that an enzyme can in principle recognize this sign. It should be noted, however, that in this example the sign can be specified only because the complementary strands are specified. If CD and C′D′ are interchanged, for example, the duplex crossover contributes −1 to the linking number rather

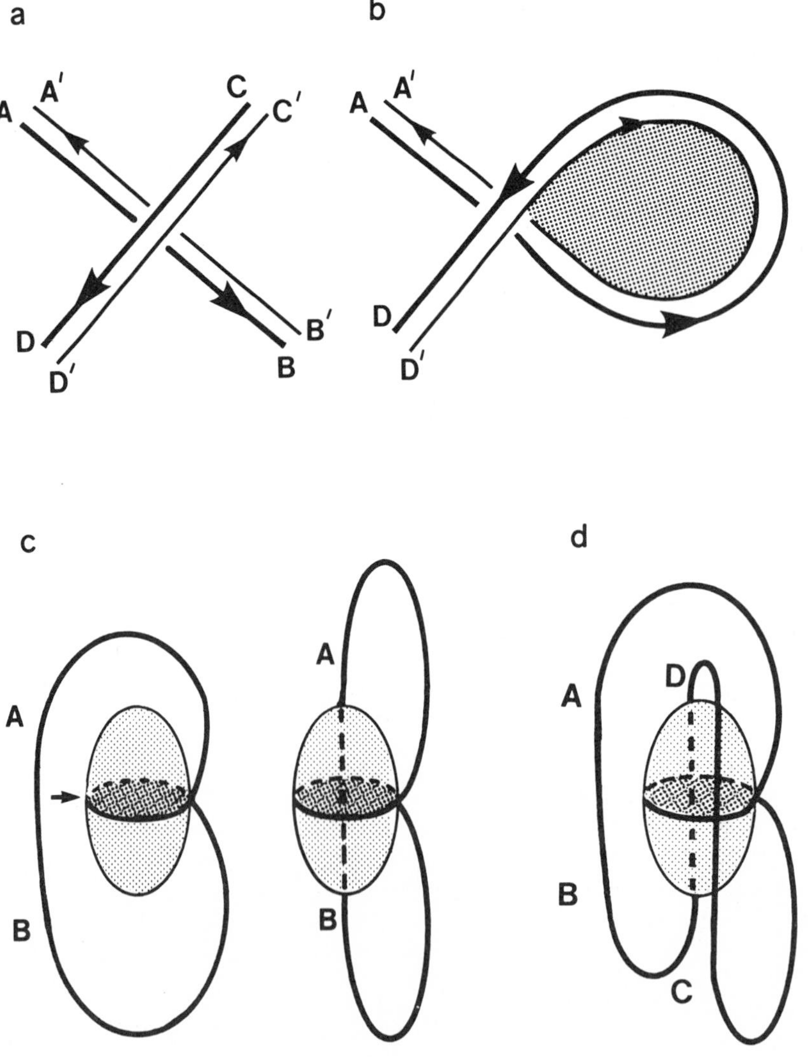

Figure 2 (See facing page for legend.)

than +1. In other words, so far as the DNA backbone chains are concerned, the enzyme can recognize the intrinsic sign of a duplex crossover only if it can sense that AB and CD are connected in the circular molecule (i.e., they are segments of the same strand).

It is possible, of course, that the enzyme is not recognizing the DNA backbone chains, but rather some unique sequences of the two duplex segments. The difficulty of a sequence-imposed sign is that the inversion of one sequence relative to the other would result in an ATP hydrolysis-coupled positive supercoiling of the DNA, which has never been observed.

The above considerations suggest that the interaction between a type-II topoisomerase and an isolated duplex crossover could be sufficient for those enzymes that can catalyze only relaxation but not supercoiling of DNA. Therefore, for gyrase, it is highly unlikely that the enzyme is recognizing an isolated crossover.

The wrapping of the DNA around the enzyme provides a solution to the directionality problem since the coiling of the DNA around the enzyme imposes a unique handedness. For example, if the crossover where duplex

Figure 2 (*a*) A crossover of two duplex segments. So far as the backbone chains are concerned, the crossover has a unique sign only if the complementary strands are recognizable (see text for details). In other words, an enzyme molecule can recognize the sign of the crossover only if it can somehow sense that AB is connected to CD and C′D′ is connected to A′B′. (*b*) The wrapping of a DNA segment around an enzyme molecule, represented by the crosshatched area, provides a way to specify the sign of the crossover. The DNA segment is shown wrapped in a right-handed way; a breakage-passage-reclosure event at the crossover converts the right-handed wrap to the left-handed wrap. (*c*) A graphic illustration of a more elaborate model involving the passage of a segment AB of a DNA ring through another segment wrapped around a gyrase molecule. The arrow indicates the position where the breakage and rejoining of the duplex segment are to occur for the passage of AB. In the drawing, the DNA is assumed to wrap around the enzyme in a right-handed way. The binding of ATP presumably shifts the equilibrium in favor of the configuration on the right. The transport of the segment AB from its position shown in the left (the outside state) to that shown in the right (the inside state), by passing through the DNA gate, reduces the linking number by 2. The hydrolysis of ATP permits the exit of the segment AB out of the enzyme without going through the DNA gate. The path of exit is unknown. In the two-gate model, AB is postulated to pass through the enzyme and exits opposite to the side it entered (see Wang et al. 1980 for a more detailed discussion). Similar models have been discussed by Mizuuchi et al. (1980) and Morrison and Cozzarelli (1981). Note that if the DNA is wrapped around the enzyme in a left-handed fashion, the transport of the DNA segment must be from the inside state to the outside state in order to reduce the linking number. The actual handedness of wrapping is yet to be determined. (*d*) A knot is introduced if the segment CD shown here, rather than the segment AB shown in *c*, crosses the DNA gate. Knotting of DNA by gyrase has not yet been observed.

passage is to occur is formed by wrapping a DNA segment around the enzyme (Fig. 2b), the correct sign of the crossover can result readily from the intrinsic handedness of the loop. Or, in a more elaborate fashion, the directional change of the linking number can result from the crossing of a duplex segment from the outside of the loop wrapped around the enzyme to the inside or vice versa (Fig. 2c, see also Wang et al. 1980). The dyadic change in the linking number by this crossing was pointed out before the discovery of gyrase (Crick 1976).

Knot Formation and the Accompanying Change in Linking Number

An elegant experiment supporting the double-stranded breakage-passage-rejoining model for type-II enzyme-catalyzed DNA topoisomerization involves the determination of the change in the linking number of the DNA substrate (Brown and Cozzarelli 1979; Liu et al. 1980). It is found that the linking number changes in steps of 2, as the topology of such a process predicts if the two ends of the transiently broken DNA molecule are not permitted to rotate freely relative to each other (Fuller 1978). This dyadic change in linking number is frequently used to test whether a topoisomerase is a type-II or type-I enzyme.

The type of model shown in Figure 2c raises a question regarding the probability of knot formation. Figure 2d illlustrates the formation of a knotted ring. If the enzyme does not dictate a unique geometry of the duplex segment that is to cross the enzyme-bound DNA segment, as one might expect from the known intermolecular catenation reaction, then the probability of knot formation would be determined by the relative statistical weights of orientations that lead to knot formation and those that do not.

Strand Passage and Enzyme Bridging

With the field of topoisomerase studies expanding rapidly, clarification of certain terms may be needed. For example, consider a type-I enzyme-catalyzed relaxation of a negatively supercoiled DNA. The basic idea is that the enzyme reversibly breaks and rejoins one strand at a time. This breakage and rejoining transiently removes the invariance of the linking number, and hence during the course of the reaction the linking number of the DNA is increased. A well-known algorithm for obtaining the linking number of two intertwined rings is based on the number of crossovers in a projection or two-dimensional representation of the intertwined rings (see, for example, Bauer et al. 1980). Therefore, it is clear that a change in linking number must be accompanied by the passage of strands. In other words, the term "strand passage" is a topological one rather than a mechanistic one.

The term "enzyme bridging" means that when a DNA strand is transiently broken by a topoisomerase, the two broken ends are held together by the enzyme. In the case of the type-II enzyme DNA gyrase, recent studies of the enzyme-DNA complex strongly support the enzyme-bridging notion (Kirkegaard and Wang 1981; Morrison and Cozzarelli 1981; Fisher et al. 1981). For the other type-II enzymes, although no direct probing of the DNA complexes of these enzymes has yet been carried out to substantiate the enzyme-bridging idea, the dyadic change in linking number by these enzymes shows that the ends of the transiently broken DNA are not free to rotate relative to each other, and hence it is likely that the ends are also held together during topoisomerization catalyzed by these enzymes.

Enzyme bridging has also been proposed for the type-I enzyme-catalyzed topoisomerizations (Brown and Cozzarelli 1981). This might be the case for the bacterial enzymes; there is at the present time no compelling evidence for enzyme bridging for the eukaryotic enzymes (for a review, see Wang 1981). In contrast to the case with the type-II enzymes, the linking number of the DNA changes in integral steps in a type-I enzyme-catalyzed relaxation reaction, and hence provides no clue as to whether the ends in the transiently broken state are bridged by the enzyme. Direct probing of the DNA-enzyme complexes is needed to provide an answer.

COVALENT COMPLEX FORMATION AND STRAND TRANSFER

One of the earliest postulates regarding DNA topoisomerases was that the breakage of a DNA backbone bond is accompanied by the formation of a covalent DNA-enzyme linkage (Wang 1971). There has been substantial support for this notion (Champoux 1977b; Gellert et al. 1977; Keller et al. 1977; Sugino et al. 1977; Champoux 1978b; Depew et al. 1978; Bean and Champoux 1980; Prell and Vosberg 1980). The recent demonstration that an active covalent complex can form between DNA and the eukaryotic type-I topoisomerase all but clinches this idea (Bean and Champoux 1981; Halligan et al. 1982). In a number of instances, the linkage has been identified as a phosphotyrosine bond (Tse et al. 1980; Champoux 1981).

The formation of an active covalent complex between DNA and the eukaryotic type-I topoisomerase is shown by a DNA strand-transfer reaction. The strand linked to the enzyme through a 3′-phosphoryl group can be transferred to a free DNA 5′-hydroxyl group (Bean and Champoux 1981; Halligan et al. 1982). This reaction differs from the topoisomerization reactions discussed earlier in that the bond reformed is chemically identical but positionally different from the bond broken.

There are several classes of enzymes that are known to catalyze DNA backbone bond exchange. One class is exemplified by the gene-*A* protein of phage ϕX174 and the gene-*2* protein of phage fd. These enzymes are involved in the initiation of synthesis of the viral DNA strand from the double-stranded replicative form. Each enzyme molecule makes a unique single-stranded cut within its site of recognition on the double-stranded circular DNA template and becomes covalently linked to the 5′-phosphoryl side of the cleaved bond.

Extension of the cleaved strand by a polymerase then commences at the 3′-hydroxyl side of the cleavage. Presumably, after one complete round of synthesis, which regenerates the enzyme recognition site, a similar cleavage by the same or a different enzyme molecule occurs. This second cleavage is accompanied by the transfer of the original enzyme-linked 5′-phosphoryl group to the newly generated 3′-hydroxyl end, yielding a single-stranded DNA ring and a nicked, circular, double-stranded DNA with an enzyme linked to the 5′-phosphoryl end ready for the next round of replication (see Chapter 8 and Kornberg 1980).

Another class is exemplified by the *int* gene product of phage λ (see Chapter 2 and Nash 1981) and the resolvases of the transposons Tn*3* and γδ (Reed 1981; Reed and Grindley 1981). These enzymes promote site-specific recombination. In each recombination event, two staggered single-chain scissions are introduced into a common core sequence of each of a pair of duplex DNA molecules, and reformation of the bonds occurs after a reciprocal strand exchange. The φX174 gene-*A* protein, fd gene-*2* protein, and λ *int* protein have been shown to exhibit DNA topoisomerase activity (Ikeda et al. 1976, 1979; Scott et al. 1977; Eisenberg et al. 1977; Meyer and Geider 1979; Kikuchi and Nash 1979).

THE DRIVING FORCES IN TOPOISOMERIZATION AND DNA STRAND EXCHANGE

In a true DNA topoisomerization reaction, the breakage and reformation of the very same phosphodiester bond are involved. Therefore, in reactions that are not coupled to ATP hydrolysis, the free energy differences of the DNA topoisomers must determine the equilibrium positions of the reactions. Of course, the free energy differences of the topoisomers are dependent on the reaction conditions. Thus, as mentioned in an earlier section, the catenation of double-stranded DNA rings can be facilitated by agents that bring about intermolecular association or aggregation of DNA, and the knotting of single-stranded DNA rings is presumably facilitated by conditions that favor intramolecular base-pairing. If the enzyme is present in stoichiometric rather than catalytic amounts in a topoisomerization reaction, then the interaction of the enzyme with the DNA topoisomers can affect the overall equilibrium provided that the energetics of the enzyme-DNA interaction are sensitive to the change of DNA topology.

Aside from the case of simple, covalently closed, double-stranded DNA rings that differ only in their linking numbers, there is a scarcity of data on the free-energy differences among DNA topoisomers. There is only a single measurement on the equilibrium constant of catenation (Wang and Schwartz 1967). No experimental data on knot formation are available, although some theoretical calculations have been carried out (Frank-Kamenetskii et al. 1975). As mentioned earlier, when knotting is involved in a duplex crossover catalyzed by a type-II topoisomerase, the accompanying change in linking number is different from that found when topoisomerizations do not involve a change of the knot type. How knotting might affect the overall thermodynamics if the DNA is also supercoiled is uncertain.

For reactions that involve strand exchange, a change in nucleotide sequences may accompany the reaction. Such a sequence rearrangement catalyzed by a site-specific enzyme would add an additional factor in determining the directionality of the reaction. In addition, an enzyme may function in a stoichiometric rather than a truly catalytic manner, and the difference in binding energies of the reactant and product sequences may contribute to the overall free energy change. The importance of these factors remains to be tested in site-specific recombination.

THE POSSIBILITY OF LOCALIZED SUPERCOILING

To date, bacterial gyrase is the only topoisomerase that catalyzes DNA supercoiling coupled to ATP hydrolysis. It is surprising that the other ATP hydrolysis-dependent DNA topoisomerases can catalyze the relaxation of either positively or negatively supercoiled DNAs, but cannot generate a supercoiled molecule from a relaxed one. It has been postulated that the phage T4 DNA topoisomerase might have a DNA supercoiling activity when the enzyme is bound to a particular DNA sequence at the origins for T4 DNA replication (Liu et al. 1979).

The general idea that an ATP hydrolysis-dependent topoisomerase might bind two sites on a DNA and catalyze the supercoiling of the DNA segment in between is an interesting one. This kind of localized supercoiling could then trigger the occurrence of a particular event within this region, such as the replication of the DNA.

For DNA gyrase, although the earlier models invoked a localized supercoiling (Liu and Wang 1978a; Mizuuchi et al. 1978; Sugino et al. 1978), the more recent ones do not. Experiments that are designed to test the idea of localized supercoiling are lacking, however.

THE BIOLOGICAL FUNCTIONS OF DNA TOPOISOMERASES

At the present time, information on the functional aspects of the topoisomerases is derived almost exclusively from studies of prokaryotic organisms. The genes *nalA* and *cou* that were so designated for being the determinants of resistance to the antibacterial agents nalidixic acid and coumermycin, have been identified as the structural genes of the subunits A and B, respectively, of DNA gyrase and are renamed *gyrA* and *gyrB*. The use of specific inhibitors and the availability of mutants in these genes have established DNA gyrase as an enzyme essential for the viability of all bacteria examined to date. Inactivation of the enzyme affects a number of cellular processes including replication and transcription (for recent reviews, see Cozzarelli 1980; Gellert 1981).

The gene *topA* encoding *Escherichia coli* DNA topoisomerase I has been identified and mapped recently (Sternglanz et al. 1980, 1981; Trucksis and Depew 1981). Several mutants defective in this gene have been isolated (Sternglanz et al. 1980, 1981). Studies on the phenotypes of these mutants are still at a preliminary stage and the in vivo functions of the enzyme are yet to be clarified. Several ideas, however, emerge from these studies.

First, the existence of mutants with the *topA* gene completely deleted shows that the enzyme is dispensable. Attempts to identify DNA topoisomerases encoded by genes other than *gyrA* and *gyrB* in the *topA* deletion strains have been unsuccessful. These results suggest that the continuous reduction in the linking number between the parental strands of a covalently closed DNA ring during replication can be accomplished by gyrase or the closely related DNA topoisomerase II′. Several groups have noted that strains deficient in DNA topoisomerase I appear to have acquired compensatory mutations in other genes (R. Sternglanz; R.E. Depew; both pers. comm.). The nature of these secondary mutations is yet to be elucidated.

A new type-I topoisomerase, designated DNA topoisomerase III, has been purified from *E. coli* (N. Cozzarelli, pers. comm.). Several mutations that appear to be compensatory to mutations in *E. coli* DNA topoisomerase I have been mapped in *gyrA* and *gyrB* (R. Sternglanz; R.E. Depew; both pers. comm.).

Second, it is almost certain that the *topA* gene is identical to *supX*, a mutation in *Salmonella typhimurium* that was originally identified by its suppression of a mutation, *leu500*, that reduces the level of transcription of the *leu* operon (Mukai and Margolin 1963; Dubnau and Margolin 1972; Dubnau et al. 1973). A likely interpretation of this phenotypical suppression is that the superhelicity of DNA in bacteria is determined by a dynamic balance of the actions of the two major topoisomerases: the negative supercoiling of the DNA by gyrase and the relaxation of the supercoiled DNA by DNA topoisomerase I. A reduction or elimination of the DNA topoisomerase I activity is therefore expected to increase the negative superhelicity of the DNA. If the transcription from the *leu500* promoter is sensitive to the superhelicity of the DNA template, as has been demonstrated for several operons in vitro and in vivo (for a review, see Gellert 1981), then the suppression of the *leu500* mutation by *topA* (*supX*) is readily explained. The effect of *topA* (*supX*) mutations on several other operons has also been reported (Dubnau and Margolin 1972; Dubnau et al. 1973).

As mentioned earlier, the functions of several phage- and plasmid-coded topoisomerases in replication and site-specific recombination are well-known. There is no indication that the initiation of replication of bacterial DNA involves an enzyme of the phage ϕX174 gene-*A* protein type. The possible involvement of bacterial topoisomerases in the breakage and rejoining of DNA strands in certain pathways of recombination is yet to be established.

In eukaryotes, no direct evidence is available on the role of the topoisomerases (for recent reviews, see Wang and Liu 1979; Cozzarelli 1980; Gellert 1981a). The involvement of a type-II DNA topoisomerase in the final segregation of progeny SV40 DNA rings following each round of replication is strongly suggested by recent studies (Sundin and Varsharvsky 1980), and it is likely that the topoisomerases are involved in many other cellular events.

The coiling of DNA in nucleosomes is sufficient to account for the underwinding of DNA in chromatin relative to free DNA. This is in contrast to the active underwinding of DNA by gyrase in prokaryotes. It is nevertheless plausible that a eukaryotic type-II DNA topoisomerase may catalytically effect DNA supercoiling under certain conditions. Liu et al. (1979, 1980) have proposed that the phage T4 type-II topoisomerase might have a replication origin-specific supercoiling activity. The eukaryotic type-II topoisomerases are similar to the T4 enzyme in the topoisomerization reactions they catalyze and may function similarly to the phage enzyme. Specific inhibitors of the eukaryotic topoisomerases or mutants in their regulatory or structural genes are clearly needed to shed some light on the in vivo functions of these enzymes.

CONCLUDING REMARKS

In the last couple of years, the topoisomerase field has experienced a new wave of excitement. A novel class of DNA topoisomerases represented by the ATP-dependent phage T4 and eukaryotic enzymes has been found; a topoisomerase activity associated with phage λ *int* protein has been established; new topoisomerization reactions such as the catenation/decatenation and knotting/unknotting of double-stranded DNA rings have been revealed; and the formation of a covalent intermediate between DNA and topoisomerase, a postulate of long standing, has finally been established.

With the discovery of the new topoisomerization reactions came the realization of a basic mechanistic difference between the two types of DNA topoisomerases: whereas the type-I enzymes break and rejoin one DNA strand at a time, the type-II enzymes break and rejoin two DNA strands in concert. The recently developed methodology of chemical and enzymatic probing at the nucleotide level has made it possible to begin examining the structures of complexes that are formed between DNA and topoisomerases. In addition, cloning techniques have facilitated the preparation of large amounts of the enzymes for detailed physical and chemical studies. Significant progress has also been made in the genetic analysis of both type-I and type-II enzymes from prokaryotes. With the rapid advancement of molecular genetics, the identification of genes coding for these enzymes in eukaryotes should be possible and studies of the functional aspects of these enzymes would in turn be greatly facilitated. There is ground for optimism that a deeper understanding of the mechanistic and functional aspects of DNA topoisomerization by these enzymes can be achieved in the near future; this deeper understanding should in turn provide new insights into the many vital cellular processes involving DNA.

ACKNOWLEDGMENTS

The bulk of the studies on DNA topoisomerases from my own laboratory has been supported by a grant from the U.S. Public Health Service (GM 24544 and its predecessor GM 14621).

REFERENCES

Baldi, M.I., P. Benedetti, E. Mattoccia, and G.P. Tocchini-Valentini. 1980. *In vitro* catenation and decatenation of DNA and a novel eucaryotic ATP-dependent topoisomerase. *Cell* **20:**461.

Bauer, W.R. 1978. Structure and reactions of closed duplex DNA. *Ann. Rev. Biophys. Bioeng.* **7:**287.

Bauer, W.R., F.H.C. Crick, and J.H. White. 1980. Supercoiled DNA. *Sci. Am.* **243:**118.

Bean, M.D. and J.J. Champoux. 1980. Breakage of single-stranded DNA by rat liver nicking-closing enzyme with the formation of a DNA-enzyme complex. *Nucleic Acids Res.* **8:**6129.

_________. 1981. DNA breakage and closure by rat liver type I topoisomerase: Separation of the half-reactions by using a single-stranded DNA substrate. *Proc. Natl. Acad. Sci.* **78:**2883.

Brown, P.O. and N.R. Cozzarelli. 1979. A sign inversion mechanism for enzymatic supercoiling of DNA. *Science* **206:**1081.

_________. 1981. Catenation and knotting of duplex DNA by type I topoisomerases: A mechanism parallel with type 2 topoisomerases. *Proc. Natl. Acad. Sci.* **78:**843.

Brown, P.O., C.L. Peebles, and N.R. Cozzarelli. 1979. A topoisomerase from *Escherichia coli* related to DNA gyrase. *Proc. Natl. Acad. Sci.* **76:**6110.

Champoux, J.J. 1977a. Renaturation of complementary single-stranded DNA circles: Complete rewinding facilitated by the DNA untwisting enzyme. *Proc. Natl. Acad. Sci.* **74:**5328.

_________. 1977b. Strand breakage by the DNA untwisting enzyme results in covalent attachment of the enzyme to DNA. *Proc. Natl. Acad. Sci.* **74:**3800.

_________. 1978. Proteins that affect DNA conformation. *Ann. Rev. Biochem.* **47:**449.

_________. 1978. Mechanism of the reaction catalyzed by the DNA untwisting enzyme: attachment of the enzyme to 3′-terminus of the nicked DNA. *J. Mol. Biol.* **118:**441.

_________. 1981. DNA is linked to the rat liver DNA nicking-closing enzyme by a phosphodiester bond to tyrosine. *J. Biol. Chem.* **256:**4825.

Champoux, J.J. and R. Dulbecco. 1972. An activity from mammalian cells that untwists superhelical DNA—A possible swivel for DNA replication. *Proc. Natl. Acad. Sci.* **69:**143.

Cozzarelli, N.R. 1980. DNA gyrase and the supercoiling of DNA. *Science* **207:**953.

Crick, F.H.C. 1976. Linking numbers and nucleosomes. *Proc. Natl. Acad. Sci.* **73:**2639.

Depew, R.E., L.F. Liu, and J.C. Wang. 1978. Interaction between DNA and *Escherichia coli* protein ω. Formation of a complex between single-stranded DNA and ω protein. *J. Biol. Chem.* **253:**511.

Dubnau, E. and P. Margolin. 1972. Suppression of promoter mutations by the pleiotropic *supX* mutation. *Mol. Gen. Genet.* **117:**91.

Dubnau, E., A.B. Lenny, and P. Margolin. 1973. Nonsense mutations of the *supX* mutant phenotype. *Mol. Gen. Genet.* **126:**191.

Eisenberg, S., J. Griffith, and A. Kornberg. 1977. ϕX174 cistron A protein is a multifunctional enzyme in DNA replication. *Proc Natl. Acad. Sci.* **74:**3198.

Fisher, L.M., K. Mizuuchi, M.H. O'Dea, H. Ohmori, and M. Gellert. 1981. Site-specific interaction of DNA gyrase with DNA. *Proc Natl. Acad. Sci.* **78:**4165.

Frank-Kamenetskii, M.D., A.V. Lukashin, and A.V. Vologodskii. 1975. Statistical mechanics and topology of polymer chains. *Nature* **258:**398.

Fuller, F.B. 1978. Decomposition of the linking number of a closed ribbon: A problem from molecular biology. *Proc. Natl. Acad. Sci.* **75:**3557.

Gellert, M. 1981a. DNA topoisomerases. *Ann. Rev. Biochem.* **50:**879.

———. 1981b. DNA gyrase and other type II DNA toposiomerases. In *The enzymes* (ed. P. Boyer), vol. XIV, p. 345. Academic Press, New York.

Gellert, M., K. Mizuuchi, M.H. O'Dea, and H.A. Nash. 1976. An enzyme that introduces superhelical turns into DNA. *Proc. Natl. Acad. Sci.* **73:**3872.

Gellert, M., K. Mizuuchi, M.H. O'Dea, and J. Tomizawa. 1977. Nalidixic acid resistance: A second genetic character involved in DNA gyrase activity. *Proc. Natl. Acad. Sci.* **74:**4772.

Gellert, M., L.M. Fisher, and M.H. O'Dea, 1979. DNA gyrase: Purification and catalytic properties of a fragment of gyrase B protein. *Proc. Natl. Acad. Sci.* **76:**6289.

Gellert, M., L.M. Fisher, H. Ohmori, M.H. O'Dea, and K. Mizuuchi. 1981. DNA gyrase: Site-specific interactions and transient double-strand breakage of DNA. *Cold Spring Harbor Symp. Quant. Biol.* **45:**391.

Halligan, B.D., J.L. Davis, K.A. Edwards, and L.F. Liu. 1982. Intra- and intermolecular strand transfer by Hela DNA topoisomerase I. *J. Biol. Chem.* **257:**3995.

Hsieh, T.S. and D. Brutlag. 1980. ATP-dependent DNA topoisomerase from *Drosophila melanogaster* reversibly catenates duplex DNA rings. *Cell* **21:**115.

Ikeda, J.-E., A. Yudelevich, and J. Hurwitz. 1976. Isolation and characterization of the protein coded by gene A of bacteriophage φX174 DNA. *Proc. Natl. Acad. Sci.* **73:**2669.

Ikeda, J.-E., A. Yudelevich, N. Shimamoto, and J. Hurwitz. 1979. Role of polymeric forms of the bacteriophage φX174 coded gene A protein in φX DNA cleavage. *J. Biol. Chem.* **254:**9416.

Keller, W., U. Muller, I. Eicken, I. Wendel, and H. Zentgraf. 1977. Biochemical and ultrastructural analysis of SV40 chromatin. *Cold Spring Harbor Symp. Quant. Biol.* **42:**227.

Kikuchi, Y. and H.A. Nash. 1979. Nicking-closing activity associated with bacteriophage *int* gene product. *Proc. Natl. Acad. Sci.* **76:**3760.

Kirkegaard, K. and J.C. Wang. 1978. *Escherichia coli* DNA topoisomerase I catalyzed linking of single-stranded rings of complementary sequences. *Nucleic Acids Res.* **5:**3811.

———. 1981. Mapping the topography of DNA wrapped around gyrase by nucleolytic and chemical probing of complexes of unique DNA sequences. *Cell* **23:**721.

Kornberg, A. 1980. *DNA replication*, p. 396. Freeman, San Francisco.

Kreuzer, K.N. and N.R. Cozzarelli. 1980. Formation and resolution of DNA catenanes by DNA gyrase. *Cell* **20:**245.

Liu, L.F. and J.C. Wang. 1978a. *Micrococcus luteus* DNA gyrase: Active components and a model for its supercoiling of DNA. *Proc. Natl. Acad. Sci.* **75:**2098.

———. 1978b. DNA-DNA gyrase complex: The wrapping of the DNA duplex outside the enzyme. *Cell* **15:**979.

Liu, L.F., R.E. Depew, and J.C. Wang. 1976. Knotted single-stranded DNA rings: A novel topological isomer of circular single-stranded DNA formed by treatment with *Escherichia coli* ω protein. *J. Mol. Biol.* **106:**439.

Liu, L.F., C.C. Liu, and B.M. Alberts. 1979. T4 DNA topoisomerase: A new ATP-dependent enzyme essential for initiation of T4 bacteriophage DNA replication. *Nature* **281:**456.

———. 1980. Type II DNA topoisomerases: Enzymes that can unknot a topologically knotted DNA molecule via a reversible double-strand break. *Cell* **19:**697.

Liu, L.F., J.L. Davis, and R. Calendar. 1981b. Novel topologically knotted DNA from bacteriophage P4 capsids: Studies with DNA topoisomerases. *Nucleic Acids Res.* **9:**3979.

Liu, L.F., L. Perkocha, R. Calendar, and J.C. Wang. 1981a. Knotted DNA from bacteriophage capsids. *Proc. Natl. Acad. Sci.* **78:**5498.

Meyer, T.E. and K. Geider. 1979. Bacteriophage fd gene II-protein. II. Specific cleavage and relaxation of supercoiled RF from filamentous phages. *J. Biol. Chem.* **254:**12642.

Miller, K.G., L.F. Liu, and P.T. Englund. 1981. A homogeneous type II DNA topoisomerase from HeLa cell nuclei. *J. Biol. Chem.* **256:**9334.

Mizuuchi, K., M.H. O'Dea, and M. Gellert. 1978. DNA gyrase: Subunit structure and ATPase activity of the purified enzyme. *Proc. Natl. Acad. Sci.* **75:**5960.

Mizuuchi, K., L.M. Fisher, M.H. O'Dea, and M. Gellert. 1980. DNA gyrase action involves the introduction of transient double-strand breaks into DNA. *Proc. Natl. Acad. Sci.* **77:**1847.

Morrison, A. and N.R. Cozzarelli. 1981. Contacts between DNA gyrase and its binding site on DNA: Features of symmetry and asymmetry revealed by protection from nucleases. *Proc. Natl. Acad. Sci.* **78:**1416.

Mukai, F.H. and P. Margolin. 1963. Analysis of unlinked suppressors of an 0° mutation in *Salmonella. Proc. Natl. Acad. Sci.* **50:**140.

Nash, H.A. 1981. Integration and excision of bacteriophage λ: The mechanism of conservative site specific recombination. *Ann. Rev. Genet.* **15**:143.

Prell, B. and H.-P. Vosberg. 1980. Analysis of covalent complexes formed between calf thymus DNA topoisomerase and single-stranded DNA. *Eur. J. Biochem.* **108:**389.

Reed, R.R. 1981. Transposon mediated site-specific recombination: A defined *in vitro* system. *Cell* **25**:713.

Reed, R.R. and N.D.F. Grindley. 1981. Transposon mediated site-specific recombination *in vitro*: DNA cleavage and protein-DNA linkage at the recombination site. *Cell* **25**:721.

Scott, J.F., S. Eisenberg, L.L. Bertsch, and A. Kornberg. 1977. A mechanism of duplex DNA replication revealed by enzymatic studies of phage φX174: Catalytic strand separation in advance of replication. *Proc. Natl. Acad. Sci.* **74:**193.

Sternglanz, R., S. DiNardo, J.C. Wang, Y. Nishimura, and Y. Hirota. 1980. Isolation of an *E. coli* DNA topoisomerase I mutant. In *Mechanistic studies of DNA replication and genetic recombination* (ed. by B. Alberts), p. 833. Academic Press, New York.

Sternglanz, R., S. DiNardo, K.A. Voelkel, Y. Nishimura, Y. Hirota, K. Becherer, L. Zumstein, and J.C. Wang. 1981. Mutations in the gene coding for *Escherichia coli* DNA topoisomerase I affect transcription and transposition. *Proc. Natl. Acad. Sci.* **78:**2747.

Stetler, G.L., G.J. King, and W.M. Huang. 1979. T4 DNA-delay proteins, required for specific DNA replications, form a complex that has ATP-dependent DNA topoisomerase activity. *Proc. Natl. Acad. Sci.* **76:**3737.

Sugino. A., N.P. Higgins, and N.R. Cozzarelli. 1980. DNA gyrase subunit stoichiometry and the covalent attachment of subunit A to DNA during DNA cleavage. *Nucleic Acids Res.* **8:**3865.

Sugino, A., C.L. Peebles, K.N. Kreuzer, and N.R. Cozzarelli. 1977. Mechanism of action of nalidixic acid: Purification of *Escherichia coli nalA* gene product and its relationship to DNA gyrase and a novel nicking-closing enzyme. *Proc. Natl. Acad. Sci.* **74:**4767.

Sugino, A., N.P. Higgins, P.O. Brown, C.L. Peebles, and N.R. Cozzarelli. 1978. Energy coupling in DNA gyrase and the mechanism of action of novobiocin. *Proc. Natl. Acad. Sci.* **75:**4838.

Sundin, O. and A. Varshavsky. 1980. Terminal stages of SV40 DNA replication proceed via multiply intertwined catenated dimers. *Cell* **21:**103.

Tse, Y.C. and J.C. Wang. 1980. *E. coli* and *Molutens* DNA topoisomerase I can catalyze catenation or decatenation of double-stranded DNA rings. *Cell* **22**:269.

Tse, Y.-C., K. Kirkegaard, and J.C. Wang. 1980. Covalent bonds between protein and DNA. *J. Biol. Chem.* **255:**5560.

Trucksis, M. and R.E. Depew. 1981. Identification and localization of a gene which specifies production of *Escherichia coli* DNA topoisomerase I. *Proc. Natl. Acad. Sci.* **78:**2164.

Wang, J.C. 1971. Interaction between DNA and an *Escherichia coli* protein ω. *J. Mol. Biol.* **55:**523.

———. 1981. Type I DNA topoisomerases. In *The Enzymes* (ed. P. Boyer), vol. XIV, p. 332. Academic Press, New York.

Wang, J.C. and K. Kirkegaard. 1981. DNA topoisomerases. In *Gene amplification and analysis,* vol. 2., *Structural analysis of nucleic acids* (ed. J.G. Chirikjian and T.S. Papas), p. 456. Elsevier/North-Holland, New York.

Wang, J.C. and L.L. Liu. 1979. DNA topoisomerases: Enzymes that catalyze the concerted breaking and rejoining of DNA backbone bonds. In *Molecular genetics* (ed. J.H. Taylor), part 3, p. 65. Academic Press, New York.

Wang, J.C. and H. Schwartz. 1967. Noncomplementarity in base sequences between the cohesive ends of coliphage 186 and λ and the formation of interlocked rings between the two DNAs. *Biopolymers* **5:**953.

Wang, J.C., R.I. Gumport, K.J. Javaherian, K. Kirkegaard, L. Klevan, M.L. Kotewicz, and Y.-C. Tse. 1980. DNA topoisomerases. *ICN-UCLA Symp. Mol. Cell. Biol.* **19:**769.

Nucleases Involved in DNA Repair

Stuart Linn
Department of Biochemistry
University of California
Berkeley, California 94720

INTRODUCTION

DNA repair has traditionally been a subject of extensive review and documentation. Therefore, this article will concentrate upon deoxyribonucleases that are directly implicated in the phenomena. Recent reviews and monographs concerning other aspects of DNA repair include Hanawalt et al. (1978, 1979), Bernstein (1981), Friedberg et al. (1981), Seeberg and Kleppe (1981), and Sutherland (1981).

Directly associating a nuclease or other protein with a role in DNA repair is a difficult task for several reasons. The difficulties are particularly true in animal cells where, for example, to date no enzyme defect has been correlated with the symptoms of any "repair-deficiency disease." In human cells, where repair-deficiency diseases are the major source of "mutants," it ought to be noted that persons with these diseases still carry out the large majority of DNA repair that is necessary for survival and development, so that in fact one might expect biochemical deficiencies in these cells to be relatively subtle. Indeed, as noted below, there are alternative pathways and enzymes for dealing with specific lesions, so that mutants lacking specific enzymes might be expected to have weak or no phenotypic response.

Another problem is that the number of loci associated with each phenotype is often inexplicably large. For example, there are more than 50 DNA repair loci in yeast and probably an equal number of such loci in *Escherichia coli*. In human cells there are at least seven complementation groups of xeroderma pigmentosum (XP) and three of ataxia telangiectasia. Trying to identify an enzyme gene product for each of these loci is difficult in part because many repair enzymes have proved to be unstable, difficult or impossible to purify, or difficult to assay for specifically.

A final problem is how to define cells that are "repair-deficient." Many possibilities exist, but each is complicated. Survival measurements are complicated by a pleiotropic lability of many mutants to perturbations of growth. Defects in the ability to excise damage from DNA are often clouded—at least in animal cells—by a remarkable ability to tolerate the maintenance of a sizable portion of damaged nucleotides for several generations. The measure of the ability to seal DNA nicks or gaps during DNA repair covers only one aspect of DNA repair. Alternatively, the estimation of DNA repair by unscheduled DNA synthesis requires the use of inhibitors to block replicative synthesis, while estimating the efficacy of repair by the extent of repair replication is often based upon the poor assumption that the amount of synthesis per lesion (patch size) is constant and is in some way related to repair efficiency. Finally, host cell reactivation appears to be a good biological criterion, but it often gives results that disagree with other methods, whereas the production of sister chromatid exchanges may be quite dependent upon the presence of bromouracil in the assay.

In summary, because of the apparent complexity of DNA repair systems and because of difficulties in designing assays, in purifying and characterizing enzymes, and in evaluating DNA repair efficiencies of particular cells, the number of DNases definitely correlated with DNA repair is relatively small.

TYPES OF DNA DAMAGE AND CAUSATIVE AGENTS

Virtually all DNA-damaging agents cause multiple types of lesions, and most lesions can be formed by several agents. In addition, the major biological response to a DNA-damaging agent might not be due to the most abundant lesion produced. For these reasons, preparing specific substrates for particular nucleases and identifying specific nucleases involved in dealing with particular damaging agents has been possible in only a few cases. The following list of types of DNA damage is intended to exemplify the above dilemma and to illustrate the need for a number of DNA repair systems; it is not an exhaustive catalog of each of the types of DNA damage that have been identified.

Base Changes and Mismatched Base Pairs

These arise from errors in replication, by recombination between nonidentical molecules to produce heteroduplex mismatches, or by spontaneous or chemically induced changes, such as those formed by deamination.

Baseless Sites

These apurinic or apyrimidinic (AP) sites arise spontaneously due to acid-catalyzed hydrolysis, by the action of DNA glycosylases (see below), by exposure to chemicals (particularly alkylating agents), or by radiation. Due to spontaneous hydrolysis, the DNA of a mammalian cell probably loses roughly 5,000–10,000 purines and 200–500 pyrimidines per 20 hours generation time (Lindahl 1979). Damaging agents increase this burden.

Cross-links

Interstrand cross-links arise after radiation, probably through the action of free radicals or baseless sites which the radiation gives rise to. Baseless sites yield cross-links by the formation of a Schiff's base between the free aldehyde group of the baseless sugar and an amino group present upon a base of the opposite strand. Finally, interstrand cross-links are formed by bifunctional alkylating agents such as nitrogen mustard or by the light-activated psoralen compounds.

Intrastrand cross-links are exemplified by pyrimidine dimers, damage by bifunctional alkylating agents, and protein-DNA cross-links induced by radiation. Generally, intrastrand cross-links would be presumed to be less difficult to repair than interstrand cross-links.

Strand Breaks, Deletions, and Duplications

DNA damaging agents cause single- and double-strand breaks by a variety of mechanisms. Often, the break has an adjacent lesion—e.g., a baseless site. Deletions or duplications are presumed to be formed during replication or recombination.

Alkylation and Bulky Adducts

Electrophiles that react with the DNA bases can be simple alkyl groups or bulky ring compounds. Each type of electrophile has its own pattern of base addition, dependent to a large extent upon whether SN_1 or SN_2 reactions occur, but also upon steric and other factors. Therefore, each agent has its peculiar biological effects (Lawley 1975; Strauss et al. 1975). The multitude of possible addition sites for each alkylating and bulky agent, some of which, like *S*-adenosylmethionine, normally occur in the cell, calls for a very large number of specific repair systems (Lindahl et al. 1979).

Oxidative Damage

Oxidative DNA damage initiated by radiation, peroxides, oxygen radicals, and various other oxidizing agents is a major source of DNA damage. The precise nature of the damage is often complex and difficult to characterize. The thymine glycol-type of damage is probably the best studied of these (Cerutti 1976).

MECHANISMS OF DNA REPAIR

The cell generally has a choice of mechanisms to repair a particular DNA lesion, and, as should be obvious from the discussion below, the consequences in terms of survival and genetic integrity depend upon the particular pathway utilized (Witkin 1975). To some extent, the choice of response is determined by where in the cell cycle the damage appears, particularly with regard to replicative DNA synthesis. Differences occur for quiescent versus growing cells as well.

A particular damaging agent often produces several types of lesions, some of which break down to yet other forms if not immediately removed. For example, a methylating agent produces many methylated base derivatives, some of which could subsequently undergo depurination; the baseless sites might then subsequently form Schiff's base derivatives through the aldehyde group of the deoxyribose (Lawley 1975). Hence, not only will different agents demand different repair functions, but a single agent often demands several enzyme systems and these demands can change with the time elapsed before removal.

Some specific repair mechanisms are discussed below.

Nucleotide Excision-repair

Nucleotide excision-repair is the classical DNA excision scheme which involves recognition of a damaged nucleotide by an incision endonuclease, the excision of the damaged region and probably some adjacent normal nucleotides, and then the filling of the resulting gap by DNA polymerase and ligase (Fig. 1). Though it was originally thought that many specific incision endonucleases exist, it now appears that these enzymes each have a somewhat general specificity.

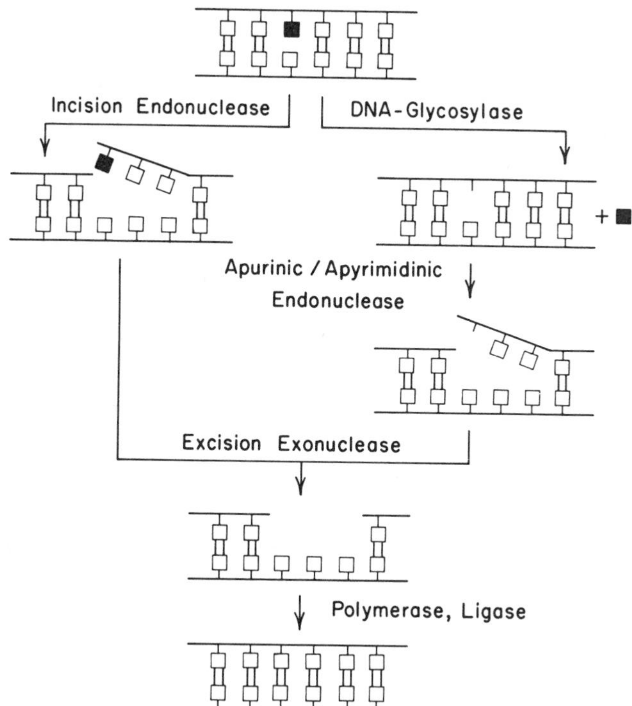

Figure 1 Two pathways of excision-repair. The left sequence is known as nucleotide excision-repair, whereas the right sequence is base excision-repair.

Some Incision Endonucleases

Although the *uvr* incision system of *E. coli* has been studied extensively in vivo, information about the specific gene products of this system has only recently been obtained (Seeberg 1981). The system acts upon UV damage, particularly pyrimidine dimers, and upon bulky chemical adducts including the psoralens. The genes, *uvrA, uvrB,* and *uvrC,* code for proteins of roughly 114,000, 84,000, and 68,000 daltons, respectively (Sancar et al. 1981a,b; Yoakum and Grossman 1981). All three genes are under the control of the *recA lexA* circuit and each of the three proteins is present individually in extracts of *E. coli*. Endonuclease activity is observed only when all three proteins are present together. Cleavage occurs near the site of damage. In the case of pyrimidine dimers, W.D. Rupp and A. Sancar (pers. comm.) have recently observed that cleavage occurs between the seventh and eighth base on the 5′ side of the dimer and between the third and fourth base on the 3′ side. Hence, the same enzyme might act both to incise and excise the damage. The *uvrA* and *uvrC* proteins bind to single-stranded DNA, and the *uvrA* protein has ATPase activity. It is unclear whether the proteins act as a complex in some sequential manner.

The repair endonucleases from human cells have not as yet been characterized, but the phenotypic properties of XP cells, coupled with the plethora of XP complementation groups, suggest the existence of an XP system in humans that is similar to the *E. coli uvr* system.

E. coli has a second enzyme, endonuclease V, that seems to have a similar, perhaps somewhat broader, specificity (Gates and Linn 1977a; Demple and Linn 1982). The enzyme prefers DNA damaged by OsO_4, UV light (but it does not act at pyrimidine dimers), acid treatment, or 7-bromomethylbenz-[*a*]anthracene. It acts most rapidly upon DNA with uracil substituted for thymine. The enzyme appears to act cooperatively and processively and makes about one double-strand break per eight single-strand nicks. Although the mechanism and specificity of the enzyme offer no simple model for its action, it is likely to serve as a back-up function for other repair systems of *E. coli*. Very similar enzymes have also been described in *Micrococcus luteus* (Hecht and Thielmann 1978) and in *Drosophila melanogaster,* an organism which appears to lack uracil glycosylase activity (A.L. Spiering and W.A. Deutsch, pers. comm.).

A surprisingly small number of other incision endonucleases have been well characterized with respect to mechanism and specificity. Endonuclease activities acting on DNA irradiated with UV light or X rays have been studied from calf thymus (Bacchetti and Benne 1975) and rat liver (Tomura and Van Lancker 1975). An endonuclease from calf thymus has been reported that acts at pyrimidine dimers (Waldstein et al. 1979), but unfortunately it was too labile for in-depth study. Finally, an endonuclease from *M. luteus* acting at 8-(2-hydroxy-2-propyl) purines in DNA has been characterized (Livneh et al. 1979a).

Some Excision "Exonucleases"

Because incision endonucleases are only now being characterized, their products are just becoming available as the appropriate specific substrates for studies of excision enzymes. Consequently, enzymes have often been designated as excision enzymes by virtue of their ability to degrade DNA containing damaged material, and genetic verification has usually been lacking. Such enzymes are sometimes single-strand specific and may recognize distortions of base-pairing or may perhaps recognize material displaced by DNA polymerase action (see Fig. 2).

The best-characterized excision activities are probably the exonucleases associated with *E. coli* DNA polymerase I. The $3' \rightarrow 5'$ exonuclease activity of this enzyme is thought to "edit" errors of polymerase action (Brutlag and Kornberg 1972), whereas the $5' \rightarrow 3'$ exonuclease is thought to act in an excision capacity (Kelley et al. 1969) by removing mono- and oligonucleotides during "nick translation" (Kelley et al. 1970; Lehman 1981) (Fig. 2). The

1. 3'→5' bacterial polymerase exonuclease (editor)

2. 5'→3' bacterial polymerase exonuclease

3. Unassociated (usually single-strand specific) exonuclease

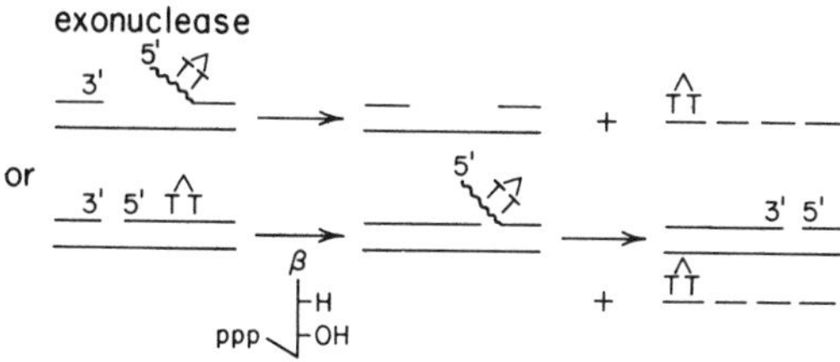

Figure 2 Some excision exonucleases. In case 2, the 5′→3′ bacterial exonuclease is shown acting in nick translation. In case 3, the distortion caused by the lack of base-pairing by a pyrimidine dimer is presumably recognized in the first example, whereas in the second example a single-strand-specific exonuclease recognizes DNA displaced by DNA polymerase action. Presumably, a single-strand-specific endonuclease could also catalyze the latter excision reaction. In any case, both examples require an enzyme capable of acting at or near a 5′ terminus. The location relative to damage of nicks by incision endonucleases has generally not been determined. A thymine dimer is arbitrarily shown as the damage.

enzyme can remove pyrimidine dimers and other types of damage, but appears to remove AP sites only reluctantly (Kelley et al. 1969; Mosbaugh and Linn 1982). 3′-AP sites are only excised slowly, while 5′-AP sites at nicks may be displaced during synthesis (see below). The exonucleases of other prokaryotic polymerases also seem able to serve in an excision capacity.

E. coli exonuclease VII (Chase and Richardson 1974a,b) and the *recBC* enzyme (Goldmark and Linn 1972; Karu et al. 1973; Tanaka and Sekiguchi 1975) also act with both a 3′→5′ and 5′→3′ polarity and can excise damaged nucleotides. Exonuclease VII is specific either for single-stranded DNA or for single-stranded termini on duplexes, whereas the *recBC* enzyme acts both on duplex and single-stranded DNA. Both enzymes form oligonucleotides and can excise pyrimidine dimers. In addition, exonuclease VII can also act from nicks formed by the *M. luteus* UV endonuclease. Deletion mutants in the structural gene of exonuclease VII, *xseA*, are fully viable, have a normal rate of thymine dimer excision, and are only somewhat UV-sensitive, presumably because of back-up by other enzymes. The mutants are also "hyper-rec." However, in cells carrying *recBC* mutations and a defective DNA polymerase I 5′→3′ exonuclease, *xseA* results in reduced excision but does not completely eliminate it (Chase et al. 1979). Evidently, exonuclease VII, the *recBC* enzyme, the 5′→3′ exonuclease of DNA polymerase I, and at least one other nuclease (perhaps the 5′→3′ exonuclease of DNA polymerase III) can each excise pyrimidine dimers in *E. coli*.

A specific 5′→3′ exonuclease (exonuclease B), as well as a second exonuclease (C), which are both capable of excising pyrimidine dimers, have been found to be induced upon phage T4 infection (Shimizu and Sekiguchi 1976). These enzymes are apparently analogous to *E. coli* polymerase-I 5′→3′ exonuclease and exonuclease VII, respectively.

In the fungi, a *Neurospora crassa* exo-endonuclease activity is capable of acting upon DNA containing damage, and certain mutants which are sensitive to UV light appear to be defective in proteolytically modifying the precursor of this enzyme (Fraser et al. 1978). A similar system would appear to exist in *Ustilago maydis* (Holloman and Holliday 1973).

In mammalian cells, DNase IV can excise pyrimidine dimers from DNA in a 5′→3′ direction. The enzyme appears to have the properties desired for an excision exonuclease (Lindahl 1971).

A so-called "correxonuclease," which has been isolated from human placenta, acts in both the 3′→5′ and 5′→3′ directions to release oligonucleotides. It can release pyrimidine dimers from DNA, notably from DNA nicked by *M. luteus* UV endonuclease (Doniger and Grossman 1976). Mammalian DNase V also hydrolyzes DNA in the 3′→5′ and 5′→3′ directions without regard to the phosphomonoester groups present. It complexes to β DNA polymerase with a 1:1 stoichiometry and forms predominantly 5′-mononucleotides (Mosbaugh and Meyer 1980). It will remove 3′-terminal baseless sugars to allow DNA polymerase β to act, and it will remove 5′-baseless termini as well (D.W. Mosbaugh and S. Linn, unpubl.). Although mammalian DNases IV and V and correxonuclease all appear capable of being excision enzymes, in each case genetic verification of their role remains to be obtained.

In conclusion, excision exonuclease capacity appears at this time to occur in several different enzymes in both prokaryotes and eukaryotes. To some extent, this apparent duplication might be due to the necessity of being compatible with the products of several different incision endonucleases. However, to some extent, the duplication may be to assure a "back up" for a very important function.

Base Excision-repair

Formation of Baseless Sites

Base excision-repair is characterized by DNA glycosylases, enzymes that recognize a specific abnormal base, then remove it from the DNA by cleaving the *N*-glycosylic bond to generate a free base and an AP site (Fig. 1). Some ten specific types of glycosylase have been reported, including those which recognize uracil (deaminated cytosine), hypoxanthine (deaminated adenine), specific alkylated purines, thymine glycols, and radiation-induced breakdown products (Lindahl 1979, 1982). At least one glycosylase, 3-methyladenine glycosylase II of *E. coli*, is inducible by exposure of cells to DNA-damaging agents (Lindahl 1982).

Cleavage by the AP Endonuclease

Baseless sites, whether formed by DNA glycosylases, DNA-damaging agents, or spontaneously, are recognized by AP endonucleases, enzymes which are ubiquitous in nature (see Friedberg et al. [1981] for a compendium of AP endonucleases from various organisms). These enzymes cleave DNA at a phosphodiester bond adjacent to the baseless sugar (see below). The resulting baseless sugar may then be excised by an exonuclease (Fig. 1) or by another AP endonuclease (see below). In either case, DNA polymerase and DNA ligase are presumed ultimately to repair the excision gap.

In studying the AP endonucleases from *E. coli* (Gates and Linn 1977b; Demple and Linn 1980; Warner et al. 1980) and human cells (Kuhnlein et al. 1976, 1978; Linsley et al. 1977; Mosbaugh and Linn 1980; Kane and Linn 1981), it has been found that each acts at AP sites in duplex DNA in one of two manners:

B_1 B_3 B_1 B_3

...P P P P... —class I→ ...P P OH + P P...

B_1 B_3

—class II→ ...P OH + P P P...

Class-I enzymes leave a baseless sugar at the 3′-terminus of the nick, whereas class-II enzymes leave a sugar-5-phosphate at the 5′-terminus. The products of the two classes of AP endonuclease can be conveniently distinguished by the fact that the 3′-terminal baseless sugar formed by class-I enzymes is not a good primer for *E. coli* DNA polymerase I (Mosbaugh and Linn 1980; Warner et al. 1980).

The two classes of AP endonuclease can act in concert at the same site to excise sugar-5-phosphate from the DNA and produce a gap of one nucleotide:

B_1 B_3 B_1 B_3

...P P P P... —class I + class II→ ...P OH + P OH + P P...

To date, no AP endonuclease has been reported that does not fall into class I or II, although we have recently noted that *E. coli* endonuclease V, which is not specific for baseless sites but does recognize them as substrate, may not follow this generalization (Demple and Linn 1982).

E. coli *AP Endonucleases*

We have classified the actions of three *E. coli* AP endonucleases. Endonuclease III is a class-I enzyme (Gates and Linn 1977b; Warner et al. 1980; Demple and Linn 1980), whereas endonuclease IV (Ljungquist 1977) and endonuclease VI (the predominant AP endonuclease of *E. coli*) (Weiss 1978) are class-II enzymes (Warner et al. 1980). In addition, T4 phage induces "endonuclease V" which exhibits class-I AP endonuclease activity (Demple and Linn 1980). Finally, Friedberg et al. (1981) have reported endonuclease VII, an activity which is specific for single-stranded polydeoxyribonucleotides containing apyrimidinic sites. The location of nicks made by this enzyme is not yet clear.

Human Fibroblast AP Endonucleases

We have also found that cultured human fibroblasts contain two types of AP endonuclease. Human AP endonuclease I passes through phosphocellulose during chromatography, whereas human AP endonuclease II is retained by phosphocellulose and elutes in a dispersed manner (Fig. 3A).

Human fibroblast AP endonuclease I is unstable and has not been extensively purified. It is a class-I enzyme (Mosbaugh and Linn 1980) and appears not to be present in cell strains of xeroderma pigmentosum complementation group D (Kuhnlein et al. 1976, 1978; Mosbaugh and Linn 1980) (Fig. 3B).

We have purified human endonuclease II from many strains of fibroblasts (Kuhnlein et al. 1976), from placenta (Linsley et al. 1977), and from HeLa cells (Kane and Linn 1981). It is a class-II enzyme (Mosbaugh and Linn 1980) and is often found to elute as multiple species during chromatography. We have observed no cell types from which it is absent.

In an attempt to determine the biological significance of the absence of AP endonuclease I from XP-D cell strains, we have studied the transfection of depurinated SV40 DNA into these cells as a function of the level of depurination (Kudrna et al. 1979). Surprisingly, the XP-D cells were considerably better than normal cells at reactivating the depurinated viral DNA, although, as expected, they were not good hosts for transfecting UV-irradiated SV40 DNA. Although the results with depurinated DNA support the presumed alteration of an AP DNA-specific factor in vivo, they do not shed light upon the significance of this change. Perhaps AP endonuclease I is particularly

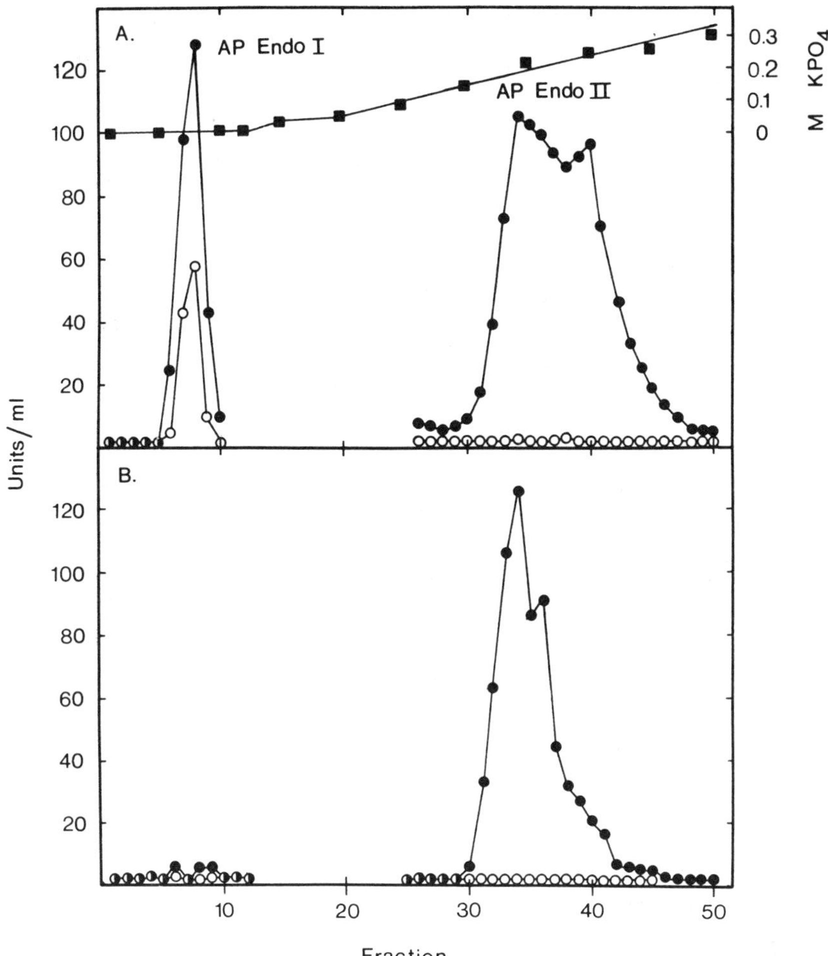

Figure 3 Elution of human AP endonuclease activity from phosphocellulose. Elution was at pH 7.5. (*A*) The extract was from normal fibroblast strain AW995. (*B*) The extract was from XP complementation group-D fibroblast strain CRL 1159 (XP5BE). (●) AP DNA substrate; (○), undamaged DNA substrate.

detrimental to transfections by DNA containing roughly 2–20 AP sites as utilized in our experiments.

Significance of Two AP Endonuclease Classes

We have no evidence that the two classes of AP endonuclease act in concert at AP sites in vivo. Should this be the case, however, a large class of DNA repair events might be overlooked during assays of in vivo repair that employ measurements of unscheduled DNA synthesis or repair replication, because the resulting patch size of essentially 1 nucleotide may not be detectable by these techniques, particularly when [^{3}H]thymidine or bromodeoxyuridine are utilized. Hence, repair mediated by DNA glycosylases and repair of other damage leading to AP sites may often have been overlooked or underestimated.

A Tripeptide AP Endonuclease Activity

The tripeptide, Lys-Tryp-Lys can cleave DNA containing apurinic sites (Behmoaras et al. 1981). Evidently, the tryptophanyl residue stacks with the bases adjacent to the apurinic sugar, thus bringing the lysine amino groups into close proximity with the baseless sugar so as to allow catalysis of β-elimination. The tripeptide thus provides a model nuclease reaction to study in detail, but also cautions that irrelevant proteins conceivably could catalyze "AP endonuclease" reactions.

Other Activities Observed to Be Associated with AP Endonucleases

Several bacterial AP endonucleases have been observed to possess other catalytic activities. The most noteworthy is the association of *E. coli* endonuclease VI (the major class-II AP endonuclease of *E. coli*) with exonuclease III (Weiss et al. 1978). The latter activity is a duplex DNA-specific, 3′→5′ exonuclease. In addition, endonuclease VI has associated DNA-3′-phosphatase and RNaseH activities (Weiss et al. 1978). The significance of the association of all of these activities on one protein is not known.

E. coli endonuclease III has a DNA glycosylase activity that acts upon 5,6-saturated thymine analogs such as dihydrodihydroxythymine and dihydrothymine (Demple and Linn 1980). Thus, the enzymes can act upon DNA containing such lesions in the following manner:

B t B → B B

...P P P P... → ...P P P P... + t

→ B B

...P P OH + P P...

We (Demple and Linn 1980; Warner et al. 1980), among others (Radany and Friedberg 1980; Seawell et al. 1980), showed that phage T4 endonuclease V cleaves DNA containing either AP sites or pyrimidine dimers in a manner first observed with the *M. luteus* "UV endonuclease" (Grossman et al. 1978). That is, the enzyme first cleaves the 5′-glycosylic bond of the pyrimidine dimer nucleotide pair, then goes on to act as a class-I AP endonuclease upon this intermediate:

B1 Py⌒Py B4 ...P P P P P... ⟶ Py B1 Py B4 ...P P P P P...

Py B1 Py B4 ...P P P P P... ⟶ Py B1 Py B4 ...P P OH + P P P...

While the glycosylase and endonuclease activities differ in their rate of heat-inactivation (Seawell et al. 1980) and sensitivity to harmane (Warner et al. 1981), they are on the same polypeptide (Nakabeppu and Sekiguchi 1981).

The *M. luteus* enzyme also carries out the above pair of reactions efficiently, but it is not reported to act as a class-I AP endonuclease at simple AP sites with the same efficiency as we have observed for the phage T4 enzyme.

The intermediate and final products of these ''UV endonucleases'' acting at pyrimidine dimers have the unique property of releasing a free pyrimidine base upon photoreversal of the dimer with UV light (Demple and Linn 1980; Radany and Friedberg 1980):

B1 Py⌒Py B4 ...P P P P P ⇌ (hν) B1 Py Py B4 ...P P P P P... ,

but,

Py B1 Py B4 ...P P P P P ⟶ (hν) B1 Py B4 ...P P P P P... + Py

and,

Py⌒Py B4 P P P... ⟶ (hν) Py B4 P P P... + Py.

Using, as a criterion, the release of free thymine from the pyrimidine dimer-containing excision products formed in vivo after subsequent UV irradiation in vitro, we have observed that neither uninfected *E. coli* (Demple and Linn 1980), nor cultured human fibroblasts (LaBelle and Linn 1982) appear to utilize a pyrimidine dimer glycosylase/AP endonuclease mechanism. *M. luteus*, on the other hand, forms excision products that release thymine base upon secondary irradiation (LaBelle and Linn, 1982) as do T4-infected *E. coli* (Radany and Friedberg 1982). These observations are consistent with the apparent lack of a pyrimidine dimer-specific excision-repair system in *E. coli* or human cells. Mutants of these cells defective in pyrimidine dimer excision—*E. coli uvr* mutants or human XP cells—are also defective in the repair of other DNA lesions. *M. luteus*, however, (and phage T4-infected *E. coli*) are unique in having large amounts of specific pyrimidine dimer "endonuclease" activity. It is curious that *E. coli* uses endonuclease III to act at dihydroxydihydrothymine-type lesions by a glycosylase/AP endonuclease mechanism, but does not use an analogous scheme for pyrimidine dimers.

The Action of DNA Polymerases upon Intermediates of Base Excision-repair

To study how *E. coli* DNA polymerase I might take part in specific processes of excision-repair, we have studied its action upon some of the products formed by the incision endonucleases discussed above (Mosbaugh and Linn 1982). We hope ultimately to answer the following questions: (1) Does the enzyme synthesize from these sites? (2) If the enzyme does synthesize from these incision sites, does it do so by nick translation or strand displacement (see Fig. 2)?

3′-AP termini formed by class-I AP endonucleases are poor primers for the polymerase (Warner et al. 1980). However, a slow rate of synthesis from these termini could be observed that increased as the reaction proceeded; this rate could be enhanced if the DNA were incubated with the polymerase before triphosphate addition (Mosbaugh and Linn 1982). After such synthesis, the newly replicated DNA was linked to the primer through an alkali-stable linkage, indicating that the polymerase first removed the baseless sugar rather than using it as a primer terminus. Indeed, free deoxyribose-5-phosphate was present after such a reaction. Similar activation by polymerase I preincubation is observed with UV-irradiated DNA which had been incised with T4 endonuclease to form 3′-AP termini and 5′-thymine dimer nucleotide termini. In summary, *E. coli* DNA polymerase I does not use 3′-baseless termini as primers, but instead slowly removes these termini, presumably with the 3′→5′ exonuclease activity, and then proceeds to synthesize DNA from the new termini.

We also have examined the action of *E. coli* DNA polymerase I at nicks formed by the class-II AP endonucleases which form 5′-sugar phosphate termini, but normal 3′-hydroxynucleotide termini (Mosbaugh and Linn 1982). Does the enzyme proceed by nick translation or strand displacement in this case? PM2 circles having an average of one incised AP site per genome were replicated with about 0.1 polymerase molecules per circle to incorporate 1000–1500 nucleotides per nick. The DNA was then examined in the electron microscope for the presence of tailed structures, which would indicate strand-displacement DNA synthesis. Tails were measured and compared with duplexes of known length to verify whether the average tail contained roughly the number of nucleotides expected from the amount of radiolabeled nucleotide incorporated.

When undamaged form-I DNA molecules were nicked by *N. crassa* nuclease to form simple 5′-phosphomonoester nicks, about 5–6% of the form-II molecules were seen to have tails; however, when molecules with apurinic sites were nicked with any of three class-II AP endonucleases, 40–80% of the form-II molecules observed had tails (Table 1). In contrast, when the 5′-terminal baseless sugar residues were removed with a class-I AP endonuclease (to generate a gap of 1 nucleotide) again only about 5–6% of the molecules had tails (Table 1). Evidently, the 5′-AP terminus, but not a simple nick or 1 nucleotide gap, prompts a strand displacement mode of synthesis.

The effect of 5′-pyrimidine dimer nucleotide termini was also studied by observing synthesis from termini formed by T4 endonuclease V acting at pyrimidine dimers (Table 1). Surprisingly, these bulky, nonbase-paired termini do not promote substantial strand displacement, regardless of whether the 3′-AP terminus formed by the T4 enzyme is removed by a class-II AP endonuclease prior to synthesis as follows:

B_1 Py⌒Py B_4 ...P–P–P–P–P... —T4 endo V→ ...P–P(B_1)–OH + P–P(Py–Py)–P(B_4)–P...

—AP Endo II→ ..P(B_1)–OH + P + P–P(Py–Py)–P(B_4)–P...

The fact that 5′-AP termini, but not 5′-pyrimidine dimer termini, promote strand displacement synthesis is curious and may be biologically significant, since strand displacement synthesis generates duplicated genetic material. Parenthetically, the 5′-AP termini can be removed by class-I AP endonucleases, a type of enzyme that we have noted to be absent from XP-D fibroblasts.

Table 1 DNA synthesis from PM2 duplex circles incised by endonuclease

DNA damage/endonuclease	Termini produced [a]	Number form-II molecules observed	% Relaxed circular molecules	
			without tails	with ≥ one tail
Undamaged				
+ *N. crassa* nuclease	3′-OH, 5′-P	117	94	6
		104	95	5
		116	95	5
Partially depurinated				
+ Human fibroblast AP endonuclease II	3′-OH, 5′-dR-P	104	58	42
+ HeLa AP endonuclease II	3′-OH, 5′-dR-P	121	51	49
+ *E. coli* endonuclease IV	3′-OH, 5′-dR-P	123	24	76
+ Fibroblast AP endonuclease II and *E. coli* endonuclease III	3′-OH, 5′-P, gap of 1	110	95	5
+ HeLa AP endonuclease II and *E. coli* endonuclease III	3′-OH, 5′-P, gap of 1	109	94	6
UV-irradiated				
+ T4 endonuclease V	3′-dR, 5′-dimer nucleotide	224	94	6
+ T4 endonuclease V and fibroblast AP endonuclease II	3′-OH, 5′-dimer nucleotide	272	88	12

DNA molecules were depurinated with acid or irradiated with UV as indicated, then treated with endonucleases as shown. Damage was at a level of approximately one apurinic site or pyrimidinic dimer per DNA molecule, respectively, and repair endonucleases were added in sufficient quantity to cleave all damaged sites. *N. crassa* nuclease was added at a level to cleave 60 – 70% of the molecules to mimic the fact that at a frequency of one damage lesion per DNA molecule, roughly 30 – 40% of the molecules sustain no damage. Indeed, in all cases, approximately 25 – 50% of the molecules were form I and are not considered for the table. DNA polymerase I was added at a level to give 1000 – 1500 nucleotides incorporated per nick, then molecules were observed by electron microscopy.

[a] Abbreviations used in the table are: 3′-OH, a 3′-hydroxynucleotide terminus; 5′-P, a 5′-terminus containing a 5′-mononucleotide; 5′-dR-P, a 5′-terminus containing deoxyribose-5-phosphate; gap of 1, a gap of 1 nucleotide formed by excision of deoxyribose-5-phosphate from an AP site with class-I and -II AP endonuclease acting in concert (see text); 3′-dR, a 3′-terminus containing deoxyribose; 5′-dimer nucleotide, a 5′-terminus containing a pyrimidine dimer attached to a deoxyribose through one glycosylic bond (see text).

The data are from Mosbaugh and Linn (1982).

Table 2 Ability of HeLa DNA polymerase to use various incision sites as primers

DNA damage/endonuclease	dNMP incorporated per nick		
	polymerase α	polymerase β	polymerase β + DNase V
Undamaged			
+ *N. crassa* endonuclease	<1	18	—
Partially depurinated			
+ *E. coli* endonuclease III	<1	2	33
+ HeLa AP endonuclease	<1	15	41
+ *E. coli* endonuclease III and HeLa AP endonuclease	<1	16	35

PM2 DNA was depurinated and treated with DNases essentially as described in the legend to Table 1. HeLa α-polymerase (fraction VII) was purified according to Fisher and Korn (1977). HeLa β-polymerase and DNase V were purified from Novikoff hepatoma as described by Mosbaugh and Meyer (1980). Polymerase assays were with 100 μM dNTPs. The data are from D.W. Mosbaugh and S. Linn (unpubl.).

Experiments have also been initiated to study the effects of baseless termini upon DNA synthesis by human (HeLa) DNA polymerases (Table 2). α-Polymerase does not act from any type of nick nor from a gap of 1 nucleotide formed by class-I and -II AP endonucleases acting at depurinated sites. Conversely, β-polymerase will act at simple nicks formed by *N. crassa* nuclease, at gaps of 1 nucleotide, or at nicks containing 5′-baseless termini with roughly equal efficiency. It will not act at nicks with 3′-AP termini, however. (The small amount of synthesis noted in Table 2 is likely due to spontaneous loss of the 3′-AP residues during the incubation.) Significantly, β-polymerase complexed to HeLa DNase V (see above) can use the 3′-AP terminus effectively, presumably by virtue of the removal of the baseless residue by the DNase. Indeed, after treatment of DNA containing either 3′- or 5′-AP sites with the complex in the presence of the four triphosphates, the resulting product can be sealed with DNA ligase, hence completing the entire excision-repair process in vitro (D.W. Mosbaugh and S. Linn, unpubl.).

Recombinational Repair

The phenomenon of postreplication repair is probably mediated to a large degree by recombinational repair mechanisms (see Chapter 2). The *recBC* type of enzyme (reviewed by Muskavitch and Linn 1981) and the single-strand-specific fungal nucleases (see Chapter 7) are noteworthy in that mutants with altered enzymes are repair defective as well as recombination defective. Unfortunately, both of these groups of enzymes exhibit several catalytic activities. The *recBC* type of enzyme has single-strand exonuclease, double-strand exonuclease, single-strand endonuclease, and double-strand DNA unwinding capabilities, whereas the fungal enzymes likely exist in multiple forms with varying specificity. At this point, it is not clear which of the activities of each group of enzyme might be specifically relevant to DNA repair.

Another difficulty in identifying the specific role of nuclease activities in recombination repair is our incomplete knowledge both of intermediate DNA structures which might be nuclease substrates and of the role of auxiliary factors such as *E. coli recA* protein or single-strand-binding protein (SSB). A model for how nucleases might act in recombination, which is based upon our current knowledge of *recA* protein and SSB, has recently been proposed (West et al. 1981), but no specific nuclease has been implicated.

Mismatch Repair

Mismatch repair is undoubtedly a very important aspect of DNA repair, certainly in bacteria and fungi where genetic studies have documented the common occurrence of this phenomenon. No specific DNases have been associated with the phenomenon, though the single-strand-specific fungal nucleases can cleave DNA at such residues in vitro (Ahmed et al. 1975; Shishido and Ando, this volume). The latter observations may be trivial, however, as they do not speak to the important question of discrimination of the incorrect from the correct nucleotide in the mismatched base pair. A hint may be provided by an elegant model which invokes methylation of the parental strand for such selection (Wagner and Meselson 1976), but direct tests of this model with appropriate heteroduplex phage DNAs have been both negative (Bauer et al. 1981) and positive (Glickman and Radman 1980; Meselson et al. 1980).

Inducible, Error-prone Transdamage Synthesis

It is becoming clear that many DNA repair systems are likely to be induced by DNA damage. In *E. coli*, for example, mutations induced by some mutagens or by UV radiation are thought to result from the action of an inducible system or systems which are at least partially controlled by induction of the *recA* and *lexA* genes (Witkin 1976). The latter genes are evidently induced by agents that interfere with replication either by directly inhibiting some replication enzyme(s) or by damaging DNA and hence blocking normal synthesis via constraints which assure fidelity of synthesis. Induction of the *recA* and *lexA* genes would then promote a mode of synthesis able to bypass these blocks, presumably in an error-prone manner. Such repair would be a desperate measure for survival when recombinational repair does not, or cannot, occur.

A popular and attractive hypothesis to explain the mutagenesis is that the inducible ("SOS"), error-prone repair system is mediated by a modified DNA polymerase activity which is able to catalyze DNA synthesis past damaged bases (Witkin 1967; Caillet-Fauquet et al. 1977). Such a modification could be mediated by *recA* protein—perhaps by its protease activity (Roberts et al. 1978). A possibility relevant to nucleases would be that the "editor" 3′→5′ exonuclease of DNA polymerase would be suppressed to allow error-prone synthesis to occur. In fact, mutator forms of the T4 polymerase with reduced exonuclease relative to polymerase have been observed to be produced by several gene-*43* mutants (Muzyczka et al. 1972).

We have observed an error-prone form of DNA polymerase I after induction of *recA*. However, it has normal 3′→5′ exonuclease activity (Lackey et al. 1982). Miyaki et al. (1977) have similarly observed increased levels of DNA polymerase II, but the associated 3′→5′ exonuclease was not studied. Evidence exists that indicates that DNA polymerase III (Bridges and Mottershead 1978), but perhaps not polymerase I (Witkin and George 1973; Barfknecht and Smith 1978), is required for the mutagenic response. Unfortunately, the complicated and still undefined—at least for DNA repair—subunit structure of polymerase III has precluded studies of DNA polymerase III and repair. Of course, forms of all three polymerases might be able to act in this function interchangeably.

In conclusion, it is likely that the molecular basis of mutagenic responses to UV radiation will be known soon and that DNase activities may be involved. The subject of DNA mutagenesis and replication in irradiated *E. coli* and mammalian cells has recently been very comprehensively reviewed by Hall and Mount (1981).

A parenthetical, related topic is the apparent necessity for an active *recA* gene to control *recBC* nuclease activity after DNA damage (Willets and Clark 1969). Indeed, Williams et al. (1981) have shown that *recA* protein strongly inhibits the single-strand endo- and exonuclease activities of the *recBC* enzyme. The duplex exonuclease activity is also inhibited when ATP-γS and oligonucleotides are present to stimulate duplex DNA binding by *recA* protein. Under these conditions, DNA is also protected from degradation by exonuclease I and λ exonuclease. These observations are subject to refinement, however, since it has been observed (K.T. Muskavitch and S. Linn, unpubl.) that SSB protein behaves synergistically with *recA* protein in these responses and that some DNA degradation by *recBC* DNase is evidently a signal for *recA* induction (presumably through the formation of oligonucleotides [Oishi and Smith 1978]). However, they will certainly ultimately be relevant to the mechanism of *recBC* DNase action in vivo.

Restoration of Normal Sequences without Phosphodiester Bond Cleavage

Whereas DNases are certainly a major part of the enzymology of DNA repair, it should be noted that at least three enzymatic reactions have been reported that could be invoked for DNA repair schemes which would not demand phosphodiester bond cleavage. Thus, photoreactivating enzymes are capable of acting to monomerize pyrimidine dimers without breaking the DNA. These enzymes have been extensively reviewed by Sutherland (1981). Likewise, O^6-methylguanine residues in DNA can be converted directly to guanine residues by transmethylation to a protein acceptor which appears in cells after adaptation to alkylating agents (Olsson and Lindahl 1980). Finally, purine residues have been observed to be reinserted specifically into un-nicked apurinic sites in DNA by fractionated extracts obtained from human fibroblasts (Deutsch and Linn 1979a,b), *E. coli* (Livneh et al. 1979b), and *D. melanogaster* (A.L. Spiering and W.A. Deutsch, unpubl.). It is not unlikely that other activities will be found that also might participate in DNase-free DNA repair.

CONCLUSIONS

The Cell Has Alternative DNA Repair Pathways at Its Disposal

Nucleotide excision-repair and base excision-repair, as depicted in Figure 1, are clearly over-generalized. For example, strand polarities must be added, it is not yet clear which exonucleases take part in excision, and the details of DNA polymerase action are not known. It is also not known to what extent AP sites might be corrected without exonuclease action—either by the joint action of the class-I and -II AP endonucleases or by base insertion. In fact, in each cell type, several schemes probably exist simultaneously.

We are also just beginning to learn some of the details about recombinational repair and about the induction of repair mechanisms. It is clear, however, that there are a series of options, each with possibly different consequences, for cellular responses to a large spectrum of damage caused by a multiplicity of individual agents. For this reason, we have only begun to understand the complexities of DNA repair; however, even after much study, our understanding of DNA damage and its repair may be possible only in probabilistic terms.

How Efficient Is DNA Repair?

A final question of interest will be to determine what is the accuracy of DNA repair. Studies with mutant forms of T4 DNA polymerase have shown that the level of synthesis fidelity could be improved by increasing the $3' \rightarrow 5'$ exonuclease activity relative to the polymerase activity (Muzyczka et al. 1972; Lo and Bessman 1976). Evidently, nature has chosen not to optimize the editing function for accuracy in the phage T4 wild type. Likewise, we have considered above the error-prone, SOS responses in bacteria. Obviously, a built-in error frequency during DNA replication and repair have potential desirability for survival from extreme levels of DNA damage and for evolution, particularly for viruses and unicellular organisms. Less subtle, however, might be the consequences for aging and development, where error frequencies could be programmed so as to provide a characteristic defined life span for each particular somatic cell type and hence ultimately for the organism as a whole

(Burnet 1974). For these reasons, studies of repair deoxyribonucleases will likely carry us toward an understanding of cellular and organismal, as well as of molecular, phenomena.

ACKNOWLEDGMENTS

The work cited from the author's laboratory was funded by grants from the United States National Institutes of Health and Department of Energy.

REFERENCES

Ahmed, A., W.K. Holloman, and R. Holliday. 1975. Nuclease that preferentially inactivates DNA containing mismatched bases. *Nature* **258:**54.

Bacchetti, S. and R. Benne. 1975. Purification and characterization of an endonuclease from calf thymus acting on irradiated DNA. *Biochim. Biophys. Acta* **390:**285.

Barfknecht, T.R. and K.C. Smith. 1978. The involvement of DNA polymerase I in the postreplication repair of ultraviolet radiation-induced damage in *Escherichia coli* K-12. *Mol. Gen. Genet.* **167:**37.

Bauer, J., G. Krämmer, and R. Knippers. 1981. Asymmetric repair of bacteriophage T7 heteroduplex DNA. *Mol. Gen. Genet.* **181:**541.

Behmoaras, T., J.-J. Toulmé and C. Hélène. 1981. A tryptophan-containing peptide recognizes and cleaves DNA at apurinic sites. *Nature* **292:**858.

Bernstein, C. 1981. Deoxyribonucleic acid repair in bacteriophage. *Microbiol. Rev.* **45:**72.

Bridges, B.A. and R.P. Mottershell. 1978. Mutagenic DNA repair in *Escherichia coli*. VIII. Involvement of DNA polymerase III in constitutive and inducible mutagenic repair after ultraviolet and gamma irradiation. *Mol. Gen. Genet.* **162:**35.

Brutlag, D. and A. Kornberg. 1972. Enzymatic synthesis of deoxyribonucleic acid. XXXVI. A proofreading function for the 3′⟶5′ exonuclease activity of deoxyribonucleic acid polymerase. *J. Biol. Chem.* **247:**241.

Burnet. 1974. *Intrinsic mutagenesis: A genetic approach to aging*. Medical and Technical Publishing Co., Ltd., Lancaster.

Caillet-Fauquet, P., M. Defais, and M. Radman. 1977. Molecular mechanisms of induced mutagenesis. Replication *in vivo* of bacteriophage ϕX174 single-stranded, ultraviolet light-irradiated DNA in intact and irradiated host cells. *J. Mol. Biol.* **117:**95.

Cerutti, P. 1976. DNA base damage induced by ionizing radiation. In *Photochemistry and photobiology of nucleic acids* (ed. S.Y. Wang), vol. 2, p. 375. Academic Press, New York.

Chase, J.W. and C.C. Richardson. 1974a. Exonuclease VII of *Escherichia coli*. Purification and properties. *J. Biol. Chem.* **249:**4545.

__________ . 1974b. Exonuclease VII of *Escherichia coli*. Mechanism of action. *J. Biol. Chem.* **249:**4553.

Chase, J.W., W.E. Masker, and J.B. Murphy. 1979. Pyrimidine dimer excision in *Escherichia coli* strains deficient in exonuclease V and VII and in the 5′⟶3′ exonuclease of DNA polymerase I. *J. Bact.* **137:**234.

Demple, B. and S. Linn. 1980. DNA *N*-glycosylases and UV repair. *Nature* **287:**203.

__________ . 1982. On the recognition and cleavage mechanism of *Escherichia coli* endonuclease V. A possible DNA repair enzyme. *J. Biol. Chem.* **257:**2848.

Deutsch, W.A. and S. Linn. 1979a. DNA binding activity from cultured human fibroblasts that is specific for partially depurinated DNA and that inserts purines into apurinic sites. *Proc. Natl. Acad. Sci.* **76:**141.

__________ . 1979b. Further characterization of a DNA-purine base insertion activity from cultured human fibroblasts. *J. Biol. Chem.* **254:**12099.

Doniger, J. and L. Grossman. 1976. Human correxonuclease. Purification and properties of DNA repair exonuclease from placenta. *J. Biol. Chem.* **251:**4579.

Fisher, P.A. and D. Korn. 1977. DNA polymerase-α. Purification and structural characterization of the near homogeneous enzyme from human KB cells. *J. Biol. Chem.* **18:**6528.

Fraser, M.J., S. Kwong, D.M. Gáler, and T.Y.-K. Chow. 1978. Regulation by proteinases of a putative *rec*-nuclease of *Neurospora*. In *DNA repair mechanisms* (ed. P.C. Hanawalt, E.C. Friedberg, and C.F. Fox), p. 441. Academic Press, New York.

Friedberg, E.C., T. Bonura, E.H. Radany, and J.D. Love. 1981. Enzymes that incise damaged DNA. In *The enzymes* (ed. P.D. Boyer), vol. 14, p. 251. Academic Press, New York.

Gates, F.T. III and S. Linn. 1977a. Endonuclease V of *Escherichia coli*. *J. Biol. Chem.* **252:**1647.

__________. 1977b. An endonuclease from *E. coli* that acts specifically upon duplex DNA damaged by ultraviolet light, osmium tetroxide, acid or X-rays. *J. Biol. Chem.* **252:**2802.

Glickman, B.W. and M. Radman. 1980. *Escherichia coli* mutator mutants deficient in methylation-instructed DNA mismatch correction. *Proc. Natl. Acad. Sci.* **77:**1063.

Goldmark, P.J. and S. Linn. 1972. Purification and properties of the *recBC* DNase of *E. coli* K-12. *J. Biol. Chem.* **247:**1849.

Grossman, L., S. Riazuddin, W. Haseltine, and K. Lindan. 1978. Nucleotide excision repair of damaged DNA. *Cold Spring Harbor Symp. Quant. Biol.* **43:**947.

Hall, J.D. and D.W. Mount. 1981. Mechanisms of DNA replication and mutagenesis in ultraviolet-irradiated bacteria and mammalian cells. *Prog. Nucleic Acids Res. Mol. Biol.* **25:**53.

Hanawalt, P.C., E.C. Friedberg, and C.F. Fox, eds. 1978. *DNA repair mechanisms*. Academic Press, New York.

Hanawalt, P.C., P.K. Cooper, A.K. Ganesan, and C.A. Smith. 1979. DNA repair in bacteria and mammalian cells. *Annu. Rev. Biochem.* **48:**783.

Hecht, R. and H.W. Thielmann. 1978. Purification and characterization of an endonuclease from *Micrococcus luteus* that acts on depurinated and carcinogen-modified DNA. *Eur. J. Biochem.* **89:**607.

Holloman, W.K. and R. Holliday. 1973. Studies on a nuclease from *Ustilago maydis*. I. Purification, properties and implications in recombination of the enzymes. *J. Biol. Chem.* **248:**8107.

Kane, C.M. and S. Linn. 1981. Purification and characterization of an apurinic/apyrimidinic endonuclease from HeLa cells. *J. Biol. Chem.* **256:**3405.

Karu, A.E., V. MacKay, P.J. Goldmark, and S. Linn. 1973. The *recBC* deoxyribonuclease of *Escherichia coli* K-12. Substrate specificity and reaction intermediates. *J. Biol. Chem.* **248:**4874.

Kelley, R.B., M.R. Atkinson, J.A. Huberman, and A. Kornberg. 1969. Excision of thymine dimers and other mismatched sequences by DNA polymerase of *Escherichia coli*. *Nature* **224:**495.

Kelley, R.B., N.R. Cozzarelli, M. Deutscher, I.R. Lehman, and A. Kornberg. 1970. Enzymatic synthesis of deoxyribonucleic acid. XXXII. Replication of duplex deoxyribonucleic acid by polymerase at a single strand break. *J. Biol. Chem* **245:**39.

Kudrna, R.D., J. Smith, S. Linn, and E.E. Penhoet. 1979. Survival of apurinic SV40 DNA in the D complementation group of xeroderma pigmentosum. *Mutat. Res.* **62:**173.

Kuhnlein, U., E.E. Penhoet, and S. Linn. 1976. An altered apurinic DNA endonuclease activity in group A and group D xeroderma pigmentosum fibroblasts. *Proc. Natl. Acad. Sci.* **73:**1169.

Kuhnlein, U., B. Lee, E.E. Penhoet, and S. Linn. 1978. Xeroderma pigmentosum fibroblasts of the D group lack an apurinic endonuclease species with a low apparent Km. *Nucleic Acids Res.* **5:**951.

LaBelle, M. and S. Linn. 1982. *In vivo* excision of pyrimidine dimers is mediated by a DNA-glyiosylase in *Micrococcus luteus* but not in human fibroblasts. *Photochem. Photobiol.* (in press).

Lackey, D., S.W. Krauss, and S. Linn. 1982. Isolation of an altered form of DNA polymerase I from *Escherichia coli* cells induced for *recA* protein. *Proc. Natl. Acad. Sci.* **79:**330.

Lawley, P.D. 1975. "Excision" of bases from DNA methylated by carcinogens *in vivo* and its possible significance in mutagenesis and carcinogenesis. In *Molecular mechanisms for repair of DNA* (ed. P.C. Hanawalt and R.B. Setlow), p. 25. Plenum Press, New York.

Lehman, I.R. 1981. DNA polymerase I of *Escherichia coli*. In *The enzymes* (ed. P.D. Boyer), vol. 14, p. 16. Academic Press, New York.

Lindahl, T. 1971. Excision of pyrimidine dimers from ultraviolet-irradiated DNA by exonucleases from mammalian cells. *Eur. J. Biochem.* **18:**407.

———. 1979. DNA glycosylases, endonucleases for apurinic/apyrimidinic sites, and base excision repair. *Prog. Nucleic Acid Res.* **22:**135.

———. 1982. DNA repair enzymes. *Annu. Rev. Biochem.* **51** (in press).

Lindahl, T., B. Rydberg, T. Hjelmgren, M. Olsson, and A. Jacobsson. 1981. Cellular defense mechanisms against alkylation of DNA. In *Molecular and cellular mechanisms of mutagenesis* (ed. J.F. Lemontt and W.M. Generoso). Plenum Press, New York.

Linsley, W.S., E.E. Penhoet, and S. Linn. 1977. Human endonuclease specific for apurinic/apyrimidinic sites in DNA. Partial purification and characterization of multiple forms in placenta. *J. Biol. Chem.* **252:**1235.

Livneh, Z., P. Elad, and J. Sperling. 1979a. Endonucleolytic activity directed towards 8-(2-hydroxy-2-propyl) purines in double stranded DNA. *Proc. Natl. Acad. Sci.* **76:**5500.

———. 1979b. Enzymatic insertion of purine bases into depurinated DNA *in vitro*. *Proc. Natl. Acad. Sci.* **76:**1089.

Ljungquist, S. 1977. A new endonuclease from *Escherichia coli* acting at apurinic sites in DNA. *J. Biol. Chem.* **252:**2808.

Lo, K.-Y. and M.J. Bessman. 1976. An antimutator deoxyribonucleic acid polymerase. I. Purification and properties of the enzyme. *J. Biol. Chem.* **251:**2475.

Meselson, M., P. Pukkila, M. Rykowski, J. Peterson, M. Radman, and R. Wagner. 1980. Methyl-directed mismatch repair—a mechanism for correcting DNA. *J. Supramol. Struct.* (Suppl.) **4:**311.

Miyaki, M., G. Sai, S. Katagiri, N. Akamatsu, and T. Ono. 1977. Enhancement of DNA polymerase II activity in *E. coli* after treatment with *N*-methyl-*N'*-nitro-*N*-nitrosoguanidine. *Biochem. Biophys. Res. Commun.* **76:**136.

Mosbaugh, D.W. and S. Linn. 1980. Further characterization of human fibroblast apurinic/apyrimidinic DNA endonucleases: The definition of two mechanistic classes of enzyme. *J. Biol. Chem.* **255:**11743.

———. 1982. Characterization of the action of *Escherichia coli* DNA polymerase I at incisions produced by repair endonucleases. *J. Biol. Chem.* **257:**575.

Mosbaugh, D.W. and R.P. Meyer. 1980. Interaction of mammalian deoxyribonuclease V, a double strand 3'⟶5' and 5'⟶3' exonuclease, with deoxyribonucleic acid polymerase-β from the Novikoff hepatoma. *J. Biol. Chem.* **255:**10239.

Muskavitch, K.M.T. and S. Linn. 1981. *recBC*-like enzymes: Exonuclease V deoxyribonucleases. In *The enzymes* (ed. P.D. Boyer), vol. 14, p. 233. Academic Press, New York.

Muzyczka, N., R.L. Poland, and M.J. Bessman. 1972. Studies on the biochemical basis of spontaneous mutation. I. A comparison of the deoxyribonucleic acid polymerase of mutator, antimutator and wild-type strains of bacteriophage T4. *J. Biol. Chem.* **248:**1417.

Nakabeppu, Y. and M. Sekiguchi. 1981. Physical association of pyrimidine dimer DNA glycosylase and apurinic/apyrimidinic DNA endonuclease essential for repair of ultraviolet-damaged DNA. *Proc. Natl. Acad. Sci.* **78:**2742.

Oishi, M. and C.L. Smith. 1978. Inactivation of phage repressor in a permeable cell system: Role of *recBC* DNase in induction. *Proc. Natl. Acad. Sci.* **75:**3569.

Olsson, M. and T. Lindahl. 1980. Repair of alkylated DNA in *Escherichia coli*. *J. Biol. Chem.* **255:**10569.

Radany, E.H. and E.C. Friedberg. 1980. A pyrimidine dimer-DNA glycosylase activity associated with the *v* gene product of bacteriophage T4. *Nature* **286:**182.

———— . 1982. Demonstration of pyrimidine dimer-DNA glycosylase activity *in vivo*: Bacteriophage T4-infected *Escherichia coli* as a model system. *J. Virol.* **41:**88.

Roberts, J.W., C.W. Roberts, and N.L. Craig. 1978. *Escherichia coli recA* gene product inactivates phage λ repressor. *Proc. Natl. Acad. Sci.* **75:**4714.

Sancar, A., N.D. Clarke, J. Griswold, W.J. Kennedy, and W.D. Rupp. 1981b. Identification of the *uvrB* gene product. *J. Mol. Biol.* **148:**63.

Sancar, A., R.P. Wharton, S. Selzer, B.M. Kacinski, N.D. Clarke, and W.D. Rupp. 1981a. Identification of the *uvrA* gene product. *J. Mol. Biol.* **148:**45.

Seawell, P.C., C.A. Smith, and A.K. Ganesan. 1980. *denV* Gene of bacteriophage T4 determines a DNA glycosylase specific for pyrimidine dimers in DNA. *J. Virol.* **35:**790.

Seeberg, E. 1981. Multiprotein interactions in strand cleavage of DNA damaged by UV and chemicals. *Prog. Nucleic Acid Res. Mol. Biol.* **26:**217.

Seeberg, E. and K. Kleppe, eds. 1981. *Chromosome damage and repair.* Plenum Press, New York.

Shimizu, K. and M. Sekiguchi. 1976. 5′⟶3′ Exonucleases of bacteriophage T4. *J. Biol. Chem.* **251:**2613.

Strauss, B., D. Scudiero, and E. Henderson. 1975. The nature of the alkylation lesion in mammalian cells. In *Molecular mechanisms for repair of DNA* (ed. P.C. Hanawalt and R.B. Setlow), p. 13. Plenum Press, New York.

Sutherland, B.M. 1981. Photoreactivating enzymes. In *The enzymes* (ed. P.P. Boyer), vol. 14, p. 482. Academic Press, New York.

Tanaka, J.-I. and M. Sekiguchi. 1975. Action of exonuclease V (the *recBC* enzyme) on ultraviolet-irradiated DNA. *Biochim. Biophys. Acta* **383:**178.

Tomura, T. and J.L. Van Lancker. 1975. The effect of a mammalian repair endonuclease on X-irradiated DNA. *Biochim. Biophys. Acta* **402:**343.

Wagner, R. and M. Meselson. 1976. Repair tracts in mismatched DNA heteroduplexes. *Proc. Natl. Acad. Sci.* **73:**4135.

Waldstein, E.A., S. Peller, and R.B. Setlow. 1979. UV-endonuclease from calf thymus with specificity toward pyrimidine dimers in DNA. *Proc. Natl. Acad. Sci.* **76:**3746.

Warner, H.R., B.F. Demple, W.A. Deutsch, C.M. Kane, and S. Linn. 1980. Apurinic/apyrimidinic endonucleases in the repair of pyrimidine dimers and other lesions in DNA. *Proc. Natl. Acad. Sci.* **77:**4602.

Warner, H.R., M.-L. Persson, R.J. Bensen, D.W. Mosbaugh, and S. Linn. 1981. Selective inhibition by harmane of the apurinic/apyrimidinic endonuclease activity of phage T4-induced UV endonuclease. *Nucleic Acid Res.* **9:**6083.

Weiss, B., S.G. Rogers, and A.F. Taylor. 1978. The endonuclease activity of exonuclease III and the repair of uracil-containing DNA in *E. coli*. In *DNA repair mechanisms* (ed. P.C. Hanawalt, E.C. Friedberg, and C.F. Fox), pp. 191. Academic Press, New York.

West, S.C., E. Cassuto, and P. Howard-Flanders. 1981. Mechanism of *E. coli recA* protein directed strand exchange in post-replication repair of DNA. *Nature* **294:**659.

Willetts, N.S. and A.J. Clark. 1969. Characteristics of some multiple recombination-deficient strains of *Escherichia coli*. *J. Bacteriol.* **100:**231.

Williams, J.G.K., T. Shibata, and C.M. Radding. 1981. *Escherichia coli recA* protein protects single-stranded DNA or gapped duplex DNA from degradation by *recBC* DNase. *J. Biol. Chem.* **256:**7573.

Witkin, E.M. 1967. Mutation-proof and mutation-prone modes of survival in derivatives of *Escherichia coli* B differing in sensitivity to ultraviolet light. *Brookhaven Symp. Biol.* **20:**17.

———— . 1975. Relationship among repair, mutagenesis and survival: Overview. In *Molecular mechanisms for repair of DNA* (ed. P.C. Hanawalt and R.B. Setlow), p. 347. Plenum Press, New York.

———— . 1976. Ultraviolet mutagenesis and inducible DNA repair in *Escherichia coli*. *Bacteriol. Rev.* **40:**869.

Witkin, E.M. and D.L. George. 1973. Ultraviolet mutagenesis in *polA* and *uvrA polA* derivatives of *Escherichia coli* B/R: Evidence for an inducible error-prone repair system. *Genet. Suppl.* **73:**91.

Yoakum, G.H. and L. Grossman. 1981. Identification of the *Escherichia coli uvrC* gene product. *Nature* **292:**171.

The ATP-dependent Restriction Endonucleases

Thomas A. Bickle
Microbiology Department
Biozentrum, University of Basel
CH-4056 Basel, Switzerland

The phenomenon of restriction and modification in bacteria was first described close to 30 years ago by Luria and Human (1952) for T-even bacteriophages and by Bertani and Weigle (1953) for phages P2 and λ. The system active against T-even phages (reviewed by Revel and Luria 1970) turned out to be unique to these phages and will not be discussed here. The effect noted by Bertani and Weigle is far more general and can be demonstrated with many different bacteriophages and hosts. The observation was that a phage stock could have different plating efficiencies when tested on closely related bacterial strains. If, however, a stock of phage was prepared on a strain that initially had grown it poorly, that stock would now grow efficiently on that strain. This acquired ability showed Lamarckian characteristics, since it was an adaptive response of the virus to the host and was lost when the virus was passaged through other hosts.

Some 10 years later, the first of a series of papers from Arber's group appeared that provided a molecular explanation for this effect (Arber and Dussoix 1962; Dussoix and Arber 1962). Physiological and genetic experiments showed that restriction and modification operated on the DNA of the infecting phage. Restriction was due to the action of a nuclease that recognized specific sequences in the incoming DNA, and modification was due to the action of another enzyme, early suspected to be a DNA methylase (Arber 1965), that modified these same specific sequences in such a way that they were no longer recognized by the nuclease. It was subsequently shown that restriction was effective against foreign DNA with little regard to how it was introduced into the cell—in a phage head, by conjugation, or by transformation (for reviews of this early work see Arber 1968; Arber and Linn 1969; Boyer 1971).

By the beginning of the 1970s, several different restriction endonucleases had been purified (Meselson and Yuan 1968; Linn and Arber 1968; Roulland-Dusoix and Boyer 1969; Smith and Wilcox 1970; Greene et al. 1974) and it was clear that they could be divided into at least two different classes based on their cofactor requirements. One of these classes comprised enzymes that require Mg^{2+}, ATP, and *S*-adenosylmethionine (AdoMet) for DNA cleavage; the second class requires only Mg^{2+}. Boyer (1971) proposed that the first class of enzymes be called type I and the second, type II. The type-II enzymes are those that have since proved to be such indispensable tools in modern molecular biology and genetics and they are dealt with in detail in Chapter 6. The ATP- and AdoMet-requiring enzymes are the subject of this chapter. It has become apparent in recent years that there are two distinct groups of ATP-requiring enzymes which are now known as types I and III (Kauc and Piekarowicz 1978). For other recent reviews see Modrich (1979), Yuan (1981), and Endlich and Linn (1981).

TYPE-I RESTRICTION ENZYMES

Occurrence

No extensive survey of bacteria for the presence of type-I (or type-III) restriction enzymes has been made. The search for type-II enzymes (Chapter 6) would not have detected the presence of ATP-requiring restriction enzymes. Apart from a single example from the genus *Haemophilus* (Gromkova and Goodgal 1976), all the restriction systems that have been shown to be of type I have been found in Enterobacteriaceae. Of these, systems of different specificity carried by many different strains of *Escherichia coli* and *Salmonella typhimurium* have been shown to be allelic and to have their structural genes mapping close to *serB* at 98.5 minutes on the *E. coli* map (Boyer 1964; Arber and Wauters-Willems 1970; Glover 1970; Bullas and Colson 1975; Bullas et al. 1980; Bachmann and Low 1980).

Genetics

The results of genetic and complementation analysis of this group led to the idea that at least three structural genes were necessary to account for all of the data. One of these genes, *hsdS* (*hsd* for *h*ost *s*pecificity for *D*NA), would be necessary in both restriction and modification for the recognition of the DNA sequence that is specific to the system. A second, *hsdM*, would be responsible, together with the *hsdS* gene product, for modification and a third gene, *hsdR*, in conjunction with both *hsdM* and the *hsdS* product, would catalyze restriction (Boyer and Roulland-Dussoix 1969; Glover and Colson 1969; Hubacek and Glover 1970). The most recent genetic analysis entirely corroborates this model. Sain and Murray (1980) have cloned the *hsd* region of *E. coli* K12 into phage λ and have shown by deletion analysis that the gene order is *hsdR*, *hsdM*, *hsdS*. They have also shown that the genes are organized into two units which are transcribed in the same direction; *hsdM* and *hsdS* are transcribed from one promoter while *hsdR* is transcribed from a second.

The Enzymes

The only type-I restriction enzymes to have been extensively purified are those from the *E. coli* strains K12 and B, *Eco*K and *Eco*B (the nomenclature is that of Smith and Nathans [1973]). These enzymes, as expected from their allelic nature, have very similar structures. Both contain three nonidentical subunits having molecular weights of about 135K, 60K and 50K (α, β, and γ), although the relative proportions of the subunits have been reported to be different for the two enzymes. For *Eco*K the structure was reported to be $\alpha_2\beta_2\gamma$ (Meselson et al. 1972) based on a determination of the molecular weight of the holoenzyme by glycerol-gradient centrifugation and SDS polyacrylamide gel electrophoretic analysis of the subunits. The situation for *Eco*B is more complex; at least three active oligomeric structures have been purified and the major form was proposed to have an $\alpha_2\beta_4\gamma_2$ subunit composition (Eskin and Linn 1972a). These differences between the two enzymes may well reflect differences in the growth conditions of the bacteria used for the different enzyme preparations or differences in the purification procedures.

Despite the fact that the purified restriction enzymes show modification activity (see later), it has been possible to purify modification enzymes that lack restriction activity from wild-type strains of *E. coli* B (Lautenberger and Linn 1972) and, more recently, from *E. coli* K12 strains lysogenic for a λ derivative carrying the *hsdK* region (B. Suri et al., unpubl.). Both of these enzymes contain the two smaller subunits found in the restriction enzyme and lack the larger subunit. This large subunit is therefore most likely the *hsdR* gene product. Analysis of deletions in the *hsd* region (Sain and Murray 1980) has shown that the smaller subunit corresponds to *hsdS* and the intermediate polypeptide to *hsdM*.

A close relationship betwen *Eco*B, *Eco*K, and the allelic *Salmonella* systems was already clear from genetic studies. More recently, the physical relationships among some of these systems have been investigated using DNA hybridization techniques (Sain and Murray 1980; Murray et al. 1982) or, in our laboratory, with antibodies raised against the purified *Eco*K, *hsdM*, and *hsdR* subunits (Murray et al. 1982). Sain and Murray showed by Southern hybridization (Southern 1975) that *E. coli* B chromosomal DNA cross-reacts well with probes derived from the *E. coli* K12 *hsd* region and that, very importantly, DNA from *E. coli* C, a strain that naturally contains no restriction-modification system, has no sequences that hybridize with the probes. This last result disposes of the argument that has often been made that type-I restriction enzymes are so complicated that they must perform some vital function other than restriction in the cell. The proponents of this theory have argued that restriction-modification mutants, or the hypothetical enzyme in *E. coli* C, are still active in this other, undefined, function. Since *E. coli* C does not contain the structural genes, it cannot produce this hypothetical enzyme. Further support of this comes from the study of *E. coli* K12 derivatives that have deleted the entire *hsd* region. These strains have no obvious phenotype other than being defective in modification and restriction (N.E. Murray, pers. comm.).

We have used our antibodies to look for the presence of cross-reacting material in *E. coli* B, *E. coli* C, and *E. coli* A, an *E. coli* strain with a restriction-modification system that is most probably type I but is not allelic to the *E. coli* B or K systems (Arber and Wauters-Willems 1970), and to the *Salmonella* SB, SP, and SQ systems (SQ is the result of a cross betwen SB and SP and has a novel specificity) (Bullas et al. 1976). As expected, the antibodies reacted well with *E. coli* B extracts and did not react at all with extracts from *E. coli* C. In fact, no differences could be detected between B and K12 strains. However, no reaction was observed with *E. coli* A, indicating that enzymes may exist that have a type-I mechanism but are physically unrelated to *Eco*K and *Eco*B. The *Salmonella* enzymes also cross-reacted with the antibodies, although the reaction with the *hsdM* subunits was much stronger than with the *hsdR*. Even with the *hsdM* antibodies, the reaction was much weaker than with the homologous enzyme or with *E. coli* B (Fig. 1). Another interesting result from these investigations was that the anti-*hsdR* sera showed some cross-reaction with the *hsdM* gene product and vice versa. Given the way in which the subunits were prepared for raising antisera (elution from a SDS-polyacrylamide gel), it is unlikely that one was contaminated with the other and it is possible that *hsdR* arose from *hsdM* by gene fusion or gene duplication. Further work is necessary to clarify this situation.

Recognition Sequences

The DNA sequences recognized by both *Eco*B (Lautenberger et al. 1978; Ravetch et al. 1978; Sommer and Schaller 1979; Lautenberger et al., 1979) and *Eco*K (Kan et al. 1979) have been determined, mainly from the analysis of

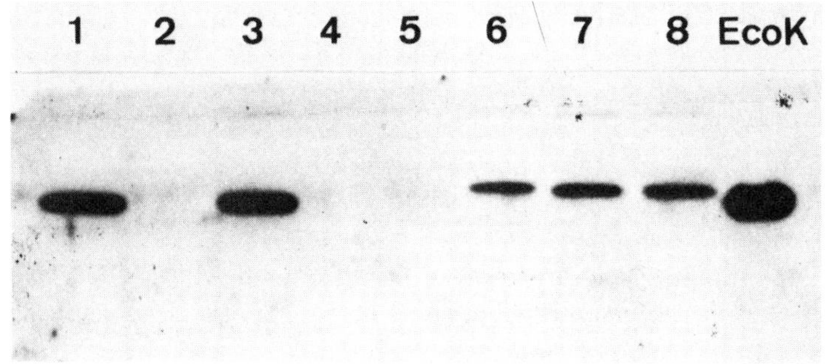

Figure 1 Physical relationships among type-I restriction enzymes. Crude extracts from about 5 × 10^7 cells were electrophoresed on an SDS-polyacrylamide gel, and the proteins were transferred electrophoretically to a nitrocellulose filter and treated with antibodies prepared against the *Eco*K *hsdM* subunit (Towbin et al. 1979). The positions of antibody binding were revealed by treatment with ^{125}I-labeled protein A followed by fluorography. The lanes contain: (1) *E. coli* B; (2) *E. coli* C; (3) a strain with the *hsdB* genes in a K12 background; (4) *E. coli* 15T$^-$; (5) a strain with the *hsdA* genes in a K12 background; (6, 7, and 8) *E. coli* strains carrying the genes for the *Salmonella* restriction systems SB, SP, and SQ, respectively. The last lane contains a sample of purified *Eco*K.

mutant sequences. Both of these recognition sites have the same, unusual overall structure in that they both consist of a specific heptameric sequence divided into a trimer and a tetramer by a spacer of nonspecific sequence of a fixed but different length for the two enzymes. Neither sequence has any element of symmetry.

The *Eco*B recognition sequence is:

5′-TGANNNNNNNNTGCT-3′
3′-ACTNNNNNNNNACGA-5′

and the *Eco*K sequence is:

5′-AACNNNNNNGTGC-3′
3′-TTGNNNNNNCACG-5′.

For both restriction systems, modification results in the methylation of adenine residues in both strands to form 6-methyladenine. For *Eco*B there is only one adenine residue in the top strand as written and this must therefore be the one that is methylated. In the lower strand, there are three adenine residues and it has been argued from the analysis of methylated oligonucleotides (Van Ormondt et al. 1973) that it is the first adenine residue in the tetrameric part of the sequence that becomes methylated.

For *Eco*K, the lower strand as written contains only one adenine residue and therefore this must be the one that is methylated. For the top strand, it is most likely that the second of the two adenine residues is methylated. One of the *Eco*K sites in phage λ DNA overlaps with a cleavage site for the type-II enzyme, *Hin*dII: GTTAAC. It is known that when the second adenine residue in the sequence is methylated, this sequence is not cleaved by *Hin*dII (Roy and Smith 1973). As the experiment shown in Figure 2 demonstrates, *Hin*dII will not cleave at this site in a DNA fragment isolated from modified λ DNA, so it is most likely that the second adenine residue in the top strand is methylated (K. Ineichen and T.A. Bickle, unpubl.).

It has been pointed out that the *Eco*K and *Eco*B recognition sequences can be arranged in such a way that, despite the differences in length of the

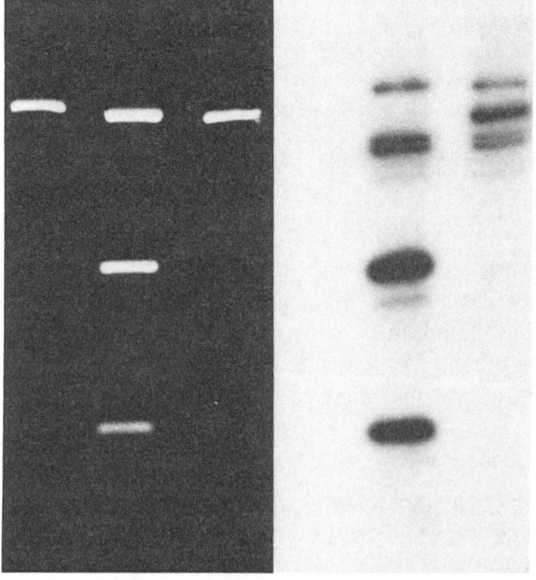

Figure 2 *Eco*K methylates the second adenine residue in the trimeric part of its specific sequence. A 260-bp *Bgl*I fragment from both K-modified and nonmodified DNA was prepared. This fragment contains an *Eco*K site whose AAC sequence forms part of a *Hin*dII site, GTTAAC. The nonmodified fragment was nick-translated (Rigby et al. 1977). The two preparations were then mixed, digested with *Hin*dII, and fractionated by agarose gel electrophoresis. In each panel, the left slot shows the modified fragment, the right the ^{32}P-labeled nonmodified fragment, and the center the *Hin*dII-digested mixture of the two. The left part of the figure shows a photograph of the stained gel and the right part an autoradiograph. The radioactive bands visible in the autoradiograph but not in the stained gel are single-stranded products of strand displacement and strand switching during nick translation (K. Ineichen and T.A. Bickle, unpubl.).

nonspecific spacer, four out of the seven base pairs are the same (Kan et al. 1979):

*Eco*B: 5′-TGANNNNNNNNTGCT-3′
*Eco*K: 5′-AACNNNNNNGTGC-3′.

The conserved positions, if this homology has any significance, include both of the methylated residues.

Reaction mechanisms

Both *Eco*K and *Eco*B have very similar reaction mechanisms and in most of what follows it can be taken that features of the mechanism determined with one enzyme are also valid for the other. These enzymes catalyze an extraordinary variety of different reactions. They are, of course, restriction endonucleases but they can also act as DNA-site-specific modification methylases and DNA-dependent ATPases and, finally, they show a DNA gyraselike action. These different reactions are discussed in detail below.

Figure 3 shows a reaction scheme derived mainly for *Eco*K. The enzyme alone shows no detectable affinity for DNA, whether it be substrate or modified. The addition of AdoMet results in a slow, allosteric conformational change to a form that is capable of binding to DNA; the AdoMet concentration dependence of this activation suggests that the enzyme has at least three binding sites for the cofactor (Hadi et al. 1975). Once the enzyme has been activated, the presence of free AdoMet is no longer required for the rest of the reactions leading to restriction (Yuan et al. 1975).

The AdoMet-activated enzyme has been shown to interact nonspecifically with DNA that contains no specific sequences because such DNA can protect

```
                E   +   AdoMet
                          ↓
                E.      AdoMet
                          ↓
                E*.     AdoMet + DNA
                          ↓
                E*.     AdoMet.DNA (non-specific)
                          ↓
                E*.     AdoMet.DNA (at recognition site)
FURTHER REACTIONS DEPEND ON THE METHYLATION STATE OF THE SITE AND ATP
     ↓                        ↓                      ↓
Modified + ATP           Half-modified         Non-modified + ATP
     ↓                      (+ ATP)
Dissociation                  ↓                      ↓
     ↓                 Full modification     Conformational change
(? inactivation)                                     ↓
                                               (? ATP hydrolysis)
                                                     ↓
                                             Tracking to cleavage site
                                                     ↓
                                                DNA cleavage
                                                     ↓
                                                ATPase activity
```

Figure 3 The reaction mechanism of type-I restriction enzymes.

the activated state (Yuan et al. 1975). These nonspecific interactions, which are probably of biological importance (Von Hippel et al. 1974; Kao-Huang et al. 1977), are succeeded by tight binding to the specific recognition sites should they be present. Surprisingly, in the absence of ATP, the enzyme binds to modified recognition sites almost as well as to nonmodified ones (Bickle et al. 1978), although the complexes can be distinguished on the basis of their stability and their sensitivity to the polyanionic glucan, heparin, a DNA analog that inhibits many nucleic acid enzymes. In the absence of ATP, these complexes with the recognition sequences are extremely stable; the half-life of the complex with a modified site is about 6 minutes whereas that with an unmodified site is about 22 (Yuan et al. 1975).

The addition of ATP to enzyme-recognition-site complexes allows further enzyme reactions that depend on the modification state of the recognition site. Gel filtration studies and observations by electron microscopy have shown that the enzyme bound to modified sites is released from the DNA and, presumably, is then free to interact with other DNA molecules (Bickle et al. 1978). It should be noted, however, that ATP has been reported to be an irreversible inhibitor of the free enzyme (Meselson et al. 1972), even in the presence of AdoMet and modified DNA.

When the restriction enzymes are bound to heteroduplex sites that are modified in only one of the DNA strands, ATP stimulates the methylation of the other strand and no DNA cleavage is observed (Vovis et al. 1974; Burckhardt et al. 1981; Suri et al. 1982). Even in the absence of ATP, these sites are relatively efficiently methylated; in our hands, these reactions show overall first-order kinetics with rate constants in the presence of ATP of 3×10^{-3} sec^{-1} and in its absence of 10^{-4} sec^{-1}. For comparison, the methylation of homoduplex DNA has a rate constant of 6×10^{-5} sec^{-1} and this reaction is inhibited about twofold by the presence of ATP (Suri et al. 1982).

Efficient methylation of heteroduplex sites is, of course, absolutely necessary for cell survival because heteroduplex sites are the normal product of semiconservative DNA replication. Should such a site remain heteroduplex until the next cycle of replication, homoduplex unmodified sites would be generated and these would be a target for restriction. In this respect, it makes good physiological sense that restriction by type-I enzymes is dependent on the presence of the methyl donor, AdoMet. If for some reason AdoMet levels are too low to support modification, they are also too low to support restriction. Lark and Arber (1970) showed that when *met*$^-$ strains of *E. coli* containing type-I restriction-modification systems were grown in the presence of the methionine analogs ethionine or norleucine, which block AdoMet synthesis but not protein synthesis, the amount of DNA in the cells more than doubled before DNA synthesis stopped and the newly synthesized DNA was completely stable. In contrast, strains containing type-II and type-III restriction-modification systems, neither of which depends on AdoMet for restriction, degraded their DNA in similar experiments (Lark 1968; Lark and Arber 1970).

The only recognition sites that can generate DNA cleavage are those that are completely unmodified. The effect of ATP on enzyme bound to these is complex (Bickle et al. 1978). The first discernible effect is a major conformational change visible in the electron microscope (perhaps the same change that inactivates the free enzyme when it is exposed to ATP) (Meselson et al. 1972). The enzyme may lose one or more subunits, but this remains to be demonstrated. This change is accompanied by an increased ability of the enzyme to bind DNA to nitrocellulose filters (Yuan and Meselson 1970), a fact that forms the basis of a very useful assay for the enzyme. *Eco*K will bind unmodified, but not modified DNA to nitrocellulose filters in the presence of AdoMet, even in the absence of ATP, but this binding is poor and somewhat irreproducible. In the presence of AdoMet and ATP, on the other hand, the binding is more reproducible and with excess DNA the amount of DNA bound is proportional to the amount of enzyme present. Both the conformational change in the enzyme and the stimulation of DNA binding to nitrocellulose filters can be elicited by nonhydrolyzable analogs of ATP. These cannot, however, substitute for ATP in the later stages of the reaction (Bickle et al. 1978).

The most surprising feature of DNA cleavage by type-I restriction endonucleases is that, although the enzyme can be shown by electron microscopy to remain bound to the recognition site throughout the reaction (Bickle et al. 1978; Rosamond et al. 1979), DNA cleavage occurs randomly at considerable distances—never less than 400 and as far as 7000 bp—from the site (Horiuchi and Zinder 1972; Adler and Nathans 1973; Bickle et al. 1978; Rosamond et al. 1979; Yuan et al. 1980a). Two somewhat conflicting accounts have appeared as to how the enzyme does this. For *Eco*B, electron microscope evidence has been obtained that the enzyme "tracks" along the DNA in one direction only, away from the trimeric part of the recognition sequence, and forms looped intermediates in which the enzyme is simultaneously bound to the recognition site and to downstream DNA sequences (Rosamond et al. 1979). A similar mechanism, again based on studies by electron microscopy, has been proposed for *Eco*K (Yuan et al. 1980a). But in this case, the enzyme could cleave on either side of the recognition sequence and with long DNA substrates, the intermediate loops were supercoiled. It was proposed that the enzyme was "pumping" the DNA past it and generating the superhelical turns in a manner similar to DNA topoisomerases (see Chapter 3). It is possible that *Eco*K and *Eco*B have slightly different reaction mechanisms in this respect, but it is most likely that the differences reflect the reaction conditions, DNA substrates, and electron microscope preparation techniques used in various laboratories.

Whatever the exact mechanism, DNA cleavage is a two-step process; the DNA is first nicked in one strand and then, several seconds later, in the second strand. In the presence of excess DNA, only single-strand breaks are found, suggesting that two enzyme molecules are required for double-strand cleavage (Meselson and Yuan 1968; Roulland-Dussoix and Boyer 1969; Lautenberger and Linn 1972b; Adler and Nathans 1973). In addition, it has been reported that *Eco*B releases about 75 bases in the form of acid-soluble oligonucleotides per cleavage event (Kimball and Linn 1976); this has not been found for *Eco*K. A further curious feature of the DNA ends produced by both *Eco*K and *Eco*B is that they are resistant to the action of polynucleotide kinase, even following phosphatase treatment (Eskin and Linn 1972; Murray et al. 1973).

Following, or possibly concomitantly with, DNA cleavage, the type-I restriction endonucleases are irreversibly transformed into potent ATPases (Eskin and Linn 1972b; Yuan et al. 1972). This transformation is accompanied by a loss of endonucleolytic ability with the net effect that the enzyme as a nuclease does not turn over: each enzyme molecule can catalyze only one DNA scission. By contrast, the ATPase activity is extremely long-lasting and goes on for hours after DNA cleavage has ceased, with each enzyme molecule hydrolyzing some 10^4 ATP molecules per minute (Boyer et al. 1971). Nonhydrolyzable ATP analogs do not allow DNA cleavage (Bickle et al. 1978) and thus it is possible that hydrolysis of the first molecule, or the first few molecules, of ATP are necessary for DNA cleavage. It has been proposed that ATP hydrolysis drives the translocation of the DNA past the enzyme bound to the recognition site (Yuan et al. 1980a). This would not explain why the hydrolysis continues for so long after DNA cleavage has finished nor why the kinetics of ATP hydrolysis should be the same during the period of DNA cleavage as at later times.

Whether the long-term ATP hydrolysis serves any biological function is still an open question; it may well be no more than a laboratory artefact. However, it is possible to envisage at least two ways in which it might be advantageous, one of them to a phage-infected cell or to a cell that has just received a conjugative plasmid, the other to the bacterial population as a whole. In order for restriction to perform its function of keeping the cellular genome pure, an invading phage DNA (or plasmid) must not only be restricted but the resulting DNA fragments must be further degraded to nucleotides by nonspecific nucleases before illegitimate recombination events can integrate these pieces into the genomic DNA. The massive ATP hydrolysis of type-I restriction enzymes may, by lowering intracellular ATP levels, decrease the frequency of illegitimate recombination and give the nonspecific nucleases a better chance. One argument against this is that in *E. coli* the major nuclease that degrades restriction fragments is the *rec*BC exonuclease (Brammar et al. 1974), which itself requires ATP (Goldmark and Linn 1972). A second way in which the ATP hydrolysis might be of benefit, this time to the population as a whole, would be if a phage-infected cell hydrolyzed so much ATP that it did not survive. This would ensure that the genetic pool of the population as a whole remained pure at the expense of a few individuals. Such an altruistic approach to bacterial population dynamics has been suggested to account for the occasional spontaneous release of colicins by colicinogenic cells (Pugsley and Rosenbusch 1981) and is somewhat reminiscent of the mechanism of interferon production.

TYPE-III RESTRICTION ENZYMES

Only three members of this class have been identified for the moment but it is probable that, as for the type-I enzymes, a systematic search would uncover many more. One member is coded by the *E. coli* prophage P1 (*Eco*P1) and a second by the resident plasmid of *E. coli* 15T$^-$ (*Eco*P15); the third is from *Haemophilus influenzae* Rf (*Hin*fIII) and is most likely coded on the chromosome (Arber and Dussoix, 1962; Arber and Wauters-Willems 1970; Piekarowicz et al. 1974). An isoschizomer of *Hin*fIII exists in *H. influenzae* Re (Piekarowicz and Kalinowska 1974; Piekarowicz 1982).

The resident plasmid of *E. coli* 15T$^-$ has a high degree of homology with the P1 prophage (Ikeda et al. 1970), and the genes specifying the *Eco*P1 and *Eco*P15 restriction systems were early shown to be allelic (Arber and Wauters-Willems 1970). When cells harboring a P1 prophage or P15 plasmid are starved for methionine in the presence of ethionine so that AdoMet synthesis is blocked, DNA synthesis continues but the newly synthesized DNA is rapidly degraded. Because cells carrying type-II restriction-modification systems show the same behavior, the type-II and type-III systems were originally grouped together (Lark 1968; Lark and Arber 1970). However, when the activity of the enzymes was examined (Meselson and Yuan 1968; Haberman 1974), it became apparent that their substrate requirements were more complicated than those of the type-II enzymes. The designation of type III for these enzymes was proposed by Kauc and Piekarowicz (1978).

Genetics

The genetics of the P1 restriction system have been complicated by the existence of two mutations, *c*2 and *c*3 (Scott 1970). These are clear-plaque mutants that fail to plate efficiently on strains carrying a P1 cryptic prophage that lacks P1 immunity but has a wild-type restriction/modification system. The *c*2 and *c*3 mutations were interpreted as being defective in modification and, since they complement each other, defining two genes involved in modification (Rosner 1973). Since at least one more gene would be required for restriction, this was taken as support for an ''at least three gene'' model for P1 restriction and modification (Reiser 1975).

Recent experiments from our laboratory have shown that there are, in fact, only two genes involved in P1 and P15 restriction and modification (S. Iida, J. Meyer, B. Báchi, T. Bickle, and W. Arber, unpubl.). The most likely interpretation of the *c*2 and *c*3 mutations is that they both lie in the same gene and that the complementation is intracistronic. The analysis of these genes by transposition mutagenesis led to the definition of two contiguous regions of the P1 genome. Transpositions into one of these regions led to a restriction$^-$modification$^+$ (Res$^-$Mod$^+$) phenotype, and define a gene that we call *res*. Transposition into the other region gave a Res$^-$Mod$^-$ phenotype defining a gene called *mod*. These regions span 3 kb and 2 kb, respectively. Both *res* and *mod* gene products are necessary for restriction, while modification requires only *mod*. When this region or the corresponding region from P15 (actually from a P1-15 hybrid phage carrying the region) were cloned into small plasmids, these plasmids conferred the corresponding modification and restriction phenotype on their host cells. Furthermore, when subclones were constructed in which most of the DNA from the first region was deleted, the strains now had the Res$^-$Mod$^+$ phenotype.

Transcriptional mapping of DNA fragments covering these regions has shown that there are two promoters, one transcribing the *mod* gene and the other the *res* gene, both directing transcription in the same direction (S. Iida et al., in prep.).

When heteroduplexes were prepared between restriction fragments carrying the P1 genes and those carrying the P15 genes, it was found that the two DNA regions are largely homologous. A 1.2-kb region of nonhomology was found within the *mod* gene and a 0.6-kb region towards the aminoterminus of the *res* gene showed partial nonhomology; thus, the gene products must be quite similar. A summary of all of these results is shown in Figure 4.

The Enzymes

Early purifications of *Eco*P1 and *Eco*P15 gave preparations containing four different subunits, only two of which had approximately the same molecular weights for the two enzymes. At that time, it was clear that these preparations were probably not homogeneous because the stoichiometry of these subunits gave a minimal molecular weight for the enzyme, considerably higher than that estimated by glycerol gradient centrifugation (Reiser and Yuan 1977). More recently, we have shown that both *Eco*P1 and *Eco*P15 each contain two subunits of molecular weights 106K and 75K (S.M. Hadi, B. Bachi, S. Iida,

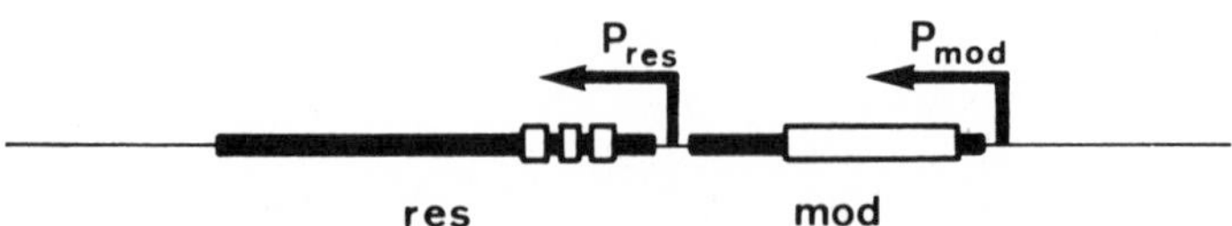

Figure 4 The organization of the genes for *Eco*P1 and *Eco*P15. The limits of the genes defined by transposition mapping are shown together with the regions of nonhomology (open boxes) between P1 and P15, the positions of the promoters, and the directions of transcription.

and T. Bickle, in prep.). As one might expect from the heteroduplex analysis of the structural genes for these enzymes, antibodies raised against one of the enzymes cross-react with both subunits of the other. The smaller of the two subunits is the product of the *mod* gene (Fig. 5) and when purified from strains containing deletions of the *res* gene is capable of modifying DNA in vitro but not of restricting it. The small subunit therefore provides both specific sequence recognition and DNA methylation functions; the larger subunit, which is practically identical for the two enzymes, must contain the endonuclease activity. The *res* gene product has also been independently identified as an approximately 100K polypeptide by transposition mutagenesis followed by infection of minicells (Heilmann et al. 1980).

Figure 5 Immunological cross-reactivity of *Eco*P1 subunits and *Eco*P15 antiserum. The figure shows the immune replica (Showe et al. 1976) made with anti-P15 antiserum of a SDS-polyacrylamide gel of (A) *Eco*P1, (B) a crude extract from cells harboring a plasmid with a deletion of the P1 *res* gene and with a res^-_{P1} mod^+_{P1} phenotype, and (C) the same strain transformed with pBR322, the cloning vehicle used. (*Top*) Stained gel; (*bottom*) immune replica.

*Hin*fIII has never been purified to homogeneity (Kauc and Piekarowicz 1978). However, it is noteworthy that the most purified preparations contain two prominent polypeptides with the same molecular weights as those of the *E. coli* enzymes.

DNA Recognition and Cleavage Sequences

The recognition sequences for all three type-III restriction enzymes have been determined. They are:

*Eco*P1 5′-AGACC-3′ (Bächi et al. 1979)
3′-TCTGG-5′

*Eco*P15 5′-CAGCAG-3′ (Hadi et al. 1979);
3′-GTCGTC-5′

*Hin*fIII 5′-CGAAT-3′ (Piekarowicz et al. 1981).
3′-GCTTA-5′

An earlier determination of the *Eco*P1 sequence showed that the last position in the top strand above could be any pyrimidine but was predominantly C (Hattman et al. 1978). However, more recent investigations give no trace of modification of the sequence 5′-AGACT-3′ or of DNA cleavage stimulated by that sequence.

DNA modification by all of the type-III restriction systems results in the introduction of methyl groups into adenine residues to give 6-methyl adenine (Brockes et al. 1972; Reiser 1975; Kauc and Piekarowicz 1978). The sequences recognized by both *Eco*P1 and *Eco*P15 contain adenosyl residues in one strand of the DNA only and thus can only be methylated on that strand within the specific sequence. In vitro modification with [^{3}H]AdoMet followed by strand separation has shown for these enzymes and for *Hin*fIII (which has adenosyl residues in both strands) that, in fact, only one strand becomes methylated (and so methylation of the other strand outside the specific sequence does not occur), in each case the top one in the sequences as shown above (Bächi et al. 1979; Piekarowicz et al. 1981). This poses a biological problem, one of many that these enzymes pose, that has yet to be resolved. When the chromosome of a cell harboring a type-III restriction-modification system replicates, the parental methyl group at a particular site will pass to one of the daughter DNA molecules, which will be fully modified. The other daughter molecule will inherit no methyl groups, will be fully unmodified, and thus ought to be a target for the restriction enzyme present in the cell, which would of course be a lethal situation. We do not yet know how the cell avoids this. A relatively trivial explanation, that the restriction enzyme might be periplasmic, has been ruled out (Bächi et al. 1979). It is conceivable that the replicating DNA is methylated immediately following replication, perhaps by modification subunits built into the complex at the replication fork. In this respect, we have shown that cells carrying a P1 prophage have, besides the restriction enzyme, which itself is capable of modifying DNA (see later), a pool of free modification subunits (S.M. Hadi et al., in prep.).

All of these enzymes cleave the DNA about 25–27 base pairs to the right of the sequence as shown to leave ends that have 2–3-base single-stranded, 5′ extensions. This considerable distance, about 9 nm, can easily be spanned by the enzyme without invoking movement along the DNA. Early experiments showed a tendency for the enzymes to cleave adjacent to thymidine residues in the DNA; this was not substantiated by more recent work (Báchi et al. 1979; Hadi et al. 1979; Yuan et al. 1980b). A likely model for the interaction is that the modification subunit recognizes and binds to the specific sequence and directs the binding of the restriction subunit to the site that will be cleaved. According to this model, the restriction subunit itself need not necessarily have a high affinity for DNA. This might explain the Res^-Mod^- phenotype of P1 and P15 mutants that have a transposon inserted in the *mod* gene but that nevertheless have an intact *res* gene together with its promoter.

Reaction Mechanisms

The reaction mechanisms of all three type-III restriction enzymes are very similar. They all require ATP for endonucleolytic activity but do not hydrolyze it. The nuclease activity is stimulated by AdoMet without its being an absolute requirement, and in the presence of both cofactors the enzymes also act as modification methylases so that restriction and modification are competing reactions. In the presence of AdoMet but without ATP, they are pure modification methylases but ATP also stimulates this reaction (Haberman 1974; Risser et al. 1974; Reiser and Yuan 1977; Kauc and Piekarowicz 1978; Yuan and Reiser 1978).

There are several experimental difficulties that complicate work with these enzymes. The first is that under optimal reaction conditions both restriction and modification are occurring. A second major problem is that these enzymes never make complete digests of the DNA, even in the absence of AdoMet, for reasons that are totally unclear. *Hin*fIII shows the most extreme behavior in this respect in that it will not cleave DNA fragments smaller than about 3000 bp (Piekarowicz and Brzezinski 1980). It should be noted that the modification reaction works efficiently with small DNA fragments. Thus, the inability of the enzyme to cleave such fragments is not due to a failure to bind to the recognition sites. A third complication, which has been demonstrated for *Hin*fIII (Piekarowicz and Brzezinski 1980) but which is probably true of the other two enzymes as well, is that the freshly purified enzymes have variable amounts of AdoMet bound to them that is lost after several weeks of storage. This, of course, complicates the analysis of the effects of AdoMet on the enzyme reactions.

Both AdoMet and ATP have been shown to serve as allosteric effectors of the enzymes in the cleavage reaction. In addition, AdoMet serves as a substrate in the modification reaction (Yuan and Reiser 1978). ATP is not detectably hydrolyzed during either restriction or modification and nonhydrolyzable ATP analogs can substitute, albeit poorly, for ATP in the nuclease reaction (Yuan et al. 1980b). Electron microscopic observations have shown that the enzyme alone, without cofactors, binds to the recognition sites in DNA, as does the enzyme-AdoMet complex (Yuan et al. 1980b). The enzymes show a difference in the apparent K_m for AdoMet in the two reactions: for restriction, the value has been reported to be 10^{-8} M whereas for modification it was 40-fold higher (Reiser and Yuan 1977).

The type-III restriction enzymes do not cleave or methylate each of their recognition sites with the same efficiency and it is not at all clear why this should be so. There is no special feature of the DNA sequences surrounding recognition sites that determines whether they are efficient or not and there is no inverse correlation, that is, a good methylation site is not necessarily a poor cleavage site and vice versa (Yuan et al. 1980b).

In conclusion, there are many aspects of the biology and reaction mechanism of the type-III restriction and modification systems that remain mysterious. Why, for example, are they such relatively inefficient nucleases in vitro, particularly under conditions that presumably reflect those inside the cell, that is, in the presence of ATP and AdoMet? One would predict from the properties of the isolated enzymes that restriction in vivo by type-III enzymes would be relatively inefficient. This is not at all the case, restriction by the P1 system, for example, is more efficient against phage λ than any of the type-I and type-II systems that can be found in *E. coli* (Arber and Dussoix 1962). Again, we have the problem that replication generates completely unmodified sites (a similar situation potentially exists for some of the type-II enzymes that recognize asymmetric sequences—*Mbo*II is an example; see Báchi et al. [1979] for a discussion of this). How does the cell avoid restricting these sites? One is left with the uncomfortable feeling that something basic is missing from the in vitro systems, perhaps an exotic cofactor or a host-coded macromolecular component.

BACTERIOPHAGE ANTIRESTRICTION MECHANISMS

In their natural environment, bacteriophages are likely to encounter a range of different hosts and, thus, potentially a number of different restriction systems. This has stimulated the evolution of ways of avoiding the worst effects of restriction in most bacteriophages that have been examined. The most striking feature of these antirestriction mechanisms is that each bacteriophage seems to have evolved a different strategy. The following discussion will be restricted to the antirestriction mechanisms of the phages of *E. coli*, although they have been detected with other organisms (e.g., see Makino et al. 1979, 1980).

Antirestriction systems have been detected in nearly all of the coliphages: T-even phages, T3 and T7, T5 (which appears to have evolved three independent systems), the single-stranded DNA-containing filamentous phages, various lambdoid phages, P1, and Mu. Of these, only the T-even phages are resistant by virtue of the fact that their DNA contains unusual bases (Sinsheimer 1954); all of the other phage DNAs contain the normal four bases.

Bacteriophage Mu

Bacteriophage Mu is not normally restricted by any of the type-I or type-III systems found in *E. coli*. However, phage mutants can be recovered that are efficiently restricted by all of these systems. These mutants define a phage gene, *mom* (for *m*odification *o*f *M*u), which provides a function that protects Mu DNA from restriction (Toussaint 1976). The *mom* gene is better expressed following induction of a lysogen than after infection and it can also provide protection to a superinfecting λ phage. The *mom* function requires a host factor, the *dam* gene product, to be effective (Toussaint 1977). The *dam* gene product is a DNA adenine methylase that methylates the sequence 5′-GATC-3′ (Hattman et al. 1978). It has now been shown that DNA from a mom^+ Mu grown on a dam^+ host has an unusual modification on 15–20% of its deoxyadenosyl residues. The modification has not yet been identified, although it is clear that it is not a methylation and that it contains a carboxyl residue (Hattman 1979). The modified residue is not found if the phage is mom^- or the host is dam^-. It is found exclusively in the dinucleotides (A, C) and (A, G) (Hattman 1980). Thus, Mu co-opts a host function, the *dam* gene product, to protect itself against restriction in a fashion that is not yet clear. The protection is conferred by a postsynthetic modification to the DNA such that it is no longer recognized by the restriction enzymes.

Phages T3 and T7

T3 and T7 are closely related phages that are both resistant to restriction by type-I restriction systems (Eskridge et al. 1967). For T7, the resistance is due to the product of the first gene to be expressed after infection, the 0.3 gene (Studier 1975), and this gene has no other known function. T3 has a gene that maps in the same position as the 0.3 gene and codes for an enzyme that hydrolyzes AdoMet to thiomethyl adenosine and homoserine (Gefter et al. 1966). This hydrolysis itself should provide an efficient antirestriction mechanism because AdoMet is an essential cofactor for the type-I restriction enzymes and indeed the SAMase gene was shown to provide the antirestriction function of T3 (Studier and Movva 1976). However, mutants have been isolated that have lost the ability to hydrolyze AdoMet but are still resistant to host restriction (Hausmann 1967). This led Krüger et al. (1977) to propose the term *ocr* for "*o*vercome *c*lassical *r*estriction" for both the T3 and T7 genes.

The antirestriction genes of T3 and T7, despite the similarities that exist in function, map position, and kinetics of expression after infection, are unrelated to each other. They show little or no homology on the level of DNA heteroduplex analysis (Davis and Hyman 1971) and the gene products have no immunological cross-reactivity and have completely different chromatographic behavior (Spoerel and Herrlich 1979; Spoerel et al. 1979; Mark and Studier 1981). However, the antirestriction mechanism of both proteins is very similar. Both of them bind stoichiometrically to type-I restriction endonucleases and the resulting complex lacks any of the measureable type-I enzyme activities (Spoerel et al. 1979; Mark and Studier 1981). Both also result in inhibition of the modification reaction.

Bacteriophage T5

Bacteriophage T5 is resistant to all of the type-I, -II and -III restriction systems commonly found in *E. coli* (Davison and Brunel 1979a). Mutants have been isolated that are sensitive to the type-II *Eco*RI restriction system but remain resistant to type-I and -III systems (Davison and Brunel 1979a). Bacteriophage T5 is peculiar in that it injects its DNA in two stages. About 8% of the DNA (the so-called First-Step Transfer [FST] DNA) is injected into the host and after a short pause the rest is injected (McCorquodale and Lanni 1964). The *Eco*RI-sensitive mutants had all acquired at different positions an *Eco*RI site in the FST DNA (Brunel and Davison 1979). This region of the genome has no *Eco*RI sites in the wild type, although six sites are found in the rest of the DNA. Revertants to *Eco*RI resistance had all lost the *Eco*RI site in the FST DNA but not those elsewhere. Evidently the FST DNA encodes a product that will protect the DNA from *Eco*RI. If the FST DNA contains an *Eco*RI site, the DNA will be digested before the antirestriction function can be expressed. The mode of action of the antirestriction function is not known. However, it has been demonstrated that DNA from both wild-type and mutant-sensitive phages grown on strains containing the *Eco*RI restriction-modification system are not protected from the action of *Eco*RI in vitro and so are not modified (Davison and Brunel 1979b).

Bacteriophage λ

Bacteriophage λ has a most unusual antirestriction system coded by an early, nonessential gene called *ral* (Debrouwere et al. 1980a; Zabeau et al. 1980). The effects of *ral* are not seen during the primary infection of a restricting host by a nonmodified phage. Restriction as reflected by the plating efficiency is the same for ral^+ or ral^- phages. However, when the progeny of the primary infection are used to reinfect the restricting host, the ral^+ progeny are found to be completely modified while the ral^- phages are only partially modified. Thus, *ral* presumably acts by stimulating the modification of completely nonmodified DNA by type-I restriction-modification systems. Apparently, nonmodified DNA is a poor substrate for modification in vivo as it is in vitro (Vovis et al. 1974; Burckhardt et al. 1981).

The *ral* gene product has also been reported to interfere with other ATP-dependent host enzymes such as the *rec*BC exonuclease. It also suppresses polarity, leading to the suggestion that it might have an antiterminator activity which, unlike the λN gene product, would not be promoter specific (Debrouwere et al. 1980b).

Bacteriophage P1

This phage has a nonessential gene called *dar* (formerly *par*) that codes for a major capsid internal protein. When this gene is deleted or inactivated by transposons, P1 becomes sensitive to restriction by type-I restriction systems. This protection is conferred upon any DNA that is packaged in a P1 head, including transduced host DNA. In a simultaneous infection with P1 and λ, the P1 DNA is protected against restriction while the λ DNA is not. The most probable scenario is that the *dar* gene product is injected into the host along with the DNA. Because other DNA simultaneously injected is not protected against restriction, *dar* inhibits restriction, in a way that still needs to be investigated, while it is still bound to the DNA (S. Iida, M. Streiff, and T.A. Bickle, unpubl.).

THE FUTURE

One other problem that remains to be investigated for many restriction systems, not only those that require ATP, is the control of expression of the structural genes for the enzymes. When bacteriophage P1 lysogenizes a cell, it is essential that the host-cell DNA be modified before the restriction enzyme gene is expressed. Experimentally, it is found that the cells do not express restriction for about 1 hour after infection and that it takes at least 3 hours for restriction activity to attain its normal level (Arber et al. 1974). In contrast, modification can be detected within a few minutes after the infection (Arber and Dussoix 1962). We have preliminary experiments that indicate that this control does not take place at the level of transcription (C. Levy and T.A. Bickle, unpubl.).

Some kind of control over the expression of restriction-modification functions must be present in all restriction systems that can move from one strain to another, whether by transduction or by conjugation. In no case has the basis for this control been established.

ACKNOWLEDGMENTS

I would like to thank Werner Arber for his thoughtful comments on the manuscript and those of my colleagues who sent me manuscripts prior to publication. Work from my laboratory has been supported by Grants from the Swiss National Foundation for Scientific Research.

REFERENCES

Adler, S.P. and D. Nathans. 1973. Conversion of circular to linear SV40 DNA by restriction endonuclease from *Escherichia coli B*. *Biochem. Biophys. Acta* **299:**177.

Arber, W. 1965. Host specificity of DNA produced by *Escherichia coli*. V. The role of methionine in the production of host specificity. *J. Mol. Biol.* **11:**247.

_________. 1968. Host controlled restriction and modification of bacteriophage. *Symp. Soc. Gen. Microbiol.* **18:**295.

_________. 1974. DNA modification and restriction. *Prog. Nucleic Acid Res. Mol. Biol.* **14:**1.

Arber, W. and D. Dussoix. 1962. Host specificity of DNA produced by *Escherichia coli*. I: Host controlled modification of bacteriophage λ. *J. Mol. Biol.* **5:**18.

Arber, W. and S. Linn. 1969. DNA modification and restriction. *Annu. Rev. Biochem.* **38:**467.

Arber, W. and D. Wauters-Willems. 1970. Host specificity of DNA produced by *Escherichia coli*. XII: The two restriction and modification systems of strain $15T^-$. *Mol. Gen. Genet.* **108:**203.

Arber, W., R. Yuan, and T.A. Bickle. 1974. Strain-specific modification and restriction of DNA in bacteria. *FEBS Proc. Meet.* **9:**3.

Bächi, B., J. Reiser, and V. Pirrotta. 1979. Methylation and cleavage sequences of the *Eco*P1 restriction-modification enzyme. *J. Mol. Biol.* **128:**143.

Bachmann, B.J. and K.B. Low. 1980. Linkage map of *Escherichia coli* K12, edition 6. *Microbiol. Rev.* **44:**1.

Bertani, G. and J.J. Weigle. 1953. Host controlled variation in bacterial viruses. *J. Bacteriol.* **65:**113.

Bickle, T.A., C. Brack, and R. Yuan. 1978. ATP-induced conformational changes in the restriction endonuclease from *Escherichia coli* K12. *Proc. Natl. Acad. Sci.* **75:**3099.

Boyer, H.W. 1964. Genetic control of restriction and modification in *Escherichia coli*. *J. Bacteriol.* **88:**1652.

_________. 1971. DNA restriction and modification mechanisms in bacteria. *Ann. Rev. Microbiol.* **25:**153.

Boyer, H.W. and D. Roulland-Dussoix. 1969. A complementation analysis of the restriction and modification of DNA in *Escherichia coli*. *J. Mol. Biol.* **41:**459.

Boyer, H.W., E. Scibiensky, H. Slocum, and D. Roulland-Dussoix. 1971. The *in vitro* restriction of the replicative form of W.T. and mutant fd phage DNA. *Virology* **46:**703.

Brammar, W.J., N.E. Murray, and S. Winton. 1974. Restriction of λ*trp* bacteriophages by *Escherichia coli* K. *J. Mol. Biol.* **90:**633.

Brockes, J.P., P.R. Brown, and K. Murray. 1972. The deoxyribonucleic acid modification enzyme of bacteriophage P1. Purification and properties. *Biochem. J.* **127:**1.

Brunel, F. and J. Davison. 1979. Restriction insensitivity in bacteriophage T5. III. Characterization of *Eco*RI-sensitive mutants by restriction analysis. *J. Mol. Biol.* **128:**527.

Bullas, L.R. and C. Colson. 1975. DNA restriction and modification systems in *Salmonella*. III. SP, a *Salmonella potsdam* system allelic to the SB system in *Salmonella typhimurium*. *Mol. Gen. Genet.* **139:**177.

Bullas, L.R., C. Colson, and B. Neufeld. 1980. Deoxyribonucleic acid restriction and modification systems in *Salmonella*: Chromosomally located systems of different serotypes. *J. Bacteriol.* **141:**275.

Bullas, L.R., C. Colson, and A. Van Pel. 1976. DNA restriction and modification systems in *Salmonella*. SQ, a new system derived by recombination between the SB system of *Salmonella typhimurium* and the SP system of *Salmonella potsdam*. *J. Gen. Microbiol.* **95:**166.

Burckhardt, J., J. Weisemann, and R. Yuan. 1981. Characterization of the DNA methylase activity of the restriction enzyme from *Escherichia coli* K. *J. Biol. Chem.* **256**:4024.

Davis, R.W. and R.W. Hyman. 1971. A study in evolution: the DNA base sequence homology between coliphages T7 and T3. *J. Mol. Biol.* **62**:287.

Davison, J. and F. Brunel. 1979a. Restriction insensitivity in bacteriophage T5. I. Genetic characterization of mutants sensitive to *Eco*RI restriction. *J. Virol.* **29**:11.

———. 1979b. Restriction insensitivity in bacteriophage T5. II. Lack of *Eco*RI modification in T5$^+$ and T5 *ris* mutants. *J. Virol.* **29**:17.

Debrouwere, L., M. Van Montagu, and J. Schell. 1980a. The *ral* gene of phage λ. III. Interference with *E. coli* ATP dependent functions. *Mol. Gen. Genet.* **179**:81.

Debrouwere, L., M. Zabeau, M. Van Montagu, and J. Schell. 1980b. The *ral* gene of phage λ. II. Isolation and characterization of *ral* deficient mutants. *Mol. Gen. Genet.* **179**:75.

Dussoix, D. and W. Arber. 1962. Host specificity of DNA produced by *Escherichia coli*. II. Control over acceptance of DNA from infecting phage λ. *J. Mol. Biol.* **5**:37.

Endlich, B. and S. Linn. 1981. Type I restriction enzymes. In *The enzymes*, 3rd ed. (ed. P.D. Boyer), vol. 14, part A, p. 137. Academic Press, New York.

Eskin, B. and S. Linn. 1972a. The deoxyribonucleic acid modification and restriction enzymes of *Escherichia coli* B. II: Purification, subunit structure and catalytic properties of the restriction endonuclease. *J. Biol. Chem.* **247**:6183.

———. 1972b. The deoxyribonucleic acid modification and restriction enzymes of *Escherichia coli* B. I: Studies of the restriction adenosine triphosphatase. *J. Biol. Chem.* **247**:6192.

Eskridge, R.W., H. Weinfeld, and K. Paigen. 1967. Susceptibility of different coliphage genomes to host-controlled variation. *J. Bacteriol.* **93**:835.

Gefter, M., R. Hausmann, M. Gold, and J. Hurwitz. 1966. The enzymatic methylation of RNA and DNA. X. Bacteriophage T3-induced *S*-adenosylmethionine cleavage. *J. Biol. Chem.* **241**:1995.

Glover, S.W. 1970. Functional analysis of host specificity mutants in *Escherichia coli*. *Genet. Res.* **13**:227.

Glover, S.W. and C. Colson. 1969. Genetics of host-controlled restriction and modification in *Escherichia coli*. *Genet. Res.* **13**:227.

Goldmark, P. and S. Linn. 1972. Purification and properties of the *rec*BC DNAse of *E. coli* K12. *J. Biol. Chem.* **247**:1849.

Greene, P.J., M.C. Betlach, H.W. Boyer, and H.M. Goodman. 1974. The *Eco*RI restriction endonuclease. *Methods Mol. Biol.* **7**:87.

Gromkova, R. and S. Goodgal. 1976. Biological properties of a *Haemophilus influenzae* restriction enzyme, *Hin*dI. *J. Bacteriol.* **127**:848.

Haberman, A. 1974. The bacteriophage P1 restriction endonuclease. *J. Mol. Biol.* **89**:545.

Hadi, S.M., T.A. Bickle, and R. Yuan. 1975. The role of *S*-adenosylmethionine in the cleavage of deoxyribonucleic acid by the restriction endonuclease from *Escherichia coli* K12. *J. Biol. Chem.* **250**:4159.

Hadi, S.M., B. Báchi, J.C.W. Shepherd, R. Yuan, K. Ineichen, and T.A. Bickle. 1979. DNA recognition and cleavage by the *Eco*P15 restriction endonuclease. *J. Mol. Biol.* **134**:655.

Hattman, S. 1979. Unusual modification of bacteriophage Mu DNA. *J. Virol.* **32**:468.

———. 1980. Specificity of the bacteriophage Mu *mom*$^+$ controlled DNA modification. *J. Virol.* **34**:277.

Hattman, S., J.E. Brooks, and M. Masurekar. 1978. Sequence specificity of the P1 modification methylase (M. *Eco*P1) and the DNA methylase (M. *Eco*dam) controlled by the *Escherichia coli dam* gene. *J. Mol. Biol.* **126**:367.

Hausmann, R. 1967. Synthesis of an *S*-adenosylmethionine-cleaving enzyme in T3-infected *Escherichia coli* and its disturbance by coinfection with enzymatically incompetent bacteriophage. *J. Virol.* **1**:57.

Heilmann, H., H.-J. Burkardt, A. Pühler, and J.N. Reeve. 1980. Transposon mutagenesis of the gene encoding the bacteriophage P1 restriction endonuclease. *J. Mol. Biol.* **144:**387.

Horiuchi, K. and N.D. Zinder. 1972. Cleavage of bacteriophage f1 DNA by the restriction enzyme of *Escherichia coli* B. *Proc. Natl. Acad. Sci.* **69:**3220.

Hubacek, J. and S.W. Glover. 1970. Complementation analysis of temperature sensitive host specificity mutations in *Escherichia coli. J. Mol. Biol.* **50:**111.

Ikeda, H., M. Inuzuka, and J. Tomizawa. 1970. P1-like plasmid in *Escherichia coli* 15. *J. Mol. Biol.* **50:**457.

Kan, N.C., J.A. Lautenberger, M.H. Edgell, and C.A. Hutchison, III. 1979. The nucleotide sequence recognized by the *Escherichia coli* K12 restriction and modification enzymes. *J. Mol. Biol.* **131:**190.

Kao-Huang, Y., A. Revzin, A.P. Butler, P. O'Conner, D.W. Noble, and P.H. von Hippel. 1977. Non-specific binding of genome-regulating proteins as a biological control mechanism: Measurement of DNA-bound *Escherichia coli lac* repressor *in vivo. Proc. Natl. Acad. Sci.* **74:**4228.

Kauc, L. and A. Piekarowicz. 1978. Purification and properties of a new restriction endonuclease from *Haemophilus influenzae* Rf. *Eur. J. Biochem.* **92:**417.

Kimball, M. and S. Linn. 1976. The release of oligonucleotides by the *Escherichia coli* B restriction endonuclease. *Biochem. Biophys. Res. Comm.* **68:**585.

Krüger, D.H., C. Schroeder, S. Hansen, and H.A. Rosenthal. 1977. Active protection by bacteriophages T3 and T7 against *E. coli* B and K-specific restriction of their DNA. *Mol. Gen. Genet.* **153:**99.

Lark, C. 1968. Studies on the *in vivo* methylation of DNA in *Escherichia coli* $15T^-$. *J. Mol. Biol.* **31:**389.

Lark, C. and W. Arber. 1970. Breakdown of cellular DNA upon growth in ethionine of strains with r^+_{15}, r^+_{P1} or r^+_{N3} restriction phenotypes. *J. Mol. Biol.* **52:**337.

Lautenberger, J.A. and S. Linn. 1972. The deoxyribonucleic acid modification and restriction enzymes of *Escherichia coli* B. I: Purification, subunit structure and catalytic properties of the modification methylase. *J. Biol. Chem.* **247:**6176.

Lautenberger, J.A., M.H. Edgell, C.A. Hutchison, III, and G.N. Godson. 1979. The DNA sequence on bacteriophage G4 recognized by the *Escherichia coli* B restriction enzyme. *J. Mol. Biol.* **131:**871.

Lautenberger, J.A., N.C. Kan, D. Lackey, S. Linn, M.H. Edgell, and C.A. Hutchison, III. 1978. Recognition site of *Escherichia coli* B restriction enzyme on ϕXsB1 and simian virus 40 DNAs: An interrupted sequence. *Proc. Natl. Acad. Sci.* **75:**2271.

Linn, S. and W. Arber. 1968. *In vitro* restriction of phage fd replicative form. *Proc. Natl. Acad. Sci.* **59:**1300.

Luria, S.E. and M.L. Human. 1952. A non-hereditary, host induced variation of bacterial viruses. *J. Bacteriol.* **64:**557.

Makino, O., H. Saito, and T. Ando. 1980. *Bacillus subtilis* phage ϕ1 overcomes host-controlled restriction by producing *Bam*Nx inhibitor protein. *Mol. Gen. Genet.* **179:**463.

Makino, O., F. Kawamura, H. Saito, and Y. Ikeda. 1979. Inactivation of restriction endonuclease *Bam*Nx after infection with phage ϕNR2. *Nature* **277:**64.

Mark, K.-K. and F.W. Studier. 1981. Purification of the gene 0.3 protein of bacteriophage T7, an inhibitor of the DNA restriction system of *Escherichia coli. J. Biol. Chem.* **256:**2573.

McCorquodale, P.J. and Y.T. Lanni. 1964. Molecular aspects of DNA transfer from T5 to host cells. I. Characterization of first step transfer material. *J. Mol. Biol.* **10:**10.

Meselson, M. and R. Yuan. 1968. DNA restriction enzyme from *Escherichia coli. Nature* **217:**1110.

Meselson, M., R. Yuan, and J. Heywood. 1972. Restriction and modification of DNA. *Annu. Rev. Biochem.* **41:**447.

Modrich, P. 1979. Structures and mechanisms of DNA restriction and modification enzymes. *Quart. Rev. Biophys.* **12:**315.

Murray, N.E., P.L. Batten, and K. Murray. 1973. Restriction of bacteriophage λ by *Escherichia coli* K. *J. Mol. Biol.* **81**:395.

Murray, N.E., J.A. Gough, B. Suri, and T.A. Bickle. 1982. Structural homologies among type I restriction-modification systems. *EMBO J.* **1**:535.

Piekarowicz, A. 1982. *Hine*I is an isoschizomer of *Hinf*III restriction endonuclease. *J. Mol. Biol.* **157**:373.

Piekarowicz, A. and R. Brzezinski. 1980. Cleavage and methylation of DNA by the restriction endonuclease *Hinf*III isolated from *Haemophilus influenzae* Rf. *J. Mol. Biol.* **144**:415.

Piekarowicz, A. and J. Kalinowska. 1974. Host specificity of DNA in *Haemophilus influenzae*: Similarity between host-specificity types of *Haemophilus influenzae* R*e* and R*f*. *J. Gen. Microbiol.* **81**:405.

Piekarowicz, A., L. Kauc, and S.W. Glover. 1974. Host specificity of DNA in *Haemophilus influenzae*: The two restriction and modification systems in strains R*b* and R*f*. *J. Gen. Microbiol.* **81**:391.

Piekarowicz, A., T.A. Bickle, J.C.W. Shepherd, and K. Ineichen. 1981. The DNA sequence recognized by the *Hinf*III restriction endonuclease. *J. Mol. Biol.* **146**:167.

Pugsley, A.P. and J.P. Rosenbusch. 1981. Release of colicin E2 from *Escherichia coli*. *J. Bacteriol.* **147**:186.

Ravetch, J.V., K. Horiuchi, and N.D. Zinder. 1978. Nucleotide sequence of the recognition site for the restriction/modification enzyme of *Escherichia coli* B. *Proc. Natl. Acad. Sci.* **75**: 2266.

Reiser, J. 1975. "The P1 and P15-specific restriction endonucleases; A comparative study." PhD thesis, Basel University.

Reiser, J. and R. Yuan. 1977. Purification and properties of the P15 specific restriction endonuclease from *Escherichia coli*. *J. Biol. Chem.* **252**:451.

Revel, H.R. and S.E. Luria. 1970. DNA-glucosylation in T-even phage: Genetic determination and role in phage-host interaction. Annu. Rev. Genet. 4:177.

Rigby, P.W.J., M. Dieckmann, C. Rhoades, and P. Berg. 1977. Labeling deoxyribonucleic acid to high specific activity *in vitro* by nick translation with DNA polymerase I. *J. Mol. Biol.* **113**:237.

Risser, R., N. Hopkins, R.W. Davis, H. Delius, and C. Mulder. 1974. Action of *Escherichia coli* P1 restriction endonuclease on simian virus 40 DNA. *J. Mol. Biol.* **89**:517.

Rosamond, J., B. Endlich, and S. Linn. 1979. Electron microscopic studies of the mechanism of action of the restriction endonuclease of *Escherichia coli* B. *J. Mol. Biol.* **129**:619.

Rosner, J.L. 1973. Modification-deficient mutants of bacteriophage P1: Restriction by P1 cryptic lysogens. *Virology* **52**:213.

Roulland-Dussoix, D. and H.W. Boyer. 1969. The *Escherichia coli* B restriction endonuclease. *Biochem. Biophys. Acta* **195**:219.

Roy, P.H. and H.O. Smith. 1973. DNA methylases of *Haemophilus influenzae* Rd. II: Partial recognition site base sequences. *J. Mol. Biol.* **81**:445.

Sain, B. and N.E. Murray. 1980. The *hsd* (host specificity) genes of *Escherichia coli* K12. *Mol. Gen. Genet.* **180**:35.

Scott, J.R. 1970. Clear plaque mutants of bacteriophage P1. *Virology* **41**:66.

Showe, M.K., E. Isobe, and L. Onorato. 1976. Bacteriophage T4 prehead proteinase II. Its cleavage from the product of gene 21 and regulation in phage-infected cells. *J. Mol. Biol.* **107**:55.

Sinsheimer, R.L. 1954. Nucleotides from $T2_r^+$ bateriophage. *Science* **120**:551.

Smith, H.O. and D. Nathans. 1973. A suggested nomenclature for bacterial host modification and restriction systems and their enzymes. *J. Mol. Biol.* **81**:419.

Smith, H.O. and K.W. Wilcox. 1970. A restriction enzyme from *Haemophilus influenzae*. I: Purification and general properties. *J. Mol. Biol.* **51**:379.

Sommer, R. and H. Schaller. 1979. Nucleotide sequence of the recognition site of the B-specific restriction modification system in *Escherichia coli*. *Mol. Gen. Genet.* **168**:331.

Southern, E.M. 1975. Detection of specific sequences among DNA fragments. *J. Mol. Biol.* **98**:503.

Spoerel, N. and P. Herrlich. 1979. Colivirus T3-coded *S*-adenosylmethionine hydrolase. *Eur. J. Biochem.* **95:**227.

Spoerel, N., P. Herrlich, and T.A. Bickle. 1979. A novel bacteriophage defence mechanism: The anti-restriction protein. *Nature* **278:**30.

Studier, F.W. 1975. Gene 0.3 of bacteriophage T7 acts to overcome the DNA restriction system of the host. *J. Mol. Biol.* **94:**283.

Studier, F.W. and N.R. Movva. 1976. SAMase gene of bacteriophage T3 is responsible for overcoming host restriction. *J. Virol.* **19:**136.

Suri, B., V. Nagaraja, and T.A. Bickle. 1982. DNA modification in bacteria. *Curr. Top. Microbiol. Immunol.* (in press).

Toussaint, A. 1976. The DNA modification function of temperate phage Mu-1. *Virology* **70:**17.

———. 1977. DNA modification of bacteriophage Mu-1 requires both host and bacteriophage functions. *J. Virol.* **23:**825.

Towbin, H., T. Staehelin, and J. Gordon. 1979. Electrophoretic transfer of proteins from polyacrylamide gels to nitrcellulose sheets: Procedure and some applications. *Proc. Natl. Acad. Sci.* **76:**4350.

Van Ormondt, H., J.A. Lautenberger, S. Linn, and A. deWaard. 1973. Methylated oligonucleotides derived from bacteriophage fd RF DNA modified *in vitro* by *E. coli* B modification methylase. *FEBS Letters* **33:**177.

Von Hippel, P.H., A. Revzin, C.A. Gross, and A.C. Wang. 1974. Non-specific DNA binding of genome regulating proteins as a biological control mechanism. I. The *lac* operon: Equilibrium aspects. *Proc. Natl. Acad. Sci.* **71:**4808.

Vovis, G.F., K. Horiuchi, and N.D. Zinder. 1974. Kinetics of methylation by a restriction endonuclease from *Escherichia coli* B. *Proc. Natl. Acad. Sci.* **71:**3810.

Yuan, R. 1981. Structure and mechanism of multifunctional restriction endonucleases. *Annu. Rev. Biochem.* **50:**285.

Yuan, R. and M. Meselson. 1970. A specific complex between a restriction endonuclease and its DNA substrate. *Proc. Natl. Acad. Sci.* **65:**357.

Yuan, R. and J. Reiser. 1978. Steps in the reaction mechanism of the *Escherichia coli* plasmid P15-specific restriction endonuclease. *J. Mol. Biol.* **122:**433.

Yuan, R., J. Heywood, and M. Meselson. 1972. ATP hydrolysis by restriction endonuclease from *E. coli* K. *Nature New Biol.* **240:**42.

Yuan, R., D.L. Hamilton, and J. Burckhardt. 1980a. DNA translocation by the restriction enzyme from *Escherichia coli* K. *Cell* **20:**237.

Yuan, R., T.A. Bickle, W. Ebbers, and C. Brack. 1975. Multiple steps in DNA recognition by restriction endonuclease from *Escherichia coli* K. *Nature* **256:**556.

Yuan, R., D.L. Hamilton, S.M. Hadi, and T.A. Bickle. 1980b. Role of ATP in the cleavage mechanism of the *Eco*P15 restriction endonuclease. *J. Mol. Biol.* **144:**501.

Zabeau, M., S. Friedman, M. Van Montagu, and J. Schell. 1980. The *ral* gene of phage λ. I. Identification of a non-essential gene that modulates restriction and modification in *E. coli*. *Mol. Gen. Genet.* **179:**63.

Type-II Restriction and Modification Enzymes

Paul Modrich
Department of Biochemistry
Duke University Medical Center
Durham, North Carolina 27710

Richard J. Roberts
Cold Spring Harbor Laboratory
Cold Spring Harbor, New York 11724

INTRODUCTION

DNA restriction endonucleases and their companion modification methylases have been identified in a wide variety of prokaryotes (Roberts 1982). Three distinct classes of restriction-modification systems have been found that differ in their cofactor requirements. The type-I and type-III enzymes have rather specific cofactor requirements and have been studied in detail, both biochemically and genetically (see Chapter 5). Although they recognize specific sequences on DNA, they cleave randomly at sites remote from the recognition sequence and so do not produce specific fragments. In contrast, the type-II restriction enzymes require only Mg^{2+} as a cofactor and they both recognize and cleave a specific sequence within the DNA molecule. It is this property that has rendered these enzymes so useful to the molecular biologists. Recombinant DNA technology and the present rapid methods for DNA sequence analysis are two important areas that have become possible because of the existence of these enzymes.

As with their type-I and type-III counterparts, the biological phenomenon of restriction and modification can be catalyzed by the type-II enzymes. However, it is important to note that for the vast majority of these enzymes there is no genetic evidence showing that they are involved in host-controlled restriction-modification, and it is quite possible that some of them have other, as yet undiscovered, roles in the cell. Whereas interest in the type-I and type-III enzymes centers around their biological properties, this is not true of the type-II enzymes. In this case, interest has focused on their use as tools with which to dissect DNA, and, for the most part, neither their biological nor their biochemical properties have been explored in detail.

This review will briefly discuss the range of specificities exhibited by the type-II restriction endonucleases and will focus on the mechanisms by which they recognize specific DNA sequences. Other recent reviews include Modrich (1979, 1982), Zabeau and Roberts (1979), and Malcolm (1981).

Purification

The first procedures used to assay restriction enzyme activity were based upon the selective degradation of foreign DNA, as opposed to host DNA. Degradation was measured as the loss of biological activity (Gromkova and Goodgal 1972; Takanami 1973) or as change in viscosity (Smith and Wilcox 1970; Middleton et al. 1972). These assays have now been totally superseded by the agarose gel assay (Aaij and Borst 1972; Sharp et al. 1973), which has the advantage of speed and permits a direct visualization of the results of a chromatographic fractionation. A key feature of this assay is that the discrete banding pattern obtained not only shows the elution profile of the enzyme, but can also reveal the presence of two different enzymes within the same bacterial strain and can indicate whether contaminating nonspecific nucleases are present.

Most purification procedures are aimed at quickly obtaining an enzyme preparation that is relatively free of contaminating nucleases and no attempt is made to obtain homogeneous protein. Only for studies of catalytic properties is homogeneous enzyme needed, or is its acquisition attempted. Because restriction enzymes are isolated from widely different bacterial sources, each containing a unique set of contaminating proteins, it would be naive to suppose that a general purification procedure exists for all restriction enzymes. Nevertheless, many of the schemes reported in the literature often represent only minor variations of that used for purifying *Hin*dII (Smith and Wilcox 1970), which was the first type-II enzyme to be isolated.

The first steps in the isolation usually involve the preparation of a high-speed supernatant of the cell lysate followed by the removal of nucleic acids by gel filtration or precipitation with either streptomycin sulfate or polyethyleneimine. Further purification is then achieved by column chromotography, and, as with most enzymes involved in nucleic acid metabolism, phosphocellulose has proved immensely useful. Other ion exchangers such as DEAE cellulose, QAE Sephadex, etc., have been used extensively, and from time to time more exotic adsorbants have been employed. In general, the enzymes are relatively easy to purify to a point where they are devoid of other nonspecific nucleases, and often one or two chromatographic steps are all that is necessary. A key factor, responsible for the success in purifying restriction enzymes, has been their remarkable stability. Indeed, many enzymes will continue to digest DNA in a linear fashion for periods in excess of 12 hours. This must reflect both their inherent stability and also the absence of significant amounts of proteases in partially purified enzyme preparations.

Since assays of crude cell extracts rarely give distinct fragmentation patterns, due to the high concentration of nonspecific nucleases, it is often difficult to quantitate the amounts of enzyme originally present. For this reason, enzyme yield is usually described in terms of the amount of enzyme finally obtained. Phage λ DNA is a commonly used substrate for monitoring cleavage, and the yield is conveniently expressed in arbitrary units, where one unit is defined as the amount of enzyme necessary to completely digest one μg of λ DNA in 1 hour at 37°C. This unit definition must of course be viewed with caution since it does not necessarily give an accurate reflection of the total amount of DNA that one might expect to cleave with a given amount of enzyme. Both the degree of purity of the enzyme and the DNA concentration can markedly influence the cleavage efficiency. Although a more rigorous definition of a unit would be desirable, the kinetic parameter necessary to establish an absolute rate can be difficult to obtain when homogeneous enzyme preparations are not available.

Characterization

The important feature that distinguishes one type-II enzyme from another is the specificity of the double-strand break, so that the most useful characteristic is the nature of the recognition sequence and the position of the cleavage sites. The most straightforward way to obtain this information involves direct sequence analysis of the ends of DNA fragment mixtures produced by restriction enzyme cleavage. The easiest approach, first used for the identification of the *Hin*dII recognition sequence (Kelly and Smith 1970) consists of labeling the 5′ ends of restriction fragments with polynucleotide kinase, followed by digestion with pancreatic DNase. This procedure generates a heterogeneous array of products, including labeled 5′ terminal di-, tri-, tetranucleotides, etc., and the nucleotide sequence common to the 5′ ends of this mixture of restriction fragments can easily be deduced (Murray 1973). Subsequent analysis of individual products with venom phosphodiesterase gives the 5′ terminal mononucleotide or, if exonuclease I is used, gives the 5′ terminal dinucleotide. For most restriction enzymes that generate either flush ends or 5′ extensions, these data will allow deduction of the recognition sequence.

An analogous approach can be used for fragments with 3′ terminal extensions. Here the 3′ ends are labeled using terminal deoxynucleotidyl transferase and α-^{32}P ribonucleoside triphosphates (Roychoudhury and Cossell 1971; Roychoudhury et al. 1976). An analogous series of experiments is then performed to deduce the sequence.

While the methods described above work well when the cleavage site is located within the recognition sequence, they are of less use when the cleavage site is located away from the sequence as occurs with enzymes like *Hga*I and *Mbo*II (Brown and Smith 1977; Brown et al. 1980). In such cases, the most useful approaches have been to employ computer programs to predict recognition sequences based on cleavage data (Fuchs et al. 1978; Gingeras et al. 1978) and then to confirm those predictions, as well as obtain the exact position of the cleavage site, using primed synthesis reactions (Brown and Smith 1977; Brown et al. 1980).

An important first step in the characterization of a new restriction enzyme involves a description of the fragment patterns that are obtained upon digestion of various substrate DNAs. Comparison of these fragment patterns with those obtained using enzymes of known specificity can indicate whether the new enyzme recognizes a new sequence or is likely to recognize a sequence identical with that of an enzyme already characterized. Enzymes from different bacterial sources that recognize the same sequence are termed isoschizomers (Roberts 1976). When two enzymes are suspected to have identical sequence specificity, this is best tested by digesting several substrate DNAs sequentially with both the new and the known enzyme and by showing that no additional DNA fragments appear. Although isoschizomers can be quickly identified in this way, one cannot assume that the location of the cleavage site within the recognition sequence will necessarily be identical. For instance, the pair of enzymes *Sma*I and *Xma*I both recognize the sequence CCCGGG, but *Sma*I cleaves to give blunt-ended fragments, whereas *Xma*I cleaves to give fragments carrying a four-base 5′ extension (Endow and Roberts 1977). Similarly, *Hha*I and *Hin*PI both recognize the sequence GCGC, with *Hha*I cutting between the second GC base pair to generate fragments with a 3′ dinucleotide extension, while *Hin*PI cleaves the first GC pair to give fragments with a 5′ terminal dinucleotide extension (Roberts et al. 1976; Shen et al. 1980).

Another interesting and often useful property of isoschizomers is that they are sometimes differentially sensitive to methylation, thus the enzymes *Mbo*I and *Sau*3A (Sussenbach et al. 1976; Gelinas et al. 1977a) both recognize the sequence GATC and cleave identically. However, *Mbo*I is unable to cleave if the A residue in the recognition sequence is modified. *Sau*3A can cleave DNA modified at the A residue but is blocked by modification at the C residue. This property of *Sau*3A is of great use in cleaving DNA that has come from *E. coli* cells because *E. coli* contains a methylase, the *dam* methylase, which methylates the sequence GATC at the A position, rendering plasmids and phages grown in wild-type *E. coli* resistant to cleavage by *Mbo*I. Another pair of isoschizomers, *Hpa*II and *Msp*I, both recognize the sequence CCGG (Garfin and Goodman 1974; I. Schildkraut, unpubl.). In this case, modification at the second C residue renders the sequence resistant to cleavage by *Hpa*II, but sensitive to cleavage by *Msp*I. This property has been exploited extensively in probing eukaryotic genomes for the presence or absence of a methyl group at this cytosine residue (Waalwijk and Flavell 1978; Bird et al. 1979).

Recognition Specificities

Among the 360 type-II restriction endonucleases that have now been characterized (see Appendix A), some 85 different specificities have already been identified, and there is no indication that the range of specificities is exhausted. Indeed, there is good reason to believe that hundreds, if not thousands, of different specificities would be found if a diligent search were conducted. One reason for this speculation is illustrated in Table 1, which shows the range of sequence patterns with which different type-II restriction endonucleases interact. The majority of these enzymes belongs to groups in which the recognition sequence contains a dyad axis of symmetry. It has been speculated that enzymes of these groups might contain subunits arranged symmetrically (Kelly and Smith 1970). The finding that the active forms of several type-II enzymes contain two identical subunits would be consistent with this proposal (see below).

Among the simple, symmetric hexanucleotide sequences, designated here as class A, more than half of the possible sequence patterns are already represented by well-characterized enzymes, as shown in Table 2. There is no reason to believe that a similar number of enzymes will not be found for the other patterns in classes B–F. Similarly, it is likely that for many of the other classes shown, the present examples are not exhaustive. Furthermore, it should be noted that, within the last year, five wholly new classes (C, D, F, N, and O) were added to this list. Examination of the specific sequences recognized within any of these classes reveals a preponderance of GC-rich sequences, and until recently there was no example of an enzyme that required only AT base pairs in its recognition sequence. This apparent anomaly has been removed by the finding of the enzyme *Aha*III, which recognizes the sequence TTTAAA (N. Brown, unpubl.).

Table 1 Sequence patterns recognized by type-II restriction enzymes

General pattern		Specific example		Known	Possible
A.	XYZZ'Y'X'	*Aha*III	TTT↓AAA	39	64
B.	XYZNZ'Y'X'	*Sau*I	CC↓TNAGG	2	64
C.	XYZNNNZ'Y'X'	*Tth*111I	GACN↓NNGTC	1	64
D.	XYZNNNNZ'Y'X'	*Xmn*I	GAANNNNTTC	1	64
E.	XYZNNNNNZ'Y'X'	*Bgl*I	GCCNNNN↓NGGC	1	64
F.	XYZNNNNNNZ'Y'X'	*Hgi*EII	ACCNNNNNNGGT	1	64
G.	XYPyPuY'X'	*Hgi*CI	G↓GPyPuCC	2	16
H.	XPyYY'PuX'	*Acy*I	GPu↓CGPyC	1	16
I.	XPuYY'PyX'	*Ava*I	C↓PyCGPuG	1	16
J.	PuXYY'X'Py	*Xho*II	Pu↓GATCPy	2	16
K.	$^{A}_{T}$XYY'X'$^{A}_{T}$	*Hae*I	$^{A}_{T}$GG↓CC$^{A}_{T}$	1	16
L.	X$^{A}_{T}$YY'$^{A}_{T}$X'	*Hgi*AI	G$^{A}_{T}$GC$^{A}_{T}$↓C	1	16
M.	XY$^{AG}_{CT}$Y'X'	*Acc*I	GT↓$^{AG}_{CT}$AC	1	16
N.	PyXXXXX	*Gdi*I	Py↓GGCCG	1	?
O.	XXXPuXX	*Tth*111II	CAAPuCA	1	?
P.	XY$^{G}_{C}$Y'X'	*Cau*II	CC↓$^{G}_{C}$GG	1	16
Q.	XY$^{A}_{T}$Y'X'	*Bbv*I	GC$^{A}_{T}$GC	3	16
R.	XXXXX	*Sfa*NI	GCATC	4	480
S.	XYY'X'	*Rsa*I	GT↓AC	8	16
T.	XYNY'X'	*Fnu*4HI	GC↓NGC	4	16
U.	XXXX	*Mnl*I	CCTC	1	240

Although for most of these enzymes the site of cleavage lies within the recognition sequence, or between the two components of the recognition sequence in cases like *Bgl*I, there are some conspicuous exceptions. For enzymes of class R and U, where the recognition sequence is asymmetric, cleavage occurs at a site remote from the recognition sequence. One clue to the nature of the protein-nucleic acid interaction involved here comes from the cleavage specificity of *Hph*I and *Mbo*II. In both cases, the sites of cleavage lie 10 bp away from the midpoint of the recognition sequence. Since 10 bp are required for one turn of the DNA helix, it therefore seems likely that the enzyme will interact with one face of the DNA by binding at one point and cleaving at a corresponding point on the next turn of the helix. This same phenomenon has recently been found for the enzyme *Bbv*I, where again cleavage occurs 8 bp away from the recognition sequence on one strand and 12 bp away on the other strand.

Several groups of enzymes display recognition specificities which are termed "relaxed." These enzymes do not distinguish between the purine or pyrimidine bases at certain positions in their recognition sequences. In other cases there is a failure to discriminate between an A or a T residue at some position in the recognition sequence, while in one case, *Acc*I, the enzyme fails to distinguish an A from a C at one position, or a T from a G at another. While

Table 2 Some restriction enzyme recognition sequences

	AATT	ACGT	AGCT	ATAT	CATG	CCGG	CGCG	CTAG	GATC	GCGC	GGCC	GTAC	TATA	TCGA	TGCA	TTAA	
			*Alu*I			*Hpa*II	*Fnu*DII		*Mbo*I *Dpn*I	*Hha*I	*Hae*III	*Rsa*I		*Taq*I			
A			*Hin*dIII		*Nsp*CI		*Mlu*I		*Bgl*II *Xho*II	*Hae*II	*Stu*I *Hae*I	*Sca*I		*Cla*I	*Ava*III		T
C			*Pvu*II	*Nde*I	*Nco*I	*Sma*I *Ava*I	*Sac*II	*Avr*II	*Pvu*I		*Xma*III *Gdi*II *Eae*I *Cfr*I			*Xho*I *Ava*I	*Pst*I	*Afl*II	G
G	*Eco*RI	*Acy*I	*Sac*I *Hgi*AI *Hgi*JII	*Eco*RV	*Sph*I *Nsp*CI	*Nae*I	*Bse*PI		*Bam*HI *Xho*II	*Nar*I *Hgi*CI *Acy*I *Hae*II	*Apa*I *Hgi*JII	*Kpn*I *Hgi*CI	*Sna*I *Acc*I	*Sal*I *Acc*I *Hin*dII	*Hgi*AI	*Hpa*I *Hin*dII	C
T							*Nru*I	*Xba*I	*Bcl*I	*Mst*I	*Bal*I *Hae*I *Eae*I *Cfr*I			*Asu*II		*Aha*III	A

nothing is known about the nature of the protein-nucleic acid interactions involved, this finding does point to the existence of a mechanism for recognizing several DNA sequences which, at first sight, might appear unrelated. It is possible that degeneracies such as these might also be a component of other protein-nucleic acid interactions, the recognition of promoters by RNA polymerases being one obvious example.

Restriction-modification systems originating in different organisms often recognize sequences that are identical or that are substantially homologous (Table 3 and Appendix A). For example, more than 20 endonucleases have been identified that recognize GATC, or a sequence in which GATC is the internal tetranucleotide (Roberts 1982). It is possible that comparison of polypeptide sequences of such enzymes may yield insight into the evolution of related systems and perhaps into those features of polypeptide structure that are important components of their specific interactions.

STRUCTURES OF TYPE-II RESTRICTION AND MODIFICATION ENZYMES

Among the few type-II enzymes that have been isolated in pure form or examined with respect to features of mechanism (Table 3), there are several differences in the mode of DNA-protein interactions. For this reason, those differences which may be of significance in the context of the sequence recognition question are emphasized below.

As summarized in Table 3, each type-II endonuclease that has been isolated in homogeneous form exhibits a single molecular-weight species upon denaturation and reduction. This suggests that each contains a single type of polypeptide chain and this has been confirmed in the case of *Eco*RI. These enzymes fall into several classes based on oligomeric states of the native proteins. The majority of type-II endonucleases exist in solution as oligomers of their constituent subunit, and, in the case of several enzymes, more than one aggregation state has been observed. There are, however, exceptions to this generality, the most notable being *Bgl*I and *Bsp*I. In these cases, only the monomer has been identified (Koncz et al. 1978; Lee and Chirikjian 1979). Since it is reasonable to assume that the mode of DNA-protein interaction tends to reflect the biologically active aggregation state(s) of the protein in question, it has been suggested that these two structural classes of type-II endonucleases may differ with respect to mechanism (Lee and Chirikjian 1979). Attempts to assign biological activity to particular oligomeric states have been successful only in the case of *Eco*RI, which exists in solution as an equilibrium mixture of dimers and tetramers (Modrich and Zabel 1976; Rubin and Modrich 1980). In this system, the tetramer-to-dimer transition is governed by an equilibrium dissociation constant of approximately 10^{-7} M, and it has been found that the dimer retains the stable form at catalytic concentrations (10^{-10} M). This finding, together with the kinetic behavior of this endonuclease (below), indicates that the dimer possesses biological activity and interacts directly with the *Eco*RI recognition sequence.

Structural analysis of type-II modification enzymes is still at an early stage. Although a number of such activities have been identified (Appendix A), only four have been isolated in pure form: *Eco*RI (Rubin and Modrich 1977), *Hpa*I (K.I. Agarwal, unpubl.), *Hpa*II (Yoo and Agarwal 1980a), and *Bsu*I (Gunthert et al. 1981a). In addition, the *dam* methylase, which is responsible for the majority of *N*-6-methyladenine residues in *E. coli* DNA (Marinus and Morris 1973), has been purified and characterized (Geier and Modrich 1979; Herman and Modrich 1981). Although it does not appear to be involved in a restriction-modification system, it will be considered here because of its structural and mechanistic similarities to the other four methyltransferases. Each of these enzymes is composed of a single polypeptide chain (30–40K), and in all five cases the stable form in solution is the monomer. Moreover, and as will be discussed below, kinetic studies indicate that the functional form of all of these enzymes is in fact the monomer. Thus, at least in the case of the *Eco*RI system, and perhaps in the *Hpa*I and *Hpa*II systems as well, the restriction endonuclease interacts with the recognition sequence as a dimer while the modification enzyme is functional as a monomer. This observation led to the suggestion that at least in the case of the *Eco*RI system, the endonuclease and methylase may differ significantly with respect to protein-nucleic acid interactions involved in specific site recognition (Rubin and Modrich 1977). This possibility might appear surprising in view of the complementary nature of the *Eco*RI enzymes as components of a restriction-modification system. Indeed, it has been established that the continued presence of an active *Eco*RI methylase is necessary to maintain cell viability in the presence of the endonuclease (Humphreys et al. 1976). This functional complementarity of two proteins that recognize the same DNA sequence led to the suggestion that the two *Eco*RI enzymes might have a common evolutionary origin (Boyer et al. 1975). In this view, the two proteins would share substantial homology within DNA-binding-site regions as well as common features of polynucleotide interactions.

Primary Sequence Data

The recent determination of the primary structures of the *Eco*RI enzymes has permitted direct assessment of structural similarities between the two proteins at the amino acid sequence level. The genes for these enzymes have been localized to a 2200-bp segment of DNA, the sequence of which has been determined in two laboratories (Jack et al. 1980; Greene et al. 1981; Newman et al. 1981). These determinations, together with partial amino- and carboxy-terminal sequence analysis of the isolated enzymes (Rubin et al. 1981), have permitted assignment of structural genes within this DNA segment, as well as

Table 3 Properties of selected type-II DNA restriction and modification systems

Microorganism	Specificity	Enzyme	M_r[a]	Recognition sequence[b]	References
B. amyloliquefaciens H	*Bam*HI	R M	(22,000)$_2$ (22,000)$_4$ n.d.	5′-G↓G-A-T-C*-C 3′-C-C*-T-A-G↑G	Wilson and Young (1975) Roberts et al. (1977) Hattman et al. (1978) Smith and Chirikjian (1979)
B. caldolyticus	*Bcl*I	R	(25,000)$_2$	5′-T↓G-A-T-C-A 3′-A-C-T-A-G↑T	Bingham et al. (1978)
B. globigii	*Bgl*I	R	(31,000)$_1$	5′-G-C-C-N-N-N-N↓N-G-G-C 3′-C-G-G-N↑N-N-N-N-C-C-G	Lee and Chirikjian (1979) Van Heuverson and Fiers (1980) Bickle and Ineichen (1980) Lautenberger et al. (1980)
B. globigii	*Bgl*II	R	(27,000)$_2$	5′-A↓G-A-T-C-T 3′-T-C-T-A-G↑A	Pirotta (1976) Bickle et al. (1980)
B. sphaericus	*Bsp*I	R M	(35,000)$_1$ ~50,000	5′-G-G↓C-C 3′-C-C↑G-G	Kiss et al. (1977) Koncz et al. (1978)
B. stearothermophilus	*Bst*I	R M	(26,000)$_1$ (26,000)$_2$ ~400,000	5′-G↓G-A-T-C-C 3′-C-C-T-A-G↑G	Catterall and Welker (1977) Levy and Welker (1978) Clarke and Hartley (1979)

B. subtilis	*Bsu*I	R M	(68,000)$_1$ (68,000)$_2$ (37,000)$_1$ (40,000)$_1$	5′-G-G↓C*-C 3′-C-C*↑G-G	Bron et al. (1975) Bron and Hurz (1980) Bron and Murray (1975) Heininger et al. (1977) Gunthert et al. (1981a)
E. coli	*dam*[c]	M	(31,000)$_1$	5′-G-A*-T-C 3′-C-T-A*-G	Geier and Modrich (1979) Herman and Modrich (1981)
D. pneumoniae	*Dpn*I[d]	R	~20,000	5′-G-A*↓T-C 3′-C-T↑A*-G	Lacks and Greenberg (1975) Lacks and Greenberg (1977) Lacks (1980)
D. pneumoniae	*Dpn*II[d]	R	~70,000	5′-G-A-T-C 3′-C-T-A-G	Lacks and Greenberg (1975, 1977)
E. coli RY13	*Eco*RI[e]	R M	(31,000)$_2$ ↑↓ (31,000)$_4$ (38,000)$_1$	5′-G↓A-A*-T-T-C 3′-C-T-T-A*-A↑G	Hedgpeth et al. (1972) Dugaiczyk et al. (1974) Greene et al. (1975) Modrich and Zabel (1976) Rubin and Modrich (1977) Bingham and Atkinson (1978)
E. coli R245	*Eco*RII	R M	(40,000)$_2$ n.d.	5′↓C-C*-A-G-G 3′-G-G-T-C*-C↑	Bigger et al. (1973) Roulland-Dussoix et al. (1974)

Table 3 Properties of selected type-II DNA restriction and modification systems (continued)

Microorganism	Specificity	Enzyme	M_r[a]	Recognition sequence[b]	References
H. aegyptius	*Hae*III	R M	n.d. n.d.	5′-G-G↓C*-C 3′-C-C↑G-G	Bron and Murray (1975) Middleton et al. (1972) Mann and Smith (1977)
H. haemolyticus	*Hha*I	R M	n.d. n.d.	5′-G-C-G↓C 3′-C↑G-C*-G	Roberts et al. (1976) Mann and Smith (1979b)
H. haemolyticus	*Hha*II	R M	$(24,000)_2$ n.d.	5′-G-A-N-T-C 3′-C-T-N-A-G	Mann et al. (1978) Mann and Smith (1979b)
H. influenzae Rd	*Hind*II	R M	~70,000 ~55,000	5′-G-T-Py↓Pu-A*-C 3′-C-A*-Pu↑Py-T-G	Smith and Wilcox (1970) Kelly and Smith (1970) Roy and Smith (1973a,b)
H. influenzae Rd	*Hind*III	R M	n.d. n.d.	5′-A*↓A-G-C-T-T 3′-T-T-C-G-A↑A*	Roy and Smith (1973a,b) Old et al. (1975)
H. parahaemolyticus	*Hph*I	R	n.d.	5′ GGTGANNNNNNNN↓ 3′ CCACTNNNNNNN↑	Kleid et al. (1976)
H. parainfluenzae	*Hpa*I	R M	$(29,000)_2$ $(38,000)_1$	5′-G-T-T↓A-A*-C 3′-C-A*-A↑T-T-G	Sharp et al. (1973) Garfin and Goodman (1974) Hines and Agarwal (1979) Agarwal (unpubl.)

H. parainfluenzae	*Hpa*II[f]	R	$(41{,}000)_2$	5′-C↓C*-G-G 3′-G-G C*↑-C	Mann and Smith (1977) Sharp et al. (1973) Garfin and Goodman (1974) Hines and Agarwal (1979) Agarwal (unpubl.) Yoo and Agarwal (1980a)
			(39,000)		
		M	(42,000)		
Moraxella nonliquefaciens	*Mno*I	R	?	5′-C↓C-G-G 3′-G-G-C↑C	Baumstark et al. (1979)
Moraxella species	*Msp*I	R	?	5′-C↓C-G-G 3′-G-G-C↑C	Roberts (unpubl.) Yoo and Agarwal (1980b)
Staphylococcus aureus 3A	*Sau*3A	R	?	5′ ↓G-A-T-C 3′-C-T-A-G↑	Sussenbach et al. (1976)

[a]In those cases where an enzyme has been isolated in pure form, stable aggregation states of the monomer are indicated. Approximate native M_r values, determined by gel filtration or sucrose gradient sedimentation, are shown for those enzymes which have been only partially purified. n.d. = not determined.

[b]Arrows indicate points of cleavage by the restriction endonuclease while bases methylated by the modification enzyme are indicated by asterisks. In all cases cleavage generates 5′-phosphoryl and 3′-hydroxyl termini, and modification yields *N*-6 methyladenine or 5-methylcytosine. In those cases where the particular base modified has been identified, both strands of the recognition site are shown as being modified. However, methylation of both DNA strands has been established only for *Eco*RI (Dugaiczyk et al. 1974), *Eco*RII (Roulland-Dussoix et al. 1974), *dam* (Geier and Modrich 1979), *Hpa*I (K.I. Agarwal, unpubl.), and *Hpa*II (Yoo and Agarwal 1980a) methyltransferases. In several cases, sites of cleavage or methylation within the recognition site have not yet been determined.

[c]As mentioned in the text, although the *E. coli dam* methylase resembles other type-II modification enzymes, it does not appear to function in restriction-modification.

[d]*Dpn*I and *Dpn*II both recognize the sequence GATC. However, *Dpn*I is novel in that it cleaves within the sequence only if the adenines on both DNA strands are modified. In contrast, *Dpn*II only attacks sites devoid of the adenine modification (Lacks and Greenberg 1975, 1977).

[e]Although there has been a claim that isolated *Eco*RI endonuclease is a monomer (Bingham and Atkinson 1978), evidence in support of this view was not provided. Moreover, this finding has not been substantiated by several other laboratories (Greene et al. 1975; Modrich and Zabel 1976).

[f]Two DNA methylases of *Hpa*II specificity, which differ in molecular weight, have been identified. Since preliminary peptide mapping suggests that the two species are related, the smaller may have been derived from the larger (Yoo and Agarwal 1980a). Similarly, two enzymes of *Bsu*I specificity have also been identified (Gunthert et al. 1981a).

evaluation of the extent to which each polypeptide is processed subsequent to biosynthesis. As summarized in Figure 1, the *Eco*RI endonuclease gene encodes a 277-residue polypeptide (M = 31,063), and processing appears limited to removal of the aminoterminal fMet. The methylase gene specifies a 326-residue protein (M = 38,048), but in this case the terminal dipeptide fMet-Ala is removed.

Comparison of the endonuclease and methylase amino acid and gene sequences has led to the identification of a single limited and imperfect DNA sequence homology that is statistically significant (Newman et al. 1981). However, the general lack of homology between the two polypeptides, and their corresponding genes, has led to the conclusion that the two proteins have a different evolutionary origin (Greene et al. 1981; Newman et al. 1981). Whereas the absence of amino acid sequence homology between the *Eco*RI enzymes does not exclude the existence of common features at higher levels of structure, the circular dichroism spectra of the two proteins indicate extensive differences at the level of secondary structure (Newman et al. 1981). Secondary structures predicted for the two proteins by the probabalistic method of Chou and Fasman (1978) are also markedly different (Newman et al. 1981, Greene et al. 1981). Thus, the dissimilar structures of the *Eco*RI restriction and modification enzymes at primary, secondary, and quaternary levels are consistent with the view that the two proteins may interact with DNA in different ways. Additional evidence in support of this idea will be considered below.

Crystallographic Studies

The availability of the primary sequences of the *Eco*RI enzymes should facilitate crystallographic analyses of these proteins. Although crystallization of the *Eco*RI methylase has not been reported, crystals of the endonuclease suitable for high-resolution structure analysis have been obtained (Rosenberg et al. 1978). The results of preliminary X-ray diffraction analysis show that the

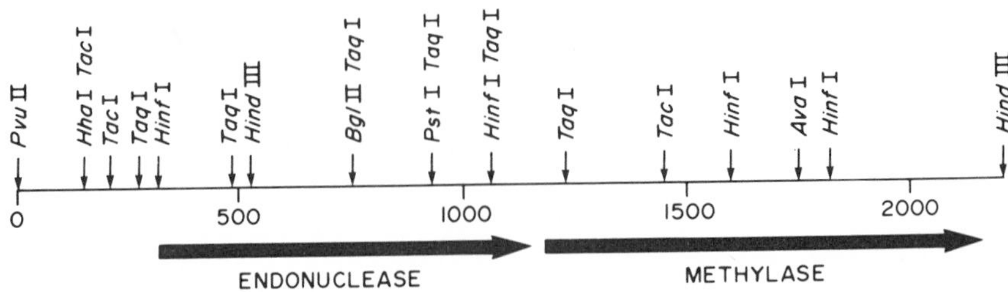

Figure 1 Organization of structural gene sequences for *Eco*RI restriction and modification enzymes (Greene et al. 1981; Newman et al. 1981). Adapted from Newman et al. (1981).

enzyme crystallizes in space group C2 with unit cell parameters of a = 209 Å, b = 129 Å, c = 50 Å, and β = 98.4°C. The crystals appear to contain four endonuclease monomers (31K) per asymmetric unit which is consistent with the proclivity of the enzyme to associate into tetramers at high protein concentrations.

An apparently identical crystalline form has been also obtained by Young et al. (1981) under different crystallization conditions. In addition, this group has successfully crystallized a complex between the endonuclease and a hexanucleotide GAATTC containing the *Eco*RI recognition sequence. The crystalline complex diffracts to a nominal resolution of 2.8 Å and is of space group $P42_12$ with $a = b =$ 183.2 Å, c = 49.7 Å, and $\alpha = \beta = \gamma$ = 90°C. The unit cell contains four endonuclease monomers plus two duplex DNA fragments in an asymmetric unit. As discussed below, this stoichiometry of one endonuclease dimer per duplex recognition sequence is consistent with the mode of interaction of the enzyme with DNA in solution. It is also interesting to note that unit cells of both the enzyme and the complex crystals have a c axial length of 50 Å. Young et al. (1981) have pointed out that this, together with the space group symmetry, suggests that the endonuclease tetramer and the complex tetramer have 50 Å as one of their dimensions.

It is apparent that the rapid acquisition of structural information on the *Eco*RI enzymes will facilitate understanding of DNA-protein interactions in this system. In particular, elucidation of crystal structures for the endonuclease and the endonuclease-DNA complex should provide substantial insight into the molecular contacts involved in DNA sequence recognition in this system, as well as protein conformational transitions that might occur during this process.

Cloned Genes

Although structural analyses of type-II restriction and modification enzymes has proceeded most rapidly in the case of the *Eco*RI system, detailed structural information from other systems is accumulating rapidly. Genes specifying the *Hha*II (Mann et al. 1978), *Eco*RII (Kosykh et al. 1980), *Pst*I (Walder et al. 1981), and *Pae*R7 (T.R. Gingeras and J.E. Brooks, in prep.) restriction and modification enzymes, as well as those specifying *Bsp*I (Szomolanyi et al. 1980) and *dam* (Herman and Modrich 1981; J.E. Brooks et al., in prep.) methylases have also been cloned. Moreover, DNA sequence analysis of the *dam* methylase gene has recently been completed (J.E. Brooks et al., in prep.) and analysis of the *Hha*II gene sequences is in progress (Hausler 1980).

DNA Cleavage by Type-II Endonucleases

Of the numerous type-II endonucleases known, the *Eco*RI enzyme has been examined in greatest detail with respect to its mechanism of DNA cleavage. In vitro restriction by type-II endonucleases, which requires only Mg^{2+} and unmodified DNA, involves hydrolysis of two phosphodiester bonds. In the case of *Eco*RI, several lines of evidence indicate that DNA cleavage occurs by the mechanism shown in Figure 2 where k_4 is rate-limiting for double-strand cleavage (Modrich and Zabel 1976; Rubin and Modrich 1978). This mechanism is based on kinetic analyses under dilute conditions where the dimer was shown to be the catalytically active species. It is not yet clear whether DNA cleavage by the *Eco*RI tetramer occurs by a similar mechanism, nor in fact whether the tetramer is catalytically active. Kinetic studies under steady-state conditions have indicated a K_m for *Eco*RI sites in the range of 1–10 nM and values for k_{cat} of 1 − 4 double-strand scissions per minute per dimer at 37°C (Modrich and Zabel 1976; Berkner and Folk 1977; Goppelt et al. 1980; W.E. Jack and P. Modrich, unpubl.).

At least some of the variation in steady-state kinetic parameters cited above may reflect differences in the DNA substrates employed. Although the occurrence of an unmodified *Eco*RI recognition sequence appears sufficient for double-strand scission, it is evident that the efficiency of cleavage is dependent on additional features of DNA structure. Thomas and Davis (1975) initially observed differences in the efficiency of double-strand cleavage of the five *Eco*RI sites of bacteriophage λ DNA. Similar results have been obtained with adenovirus DNA by Forsblum et al. (1976), who attributed differences in double-strand cleavage efficiency to K_m effects. In both studies, conditions known to stabilize DNA secondary structure were found to enhance differences in the efficiency of double-strand cleavage between substrate sites. In this respect, it is noteworthy that cleavage of the *duplex* form of TGAATTCA by the endonuclease at 15°C is governed by the same V_{max} as cleavage of SV40 DNA (Greene et al. 1975). However, the K_m for the oligonucleotide duplex is 200-fold greater than that for the viral molecule. These observations also indicate that interaction of the endonuclease with its recognition site is modulated by sequences external to the *Eco*RI sequence.

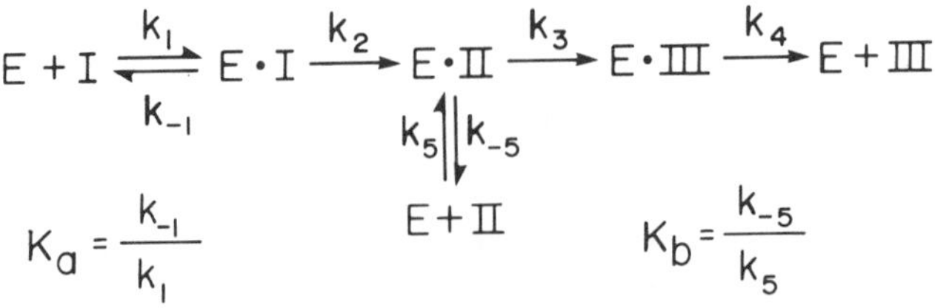

$$E + I \underset{k_{-1}}{\overset{k_1}{\rightleftharpoons}} E \cdot I \xrightarrow{k_2} E \cdot II \xrightarrow{k_3} E \cdot III \xrightarrow{k_4} E + III$$

$$E \cdot II \underset{k_{-5}}{\overset{k_5}{\rightleftharpoons}} E + II$$

$$K_a = \frac{k_{-1}}{k_1} \qquad K_b = \frac{k_{-5}}{k_5}$$

Figure 2 Proposed mechanism of DNA cleavage by *Eco*RI endonuclease. I, II, and II represent, respectively, an intact *Eco*RI sequence; an intermediate species that has been subject to cleavage in only one DNA strand; and a site that has suffered double-strand hydrolysis. Rate constants shown were determined at 30°C ($K_2 \geq 40$ min^{-1} at 30°C, $K_3 = 14$ min^{-1}, $K_{cat} = 1$ min^{-1}; Modrich and Zabel 1976). Other details are provided in the text. Adapted from Rubin and Modrich (1978).

DNA structure external to the *Eco*RI site also affects the mechanism of double-strand cleavage by the endonuclease. Double-strand hydrolysis of ColE1, G4 RFI, and pBR322 DNA substrates at 37°C has been shown to proceed without significant dissociation of the E·II intermediate (Fig. 2). Under these conditions, the form-II intermediate remains enzyme-bound, and the two single-strand events required for double-strand cleavage are coupled (Modrich and Zabel 1976; Rubin and Modrich 1978; Langowski et al. 1980). These findings indicate that the endonuclease can interact with both DNA strands during a single binding event under these conditions. Given that the *Eco*RI sequence has twofold symmetry and that the active form of the endonuclease is a dimer of a single polypeptide chain, these results have been interpreted in terms of symmetry of the complex between endonuclease and the *Eco*RI sequences (Rubin and Modrich 1978).

In contrast to the mode of hydrolysis of the substrates cited above, *Eco*RI cleavage of SV40 or pMB9 has been found to proceed with substantial dissociation of the E·II intermediate (Ruben et al. 1977; Rubin and Modrich 1978; Halford et al. 1979). As in the case of differential efficiency of double-strand cleavage, the substrate dependence of mechanism has been attributed to outside sequence effects (Rubin and Modrich 1978), although it is not clear whether the two phenomena are related. Since *Eco*RI interacts with about 10 bp (below), such phenomena may simply reflect nucleotide sequences flanking the *Eco*RI sites in question. Indeed the substrate dependence of the mechanism of cleavage by *Eco*RI correlates with the extent to which purine-pyrimidine symmetry persists beyond the *Eco*RI site (Rubin and Modrich 1978). However, factors other than nearest neighbor effects cannot be excluded.

Although mechanistic studies of other type-II endonucleases are still at an early stage, the available information suggests that DNA cleavage may occur by a mechanism similar to that for *Eco*RI (Fig. 2). Thus, *Bam*HI (Hinsch et al. 1980), *Bsp*I (Koncz et al. 1978), *Hpa*I, and *Hpa*II (P. Dwyer-Hallquist and K.I. Agarwal, unpubl.) endonucleases have been observed to turn over in vitro. Furthermore, evidence for a cleavage intermediate has been presented in the cases of *Bam*HI (Smith and Chirikjian 1979), *Hin*dIII and *Sal*I (Halford et al. 1979), and *Hpa*II (Ruben et al. 1977). In the case of *Cla*I cleavage, it has been concluded that the rate-limiting step is dissociation from product (Hinsch and Kula 1981), as is the case for *Eco*RI (Fig. 3).

Thus, although limited, current information suggests that an intermediate containing a scission in only one DNA strand is a common feature of DNA cleavage by type-II endonucleases. It is evident that in such a mechanism interaction between endonuclease and intermediate is necessarily asymmetric. However, it is also possible that asymmetry may occur during the introduction of the first single-strand break. Indeed, strand preferences have been observed during cleavage of short DNA duplexes by *Hpa*II (Baumstark et al. 1979) and *Msp*I (Yoo and Agarwal 1980b). In both studies, cleavage of a 9-bp DNA duplex was monitored:

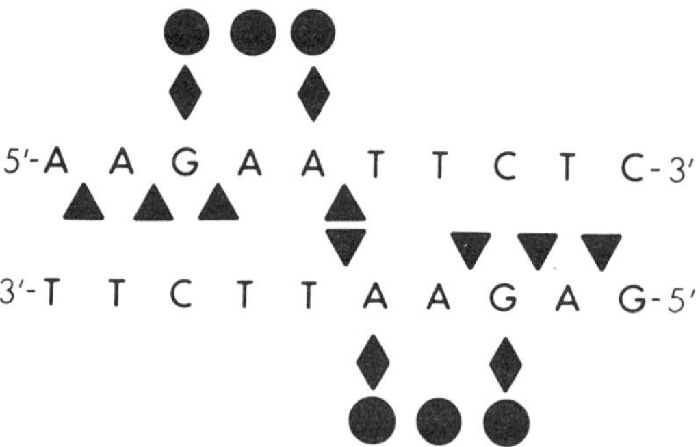

Figure 3 Results of alkylation protection and alkylation interference analysis on specific complexes between *Eco*RI endonuclease and the *Eco*RI sequence of pBR322 DNA (Lu et al. 1981). (●) Purines protected against alkylation by *Eco*RI endonuclease; (◆) purines, the methylation of which interferes with specific complex formation; (▲) phosphates, the ethylation of which interferes with specific binding.

5′ pGp Ap Ap Cp Cp Gp Gp Ap Gp A
3′ Tp Tp Gp Gp Cp Cp Tp Cp Tp.

In the case of *Hpa*II, the rate of cleavage within the lower pyrimidine-rich strand was found to be about three times that for cleavage of the upper strand. *Msp*I exhibits the opposite preference, cleaving the upper purine-rich strand about twice as rapidly as the other. In contrast, both strands were attacked at an equal rate by another isoschizomer, *Mno*I (Baumstark et al. 1979). Since these three endonucleases all recognize the internal CCGG sequence, the strand preferences of *Hpa*II and *Msp*I presumably reflect differential effects of nucleotides surrounding this sequence. These findings therefore suggest that these two endonucleases interact with nucleotides beyond the required tetranucleotide, but do so in different ways. They would also indicate that while the proteins may interact equivalently with the CCGG sequence within the two DNA strands, a certain degree of asymmetry may exist in protein-DNA complexes by virtue of interaction with outside nucleotides.

The biological phenomenon of restriction-modification relies upon unmodified recognition sites being subject to cleavage, while sites methylated on both DNA strands are resistant. The interaction of an endonuclease with recognition sequences methylated on only one DNA strand is of interest for several reasons. First, such sites are of biological importance, since they will result from semiconservative DNA synthesis. Second, the efficacy of hemimethylated sites as substrates for an endonuclease provides insight into the mode of enzyme-DNA interaction in a given system. Type-II endonucleases examined in this respect fall into two classes. Restriction endonucleases *Eco*RI (Rubin and Modrich 1977), *Dpn*I, and *Dpn*II (Vovis and Lacks 1977) have been shown not to cleave hemimethylated recognition sites. Hemimethylated duplexes have also been shown to be resistant to double-strand cleavage in the case of the *Eco*RII system (Roulland-Dussoix et al. 1974), but the possibility of single-strand cleavage has not been examined. In contrast, it has been found that recognition sites methylated on only one DNA strand are subject to single-strand scission within the unmodified strand in the *Bsp*I system (Koncz et al. 1978). Similarly, a duplex GATC site containing 5-methylcytosine in only one strand is subject to single-strand cleavage in the unmethylated strand by *Sau*3A (Streeck 1980). The basis for these distinct responses to hemimethylated sites is not yet known. However, it can be noted that *Eco*RI and *Bsp*I, which have been obtained in pure form, differ with respect to their stable oligomeric states in solution (Table 3).

METHYL TRANSFER BY TYPE-II MODIFICATION ENZYMES

Only five methylases, *Bsu*I (Gunthert et al. 1981a,b), *Eco*RI (Rubin and Modrich 1977), *Hpa*I (K.I. Agarwal, unpubl.), *Hpa*II (Yoo and Agarwal 1980a), and *dam* (Herman and Modrich 1981) have been isolated in pure form. Each of these enzymes is kinetically simple. The rate of methyl transfer is first order with respect to enzyme concentration and Michaelis-Menton behavior is obeyed with respect to both AdoMet and DNA substrates. Since the stable form of each enzyme is the monomer (Table 1), these findings have led to the conclusion that each methyltransferase functions as a monomer. Like type-II endonucleases, these modification enzymes have very low turnover numbers of 3 (*Eco*RI), 19 (*dam*), and 1 (*Hpa*I) methyl transfers per minute per monomer at 37°C (Rubin and Modrich 1977; Herman 1981; P. Dwyer-Hallquist and K.I. Agarwal, unpubl.).

In the case of the *Eco*RI and *dam* enzymes, the mechanism of methyl transfer is consistent with the monomer being the functional form of these proteins. To determine whether one or two methyl groups are transferred to an unmodified site per DNA binding event, Rubin and Modrich (1977) examined the conversion of *Eco*RI sites to an endonuclease-resistant form as a function of the average extent of methylation. In similar experiments, Herman and Modrich (1981) monitored the conversion of GATC sequences to a form sensitive to *Dpn*I, an endonuclease that only cleaves methylated DNA, as a function of the extent of *dam* methylation. These studies indicate that both modification enzymes transfer a single methyl group to a recognition site per binding event and, in the case of the *Eco*RI enzyme, also demonstrate that kinetic parameters for methylation of a hemimodified site are not more favorable than those for methyl transfer to the unmodified sequence (Rubin and Modrich 1977). In the case of the *Bsu*I and *dam* methylases, a slight preference for methyl transfer to hemimethylated sites has been observed

(Gunthert et al. 1981b; Herman and Modrich 1981). These findings are in contrast to the reaction mediated by type-I modification enzymes for which a marked preference for half-modified sites has been demonstrated (Vovis et al. 1974).

These five methylases are novel in the sense that they represent the first examples of proteins which interact with twofold symmetrical DNA sequences in the monomeric state. In the absence of internal symmetry within the protein monomers, these findings could imply that complexes between such enzymes and their recognition sequences are necessarily asymmetric. For the *Eco*RI system, this work, together with comparable studies on the endonuclease (above), demonstrates that the two enzymes interact with their common recognition sequence in different ways (Rubin and Modrich 1977).

FIDELITY OF TYPE-II RESTRICTION AND MODIFICATION ENZYMES

Analysis of a variety of sequence-specific proteins has demonstrated that they typically bind to DNA molecules lacking recognition sites, albeit with reduced affinity (von Hippel and McGhee 1972; von Hippel 1979). In principle, the spectrum of such interactions may range from those that are truly nonspecific to those in which a significant portion of the interaction energy is dictated by the DNA sequence. The latter class is expected to include interactions with a hierarchy of DNA sites that are to some extent homologous with the recognition sequence of the protein in question (e.g., von Hippel 1979). In the case of several type-II enzymes, it has been possible to monitor a presumably related phenomenon, namely catalysis characterized by reduced sequence specificity in vitro.

The frequency with which type-II endonucleases cleave endogenous cellular DNA in vivo appears to be very low. For example, it has been demonstrated that in vivo *Eco*RI restriction of bacteriophage λ DNA destroys biological activity of the viral molecule with a probability of 0.8–0.9 per *Eco*RI site present (Murray and Murray 1974; Thomas et al. 1974). If one assumes that double-strand cleavage within the *E. coli* chromosome (4×10^6 bp) would confer lethality with a similar probability, it is evident that in vivo restriction cleavage errors would be quite rare. It must be noted, however, that the error rate of in vivo restriction may not solely reflect fidelity of sequence recognition by the restriction enzyme. It can be argued that the restriction endonuclease and chromosome are differentially compartmentalized within the cell, or alternatively, that secondary sites subject to low-efficiency restriction are also subject to modification and thus rendered resistant to the endogenous endonuclease. Indeed, some evidence for in vivo modification of secondary sites by *Eco*RI methylase has been obtained (Woodbury et al. 1980a).

The analysis of in vitro fidelity of type-II enzymes is still at an early stage and has not yet been assessed in a systematic way in any system. Moreover, caution must be exercised in attributing in vitro restriction or modification events at secondary sites to the same enzyme species that mediate activity at canonical sequences, particularly in those cases involving poorly characterized enzyme preparations. Since it is well known that proteins can be subject to chemical alteration during isolation, the specter of chemically distinct enzymatic species in a purified preparation must be carefully considered in fidelity experiments. This is not a trivial point since, as will be detailed below, evidence for such effects already exists in the case of DNA cleavage by the *Eco*RI endonuclease (Bishop 1979; Bishop and Davies 1980).

Although double-strand cleavage by type-II endonucleases exhibits a strict requirement for established recognition sequences under optimal reaction conditions, two modes of DNA cleavage characterized by reduced specificity have been observed in vitro. Several laboratories have found that under optimal reaction conditions certain DNA sites can be subject to single- but not double-strand cleavage by type-II endonucleases (Bishop 1979; Bishop and Davies 1980; Maxam and Gilbert 1980). Sequences ''nicked'' under these conditions typically are largely homologous to the canonical recognition site, differing by one base pair. This phenomenon has been analyzed in greatest detail in the case of *Eco*RI. Bishop (1979) initially reported that the sequence $\genfrac{}{}{0pt}{}{\mathrm{d(G{\downarrow}A\text{-}A\text{-}T\text{-}T\text{-}A)}}{\mathrm{d(C\text{-}T\text{-}T\text{-}A\text{-}A\text{-}T)}}$ is subject to single-strand cleavage by *Eco*RI and suggested that the occurrence of the sequence GAATT is sufficient for cleavage within a DNA strand. However, subsequent studies (Bishop and Davies 1980) demonstrated that this reaction is mediated only by certain preparations of the endonuclease. Hence, the protein catalyzing the nicking activity, designated EcoRI^{n}, is distinguishable from the enzymatic species that catalyzes double-strand cleavage within the canonical site.

"Star" Activities

A second type of variation in the specificity of type-II enzymes has been observed under conditions of altered reaction environment. These effects, which generally involve ''relaxation'' of specificity, have been documented for a number of endonucleases, including *Eco*RI (Polisky et al. 1975; Hsu and Berg 1978; Tikchonenko et al. 1978; Malyguine et al. 1980; Woodbury et al. 1980b), *Bam*HI (George et al. 1980; Malyguine et al. 1980), *Bst*I (Clarke and Hartley 1979), *Bsu*I (Heininger et al. 1977), *Hha*I (Malyguine et al. 1980), and *Hin*dIII (Hsu and Berg 1978). This relaxation of specificity is typically observed under conditions of high endonuclease concentration and has been promoted by variation of pH and ionic strength, replacement of Mg^{2+} by Mn^{2+}, and supplementation of reactions with glycerol, dimethylsulfoxide, or other water-miscible organic solvents.

Polisky et al. (1975) initially demonstrated a reduction in the specificity of highly purified *Eco*RI under conditions of elevated pH and low ionic strength. Analysis of DNA termini generated by this activity (designated *Eco*RI*) suggested that cleavage was occurring within the tetranucleotide sequences

X↓AATTY. However, even under *Eco*RI* conditions DNA cleavage was observed to occur much more rapidly at the canonical *Eco*RI sequence, with cleavage at alternate sites exhibiting a gradient of preference for nucleotide X of dG (0.59) > dA (0.26) > dT (0.14) >> dC (0.006).

In related studies, Hsu and Berg (1978) and Tikchonenko et al. (1978) demonstrated a similar reduction in specificity in the presence of Mn^{2+} or organic solvents (glycerol, dimethylsulfoxide, dimethylacetamide, and dimethylformamide). It would appear that a similar set of DNA sites is subject to cleavage under the several conditions of reduced specificity (Tikchonenko et al. 1978; Woodbury et al. 1980b), and synergistic effects have been observed (Tikchonenko et al. 1978). Although it is difficult to generate limit *Eco*RI* digests under conditions of low ionic strength and elevated pH, or at moderate salt concentrations in the presence of Mn^{2+}, near-limit digests have been observed under the former conditions in the presence of Mn^{2+} (Tikchonenko et al. 1978).

More recently von Hippel and colleagues (Woodbury et al. 1980b) have exploited the availability of the total DNA sequence of ϕX174 (Sanger et al. 1978), which is devoid of *Eco*RI sites, to examine the sequences cleaved under *Eco*RI* conditions. In these experiments, *Eco*RI* sites within the duplex form of the DNA were localized by restriction mapping and corresponding regions of the viral DNA sequence inspected for sequences homologous to the canonical *Eco*RI site. This analysis suggested that the most labile *Eco*RI* sites within this DNA are GGATTT, AAATTT, GAATTT, and GAATTA. However, the observed restriction patterns of ϕX174 DNA were not consistent with *Eco*RI* cleavage at all AAATTT sequences. In addition, GGATTT sites, which were identified as the most labile sequences in the ϕX174 molecule, appear refractory to cleavage in SV40 DNA (Hsu and Berg 1978; Woodbury et al. 1980b). Although these discrepancies may reflect the nature of flanking sequences, it is evident that those elements of a DNA sequence that govern the reduced specificity of cleavage by *Eco*RI require further definition.

Relaxation of specificity of type-II endonucleases has generally been attributed to the effects of altered reaction environment on the nature of complexes formed between the substrate and a chemically authentic form of the endonuclease. This explanation is consistent with the presence of secondary cleavage activity in highly purified and, in several cases, homogeneous endonuclease preparations (Polisky et al. 1975; Hsu and Berg 1978; George et al. 1980; Malyguine et al. 1980; Woodbury et al. 1980), cofractionation of *Eco*RI and *Eco*RI* activities (Tikchonenko et al. 1978; P.J. Greene, unpubl.), and similar effects of Mn^{2+}, glycerol, and other solvents on several endonucleases of differing specificity. In this case, the hierarchy of susceptibilities of secondary sites would be expected to bear on the mechanisms of sequence recognition by type-II enzymes, and in the case of *Eco*RI, a model for recognition based on the specificity of *Eco*RI* cleavage has been proposed (Woodbury et al. 1980b).

In contrast, Tikchonenko et al. (1978) have suggested that the *Eco*RI and *Eco*RI* activities may reflect distinct enzymatic species. They have reported that *Eco*RI and *Eco*RI* activities of purified preparations are differentially labile during prolonged storage, that *Eco*RI endonuclease isolated in the presence of nonionic detergents contains more *Eco*RI* activity than enzyme prepared in their absence, and that *Eco*RI* activity is abolished by parachloromercuribenzoate while *Eco*RI activity is unaffected by this reagent. Although these observations do not prove that *Eco*RI and *Eco*RI* cleavage reactions are mediated by distinct enzymatic species, it is apparent that the origin and nature of *Eco*RI* activity have not yet been completely resolved.

Reduction in specificity of methyl transfer by the *Eco*RI modification enzyme has also been observed at high enzyme concentrations or in the presence of glycerol (Berkner and Folk 1978; Woodbury et al. 1980a). Under these conditions, methyl transfer still occurs exclusively to adenine residues (Berkner and Folk 1978). Since $d(AT)_n \cdot d(AT)_n$ accepts methyl groups slowly under such conditions while $d(A)_n \cdot d(T)_n$ does not, it has been suggested that the minimal substrate requirement for the methylase is an $^{\mathrm{AT}}_{\mathrm{TA}}$ duplex (Woodbury et al. 1980a). Furthermore, the fact that natural DNA methylated under reduced specificity conditions is largely resistant to *Eco*RI* cleavage indicates that an overlapping set of sites is recognized by the reduced specificity forms of the two enzymes. Indeed, it has been proposed that reduced specificity modification may function physiologically to protect noncanonical sites against *Eco*RI* cleavage in vivo (Berkner and Folk 1978). That in vivo modification of secondary sites may occur is suggested by the finding that DNA isolated from a strain bearing the *Eco*RI restriction-modification system is a poor acceptor for noncanonical methylation in vitro, as compared with DNA isolated from *E. coli* strains that do not carry this restriction-modification system (Woodbury et al. 1980a). However, a comparative analysis of the sensitivity of such DNA preparations to in vitro *Eco*RI* cleavage has not been reported.

Cleavage of Single-stranded DNA

The specificity of type-II enzymes with respect to polynucleotide conformation has also received substantial attention. Most of this work has dealt with the activity of such enzymes on single-stranded as opposed to duplex DNA. It has been found in several laboratories that single-stranded circular DNAs isolated from virions of f1, M13, and ϕX174 are subject to cleavage at reduced rates by some, but not all, type-II endonucleases (Blakesley and Wells 1975; Horiuchi and Zinder 1975; Godson and Roberts 1976). Moreover, the pattern of DNA fragments obtained from cleavage of the single-stranded molecules is the same as that obtained from restriction of the duplex forms. The mechanism of this single-strand cleavage reaction has been assessed in the cases of *Hae*II, *Hha*I (Blakesley et al. 1977), and *Msp*I (Yoo and Agarwal 1980b) and has led to slightly different interpretations.

Several lines of evidence led Blakesley et al. (1977) to conclude that cleavage of ϕX174 single-strands by *Hae*III or *Hha*I does not involve cleavage of recognition sequences in the single-stranded state, but rather occurs within regions of duplex helical structure, which may be stable, at least transiently. Cleavage of single strands by these two endonucleases was exquisitely sensitive to inhibition by low levels of actinomycin D and netropsin, both of which bind only to duplex DNA. Second, cleavage of single-stranded and duplex forms of ϕX174 were differentially sensitive to temperature. For example, *Hae*III hydrolysis of single strands exhibited a much lower temperature optimum (47°C) than did the reaction with duplex circles (72°C). Moreover, near the respective temperature optima certain sites in the single-stranded substrate were rendered resistant to hydrolysis while all sites within the double-helical substrate remained sensitive. These findings strongly suggest that *Hae*III and *Hha*I sites within ϕX174 single strands are located within regions of secondary structure that are differentially stable (Blakesley et al. 1977).

In more recent experiments, Yoo and Agarwal (1980b) analyzed cleavage of the complementary deoxyoligonucleotides GAACCGGAGA and TCTCCGGTT by *Msp*I. While the 9-bp duplex formed between the oligonucleotides was readily cleaved by the endonuclease, it was demonstrated that the individual oligonucleotides were also subject to cleavage at reduced rates by the enzyme. Since the self-complementarity of each oligonucleotide is limited to the CCGG recognition sequence, it is unlikely that stable duplex structures form at the temperature used. Furthermore, examination of DNA sequences surrounding CCGG recognition sites within the ϕX174 molecule indicated that a 6-bp duplex is the longest region of secondary structure involving an *Msp*I site that can be formed by intramolecular base-pairing within the viral single strand. Hence, it is highly unlikely that *Msp*I cleavage of ϕX174 single strands requires preexisting regions of stable duplex structure.

While these findings may indicate that cleavage occurs within single-stranded regions by *Msp*I, the rates of cleavage of the individual oligonucleotide strands mentioned above were found to decrease markedly with increasing temperature in the range 4°C to 37°C, suggesting a more complex mechanism. Based on these observations and the properties of other type-II endonucleases, Yoo and Agarwal (1980b) proposed the following model for *Msp*I action: A form of the endonuclease characterized by twofold symmetry (postulated to be the dimer) is capable of interaction with CCGG sequences in either the single-stranded or duplex conformation, with complexes involving duplex sites possessing elements of two-fold symmetry. When the substrate is single-stranded, it is proposed that two identical DNA-protein complexes interact to yield an intermediate similar to that resulting from interaction of the enzyme with preformed duplex substrate. It is argued that in either case, enzyme-induced transient complexes so formed are rapidly converted to an equimolar complex of recognition sequence and enzyme subunit, with interactions among subunits and between subunits and recognition sequence occurring in a concerted manner to determine the rate of DNA cleavage. Thus, the distinction between this model and that of Blakesley et al. (1977) is that the latter involves endonuclease recognition only within regions of *preformed* secondary structure. It should be emphasized that the model of Yoo and Agarwal (1980b) is based on the assumption that duplex *Msp*I sites could not have preexisted in their experiments. Although the formation of stable regions of secondary structure was excluded in this work, the authors note that transient formation of duplex recognition sites, which could be subsequently stabilized by interaction with enzyme, was not ruled out.

In the case of those endonucleases that are inactive on natural DNA in the "single-stranded" configuration, it could be argued a priori that failure to observe cleavage reflects inhibition by the high levels of nonspecific single-stranded sequence present. However, analysis of oligonucleotide cleavage by *Eco*RI and *Hpa*II endonucleases indicates that at least these enzymes cleave only the duplex form of their recognition sites. In the case of *Eco*RI, Greene et al. (1975) demonstrated cleavage of the self-complementary octanucleotide TGAATTCA, but the dependence of the rate of cleavage on oligonucleotide concentration was greater than first order. Since the duplex and single-stranded forms of the octanucleotide were in equilibrium at reaction temperature, it was argued that the anomalous rate dependence reflected a requirement for duplex structure. Analysis of the kinetic data in terms of the equilibrium constant governing octanucleotide association was consistent with this view. Similarly, Baumstark et al. (1979) have demonstrated that while the 9-bp duplex formed between GAACCGGAGA and TCTCCGGTT is readily cleaved by *Hpa*II, the individual single-strands are not. It is interesting to note that, as discussed above, these oligonucleotides are those which were subject to cleavage by *Msp*I in the "single-stranded" form. Thus, *Hpa*II and *Msp*I differ in this respect despite their common recognition sequence.

Methylation of Single-stranded DNA

*Hae*III, *Hpa*II, and the *dam* modification enzymes have been found to methylate denatured salmon sperm or T7 DNAs, albeit at rates somewhat less than that for duplex substrates (Mann and Smith 1979a; Herman and Modrich 1981). However, the possibility that methylation of single-stranded DNA by these enzymes occurs within regions of secondary structure has not been explored. In contrast, *Eco*RI and *Hpa*I modification enzymes do not methylate denatured DNA (K.I. Agarwal, unpubl.; R.A. Rubin and P. Modrich, unpubl.). Perhaps the most intriguing observations in this respect have been made with the *Hha*I methylase (Mann and Smith 1979a). This enzyme has been found to methylate a random copolymer of $d(G,C)_n$ to an extent consistent with the expected frequency of occurrence of the recognition sequence, GCGC. While modification of this polymer could reflect local secondary structure, the enzyme also catalyzes methyl transfer to random $d(N\text{-}AcG,C)_n$ and $d(X,C)_n$. Since these latter polymers were devoid of Watson-Crick secondary structure, as judged by the absence of thermal melting transitions, these observations suggest that the *Hha*I enzyme can recognize and methylate a recognition site in the single-stranded conformation. However, both rate and extent of methyl transfer to denatured *M. luteus* DNA were found to be only 30% of those observed with the duplex form of this substrate (Mann and Smith 1979a). The reduced extent of methyl transfer by the *Hha*I enzyme to single-stranded natural DNA suggests that at least some sites are not subject to modification in this form.

Cleavage of DNA-RNA Hybrids

The potential for cleavage of DNA-RNA hybrid duplexes by type-II endonucleases has also been examined in one study (Molloy and Symons 1980). RNA-DNA hybrids prepared by reverse transcription were subjected to hydrolysis by a number of type-II enzymes, with cleavage of the isotopically labeled DNA product being monitored. *Eco*RI, *Hae*III, *Hha*I, *Hin*dII, and *Msp*I were found to cleave a substantial fraction of the labeled DNA and the size of products was consistent with the location of recognition sequences within the RNA template used for reverse transcription. Unfortunately, neither the substrates nor the products of endonuclease action were characterized with respect to DNA-RNA hybrid structure. It is possible that DNA duplexes represented a portion of the reverse transcription product, or alternatively that hydrolysis of the RNA component of RNA-DNA duplexes (by RNase contamination of restriction enzyme preparations) resulted in DNA single strands that could have been cleaved by several of the endonucleases tested.

TYPE-II ENZYME-DNA INTERACTION: THERMODYNAMIC AND KINETIC PARAMETERS

The analysis of structures and catalytic properties of type-II endonucleases and methyltransferases outlined above has provided a crude picture of the modes of interaction of such proteins with their recognition sites. However, they have not indicated whether sequence-specificity is manifested at the level of DNA binding or catalysis. Furthermore, alternative methods are required to delineate contacts between protein and polynucleotide that are important in establishing specificity.

The binding of type-II enzymes to DNA has been monitored by several methods. In the absence of a divalent cation, conditions under which DNA cleavage does not occur, *Eco*RI endonuclease forms site-specific complexes with DNA that are retained on nitrocellulose membranes (Modrich and Zabel 1976; Modrich 1979; Halford and Johnson 1980; Jack et al. 1980; Rosenberg et al. 1980). Such complexes are site-specific since their formation is markedly reduced if the DNA bears the *Eco*RI modification (Modrich 1979; Jack et al. 1980; Rosenberg et al. 1980) or if the *Eco*RI recognition sequence has been selectively deleted from the molecule (Halford and Johnson 1980; Jack et al. 1980). The only other type-II endonuclease that has been examined using this assay is *Bgl*I (Lee and Chirikjian 1979). While this enzyme forms complexes that can be retained on nitrocellulose with SV40 DNA in the absence of Mg^{2+}, these complexes have not been shown to be site specific.

In the case of the *Eco*RI enzyme, specific binding as monitored by this method is hyperbolic and exhibits the requisite criteria for an equilibrium process. Furthermore, the simple nature of specific binding observed during titration of DNA with increasing endonuclease indicates that oligomerization of enzyme is apparently not required for specific complex formation. Since the stable form of the *Eco*RI enzyme is the dimer in the concentration range employed in these experiments, these findings demonstrate that the dimer interacts directly with the *Eco*RI sequence.

At 37°C and ionic strengths of 0.07–0.15 M, the apparent equilibrium dissociation constant for specific complexes between *Eco*RI endonuclease and plasmid pBR322 (one *Eco*RI site) lies between 10^{-11} and 10^{-10} M (Modrich 1979; Jack et al. 1980; Rosenberg et al. 1980). Moreover, binding curves obtained under conditions of invariant endonuclease concentration have demonstrated that a large fraction of the isolated endonuclease (as the dimer) is active in site-specific binding (Jack et al. 1980). Since such complexes undergo rapid cleavage upon addition of Mg^{2+} (Modrich and Zabel 1976; Jack et al. 1982), this implies that a large fraction of the purified *Eco*RI endonuclease is catalytically active. At 22°C and comparable ionic strengths, similar studies with derivatives of bacteriophage λ DNA containing only one *Eco*RI site yielded apparent dissociation constants on the order of 10^{-9} M (Halford and Johnson 1980). It should be noted that these binding constants have generally been estimated assuming a simple two-state model in which enzyme not involved in specific complex formation is presumed to be free in solution. However, the endonuclease also binds to DNA sites other than at its recognition sequence, GAATTC. Since the effects of this alternative type of interaction are neglected in the two-state model, these dissociation constants should be regarded as only approximate.

The interaction of *Eco*RI with nonspecific DNA has been examined by several techniques. Woodhead and Malcolm (1980) have demonstrated that nonspecific DNA sequences can protect *Eco*RI against inactivation by methyl acetimidate. Using this protection assay, equilibrium dissociation constants were estimated for several DNA species. At 20°C in the absence of Mg^{2+} and at an ionic strength of 0.2 M, this method yielded equilibrium dissociation constants of 1.2 μM (ϕX174 RF DNA), 0.7 μM (SV40 DNA), 15 μM (λ DNA), 290 μM (d[GC]$_n$), and 300 μM (d[AT]$_n$), with concentrations expressed in terms of base-pair equivalents. Although both SV40 and λ DNA contain *Eco*RI sequences, protection afforded by these molecules was largely due to interaction with other DNA sequences since endonuclease was present in large molar excess over recognition sites. Interaction of *Eco*RI with sequences other than the *Eco*RI site has also been monitored by the nitrocellulose membrane assay. Equilibrium competition experiments, in which specific complex formation with pBR322 was competed by a derivative of this plasmid lacking the *Eco*RI sequence, have also indicated a nonspecific binding constant in the μM range (B.J. Terry and P. Modrich, unpubl.).

Competitive inhibition of DNA cleavage has demonstrated that *Eco*RI also binds nonspecifically to a variety of polynucleotides in the presence of Mg^{2+} (Langowski et al. 1980), although enzyme affinity for nonspecific sites may be reduced slightly by this cation (Halford and Johnson 1980). Other than natural DNA containing *Eco*RI sites, the highest-affinity inhibitors identified in these studies were several alternating DNA copolymers and tRNA. Hence, as in the case of *lac* repressor (Jovin 1976; von Hippel 1979), nonspecific interaction with polynucleotides is a characteristic of *Eco*RI.

The nature of the nonspecific interactions between *Eco*RI and DNA are not yet understood. Woodhead and Malcolm (1980) have suggested that nonspecific interactions between endonuclease and DNA primarily reflect binding to *Eco*RI* sequences. As yet, no evidence to support this contention is available. Indeed, it has been shown that nonspecific interactions are not affected by extensive methylation of the DNA by *Eco*RI modification enzyme under conditions of reduced specificity methyl transfer (Jack et al. 1980), which is known to protect at least some secondary sites against *Eco*RI* cleavage (Berkner and Folk 1978; Woodbury et al. 1980a). It could be argued, however, that in contrast to the effects of modification on the binding of the endonuclease to the canonical *Eco*RI sequence, methylation of *Eco*RI* sites may not reduce their affinity for the endonuclease.

In addition to their high thermodynamic stability, specific complexes between *Eco*RI endonuclease and DNA formed in the absence of Mg^{2+} are also kinetically very stable. Specific complexes between the enzyme and plasmid pBR322 DNA dissociate with a first-order rate constant of 7×10^{-4} sec^{-1} (at 37°C and ionic strength of 0.07) corresponding to a half-life of 16 minutes (Jack et al. 1980, 1982). Surprisingly, the kinetic stability of such complexes has been found to be markedly dependent on the chain length of the DNA in which the *Eco*RI site is embedded. Jack et al. (1982) have analyzed dissociation kinetics of specific complexes formed with nine linear DNA fragments ranging in size from 34 bp to 6200 bp, all of which were derived from pBR322 and contained the *Eco*RI site of this plasmid in a centrally located position. These experiments demonstrated an increase in the first-order rate constant governing dissociation from a value of 8×10^{-5} sec^{-1} ($t_{1/2}$ = 140 min) for the 34-bp molecule to 7×10^{-4} sec^{-1} for the longest linear duplex. Therefore, DNA sequences outside the *Eco*RI site kinetically destabilize specific complexes.

In contrast to the marked dependence of dissociation rate on DNA chain length, competition experiments demonstrated that the intrinsic equilibrium constant governing interaction of the endonuclease with the *Eco*RI sequence is independent of DNA chain length. The equilibrium ratio of specific complexes formed with long and short fragments was 1:1 when the two DNAs were present at equimolar concentrations in the same reaction (Jack et al. 1982). Furthermore, as predicted by microscopic reversibility, the rate of specific complex formation was also enhanced by DNA chain length. Thus, in reactions containing equimolar long and short molecules, specific complex formation by the endonuclease favored the longer DNA chain during the initial stages of the reaction, with an equilibrium ratio of 1:1 being achieved upon further incubation. The magnitude of the initial preference for long DNA molecules (ratio of long:short DNA molecules bound) in such reactions reflects the relative rate constants governing association of enzyme with the two sizes of DNA molecule. Analyses of seven different pairwise combinations of DNA fragments demonstrated that the initial preference for binding to the longer DNA was always identical to the ratio of first-order rate constants governing *dissociation* of enzyme from the fragments in question. In other words, the ratio of association rate constants for two DNA fragments was always found to be equivalent to the ratio of their dissociation rate constants, a finding consistent with the observed lack of dependence of the equilibrium constant on DNA chain length. Since longer DNAs are also subject to preferential cleavage in the presence of Mg^{2+} (Jack et al. 1982; W.E. Jack and P. Modrich, unpubl.), it would appear that similar effects are operative under catalytic conditions.

These findings imply that DNA sequences outside the recognition site are involved in the major kinetic path by which *Eco*RI endonuclease locates and leaves its recognition sequence. The results cited above are consistent with theoretical treatments (Richter and Eigen 1974; Schranner and Richter 1978; Berg et al. 1981) for a unidimensional facilitated diffusion mechanism in which a sequence-specific protein initially binds to nonspecific sites and then locates its recognition sequence by random diffusion along the DNA helix. In particular, results obtained with *Eco*RI suggest that the average distance scanned by the random walk mechanism is about 1000 bp per DNA binding event under the experimental conditions employed. It is important to note that evidence for the involvement of such a mechanism in specific site location by a protein is not limited to *Eco*RI. von Hippel and colleagues (Winter et al. 1981) have examined the effects of DNA chain length on kinetics in the case of the *lac* repressor and have drawn similar conclusions.

Despite the ability of the *Eco*RI endonuclease to form site-specific complexes in the absence of Mg^{2+}, attempts to detect specific binding by the *Eco*RI methylase in the absence of AdoMet have not yet been successful (Modrich 1979; Jack et al. 1980; Woodhead and Malcolm 1980). While titration of DNA with methylase can lead to retention of a large fraction of input DNA on nitrocellulose membranes, a large molar excess of methylase is required. Similarly, titration of a fixed concentration of methylase with DNA leads to the retention of DNA-protein complexes on filters, but only at a level of 10–20% that expected based on input enzyme. Lastly, formation of filter-retained complexes in either case occurs with DNA containing unmodified or modified *Eco*RI sites, or with DNA devoid of the *Eco*RI sequence (Woodhead and Malcolm 1980; R.A. Rubin and P. Modrich, unpubl.). These findings could reflect formation of nonspecific DNA-methylase complexes that are retained on nitrocellulose with low efficiency, or perhaps less likely, the presence of a low-level DNA-binding contaminant in purified preparations of the enzyme. This failure to detect specific methylase-DNA complexes may be a consequence of the assay procedures employed. For example, it is possible that specific complexes are not retained by nitrocellulose membranes. Alternatively, as in the case of the type-I enzymes (Chapter 5), site-specific binding of the methylase may require the presence of AdoMet. This would imply an ordered addition of substrates during the methyl transfer reaction and would be consistent with the ability of the methylase to transfer only one methyl group per DNA binding event. Although such a mechanism has not yet been established, this possibility has also been suggested for the *Bsu*I methyltransferase. Both this methylase and the *Eco*RI enzyme bind AdoMet in the absence of DNA (Gunthert et al. 1981b; D. Albergo and P. Modrich, unpubl.)

TYPE-II ENZYME-DNA INTERACTION: DNA DETERMINANTS IMPORTANT IN SPECIFIC RECOGNITION

Two approaches have been used in attempts to define DNA contacts important in specific interactions between type-II enzymes and their recognition sequence. One approach has been to examine the effects of base-analog substitution, within the recognition sequence, on specific interactions. A number of analogs are available that form good Watson-Crick base pairs and differ in simple ways from the common bases. Interaction of type-II enzymes with analog-substituted sites can be examined under catalytic, as well as noncatalytic, conditions to define the effects of DNA substituents on parameters governing specific recognition and catalysis.

An alternative approach, which was developed by Gilbert et al. (1976), relies on partial chemical alkylation of DNA fragments and is related to the chemical method for DNA sequence analysis (Maxam and Gilbert 1980). In one variation, the alkylation protection method, an end-labeled DNA fragment is subjected to partial dimethylsulfate alkylation in the presence and absence of a DNA-binding protein. Subsequent depurination and strand cleavage permit identification of those purines whose reactivities toward the reagent are altered by the bound protein. Since this reagent methylates the N^7 of dG (major groove) and the N^3 of dA (minor groove) (Singer 1975), this procedure is used to infer the presence of polypeptide in the vicinity of affected nitrogens.

In the second variation, alkylation interference (Siebenlist et al. 1980), an end-labeled DNA fragment is partially alkylated prior to interaction with the protein. After addition of the protein, the DNA is separated (by filtration through a nitrocellulose membrane) into a fraction that is capable of binding to the protein and one that is not. Subsequent analysis permits identification of those nucleotides that, in the alkylated form, interfere with DNA-protein interaction. In addition to the use of dimethylsulfate, alkylation interference studies have also employed ethylnitrosourea as an ethylating agent for backbone phosphates (Sun and Singer 1975; Siebenlist et al. 1980). The alkylation interference technique, which is analogous to base-analog substitution, thus allows identification of N^3 of dA, N^7 of dG, and phosphates which may provide contacts with the protein.

Although covalent alteration of a nucleotide determinant used as a contact during sequence discrimination would be expected to interfere with recognition by a protein, the possibility of indirect effects must also be considered. The available information concerning the dependence of DNA secondary structure on nucleotide sequence is incomplete, but in at least one model system, a major effect on helix conformation as a consequence of extensive 5-methylcytosine substitution has been documented (Behe and Felsenfeld 1981). The possibility of conformational effects has generally not been excluded in the analog and alkylation experiments discussed below. Similarly, indirect steric effects are also possible in alkylation interference studies or in experiments employing analogs with bulky substituents. For example, modification of a dC or dA residue by a type-II methylase (Table 3) would directly affect the interaction between the corresponding restriction enzyme and the base determinant methylated (N^6 of dA or C^5 of dC). However, analysis of molecular models indicates that the presence of such methyl groups may also interfere sterically with protein bonding to other determinants of the base pair modified and may even interfere with adjacent base pairs.

*Eco*RI Restriction-modification Enzymes

Kaplan and Nierlich (1975) and Berkner and Folk (1977) have examined *Eco*RI cleavage and methylation of several bacteriophage DNAs which, in their native state, contain analogs of dT and dC, with substituent variation at the 5 position of the pyrimidine ring. Since this position is exposed in the major groove of B-DNA, the simplest interpretation of substituent effects at this position is to define major groove contacts between protein and polynucleotide. Bacteriophage T4 DNA, which contains glucosylhydroxymethylcytosine in place of cytosine, is resistant to cleavage by *Eco*RI (Kaplan and

Nierlich 1975; Berkner and Folk 1977). Since nonglucosylated T4 DNA is hydrolyzed by the enzyme (Berkner and Folk 1977), this resistance evidently reflects steric hindrance by the presence of glucosyl residues in the major groove. Furthermore, the fact that nonglucosylated DNA is a substrate for the enzyme implies that the presence of a hydroxymethyl substituent at the 5 position of dC does not block cleavage. However, since cleavage of the nonglucosylated substrate was only analyzed qualitatively, the possibility of some interference with recognition or catalysis by this substituent cannot be excluded. Results obtained with this and other analog-containing substrates are summarized in Table 4.

In related experiments, Berkner and Folk (1977) demonstrated that phage PBS2 DNA is hydrolyzed at normal rates by the *Eco*RI enzyme. Since this DNA contains uracil in place of thymine, this finding implies that the 5-methyl of thymine is not required for recognition or hydrolysis by the enzyme. In contrast, DNA from bacteriophage ϕe, which contains hydroxymethyluracil in place of thymine, was found to be an extremely poor substrate for *Eco*RI endonuclease. This effect appears to reflect a block at the level of DNA cleavage rather than recognition since the K_m for hydrolysis of ϕe DNA is similar to that for λ DNA, with the defect being evident in the V_{max}. It is noteworthy that bromouracil-containing λ DNA has also been found to be subject to cleavage at reduced rates (Marchionni and Roufa 1978), although the kinetic basis for this effect has not been determined. Thus, while the 5-methyl of thymine is not essential for recognition or catalysis by the endonu-

Table 4 Potential DNA contacts within the *Eco*RI sequence

Base	Position	Groove	Methylase contact	Endonuclease contact	Reference
G	N^7	major	?	yes	Lu et al. (1981)
	$2\text{-}NH_2$	minor	yes	no	Modrich and Rubin (1977)
A	N^3	minor	?	no	Lu et al. (1981)
A	$6\text{-}NH_2$	major	yes	yes	Dugaiczyk et al. (1974)
	N^3	minor	?	yes	Lu et al. (1981)
T[a] T	$5\text{-}CH_3$	major	yes	no	Berkner and Folk (1977)
C	5-H	major	no	no	Berkner and Folk (1977) Kaplan and Nierlich (1975)

Positive assignment as a contact is subject to the reservations cited in the text.

[a] Assignment of the $5\text{-}CH_3$ of dT residues as a methylase contact is based on analysis of PBS2 DNA which contains dU in place of dT (Berkner and Folk 1977). Since both dT residues are presumably substituted within *Eco*RI sites of this viral molecule. It is not clear whether the $5\text{-}CH_3$ of one or both nucleotides is required.

clease, its replacement by bromo- or hydroxymethyl groups substantially reduces the rates of DNA cleavage. Reduced activities with the latter substituents could be the consequence of a conformational transition in the substrate, or alternatively may be due to electronic or steric effects reflecting their increased polarity or size.

To minimize the possibility of conformational perturbation of the DNA substrate in analog studies, Modrich and Rubin (1977) selectively replaced the dG residues within the *Eco*RI site of ColE1 DNA with deoxyinosine (dI) without altering other dG residues in the molecule. This substitution, which is equivalent to the loss of the 2-amino group of dG exposed in the minor groove of B-form DNA, was without effect on the rate of cleavage by *Eco*RI. Hence, this DNA functional group is not required for recognition or catalysis by the enzyme.

Although the analog studies summarized above have not identified DNA determinants important in sequence recognition by the *Eco*RI endonuclease, application of the alkylation protection and alkylation interference methods has suggested DNA contacts important in this process (A.L. Lu and P. Modrich, unpubl.). Results of these studies are summarized in Figure 3 and Table 4. It should be reiterated that application of these techniques is dependent on the formation of stable and isolable complexes between protein and DNA. Hence, such experiments with *Eco*RI endonuclease have been performed in the absence of Mg^{2+}. As shown in Figure 3, the N^7 position (major groove) of the dG residue and the N^3 position (minor groove) of both dA residues within the *Eco*RI sequence were protected against dimethylsulfate alkylation by the endonuclease. In the absence of substantial helix perturbation upon endonuclease binding, these results suggest the presence of polypeptide in both major and minor grooves in the vicinity of these purine nitrogens. Moreover, interaction in the minor groove outside the *Eco*RI site was suggested by a weak protection of N^3 of a dA residue two base pairs 5′ to the recognition sequence (not shown in Fig. 3).

Complementary alkylation interference analyses demonstrated that methylation of the N^7 of the *Eco*RI site dG residue or the N^3 of the dA adjacent to the dyad axis resulted in a marked reduction in specific complex formation, suggesting that these purine nitrogens may be used as contacts by the endonuclease (Lu et al. 1981). Although the 5′ penultimate dA in the *Eco*RI sequence was protected against chemical alkylation in the presence of the enzyme, methylation of this purine did not interfere with specific complex formation. Therefore, the N^3 of this dA residue does not interact with the enzyme. Methylation of the *Eco*RI sequence by the modification enzyme represents a natural example of methylation interference and the simplest explanation for the resulting block to endonuclease recognition is that the 6-amino group (major groove) of the central dA is an essential contact for the enzyme.

Ethylation interference studies have suggested that four phosphates in each DNA strand are important in specific DNA endonuclease complexes (Fig. 3; Lu et al. 1981). It is interesting that two of these four residues are 5′-external to the *Eco*RI sequence. If it is assumed that these phosphates are involved in ionic bonding to the protein, this finding implies that a substantial fraction of the electrostatic component of the binding energy is due to interaction with backbone phosphates outside the recognition site. The third phosphate, the alkylation of which markedly suppresses specific binding, is that centrally located within the *Eco*RI sequence. The last of the four phosphates implicated in this study is the one involved in the phosphodiester bond that is hydrolyzed by the enzyme. Surprisingly, ethylation of this phosphate did not inhibit specific binding as severely as did alkylation of the other three. While the significance of this observation is not yet clear, it is likely that this phosphate is coordinated to Mg^{2+} in the cleavage complex. Consequently, the nature of interaction between polypeptide and this phosphate may well differ from the mode of bonding to other phosphate residues.

A representation of the spatial distribution of endonuclease "contacts" along the helix can be obtained by projecting the B-DNA surface onto a cylinder and then unwrapping the cylindrical projection (Siebenlist et al. 1980) (Fig. 4). Inspection of this projection, or of space-filling models, illustrates several interesting features concerning the placement of DNA determinants implicated in specific recognition. For example, potential endonuclease contacts are accessible primarily from one side of the helix. Also evident are potential contacts within the major groove between the points of cleavage of the two DNA strands. This representation also demonstrates the proximity (4 Å) of the 6-amino substituents of the dA residues modified by *Eco*RI methylase.

Several lines of evidence considered above led to the suggestion that the *Eco*RI endonuclease dimer interacts with its recognition sequence to yield a complex possessing elements of twofold symmetry. The alkylation experiments summarized in Figures 3 and 4 confirm the presence of symmetry at the fine-structure level and establish the equivalence of interaction with the two strands of the 6-bp *Eco*RI sequence. However, this work has also shown that the DNA-binding site of the enzyme encompasses about 10 bp. This additional interaction with nucleotides external to the recognition sequence could impose a degree of asymmetry on the specific endonuclease-DNA complexes.

Since specific complexes between *Eco*RI methylase and DNA have not been demonstrated under noncatalytic conditions, attempts to elucidate the DNA contacts made by this enzyme have involved analysis of analog effects on the kinetics of methyl transfer. The results of these studies, which are summarized in Table 4 and Figure 4, are in marked contrast to those obtained with the endonuclease. Thus, bacteriophage T4 DNA, which is resistant to cleavage by the endonuclease, was found to be an excellent substrate for the modification enzyme (Berkner and Folk 1977). This surprising finding has demonstrated that the presence of the bulky glucosylhydroxymethyl moiety in

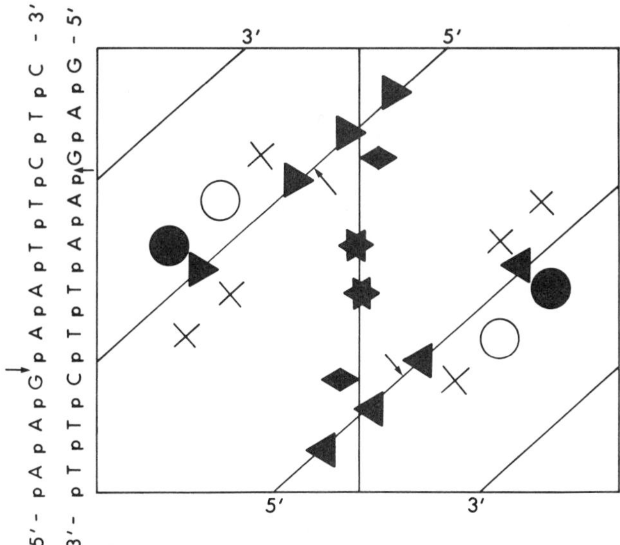

Figure 4 Potential contacts between *Eco*RI enzymes and their recognition sequence. The representation shown is based on a cylindrical projection of the DNA surface which has been opened along a line parallel to the helix axis (Siebenlist et al. 1980). Major and minor grooves are oriented diagonally. The helix repeat was assumed to be 10.5 bp per turn. Points of endonuclease cleavage are indicated by arrows. Endonuclease contacts: (▲) Phosphates identified by virtue of ethylation interference; (●) N^3 of dA residues within the *Eco*RI site identified on the basis of methylation interference; (○) N^3 of dA residues which are protected against dimethylsulfate alkylation by the endonuclease, but which do not interfere with specific binding when methylated; (◆) N^7 of dG residues, the alkylation of which interferes with binding; (★) 6-NH_2 of dA residues which are methylated by *Eco*RI modification enzyme. Methylase contacts identified by base analog substitution are indicated by Xs. These are not required by the endonuclease. Those in the minor groove correspond to the 2-NH_2 of dG, those in the major groove to the 5-CH_3 of dT. Although the 5-CH_3 of both dT residues in each strand are indicated, only one may be required (text). The 6-NH_2 of dA residues (★) are sites of modification by the methylase and, hence, are clearly required by the enzyme.

the major groove does not interfere with interaction between this enzyme and its recognition site. The methylase also differs from the endonuclease on dU-containing PBS2 DNA, which, in the case of the modification enzyme, is an extremely poor substrate (Berkner and Folk 1977). However, it has not been determined whether the replacement of thymine by uracil in this DNA affects site-specific binding or subsequent catalysis. Nevertheless, this result suggests that the 5-methyl group of one or both dT residues within the *Eco*RI site has an important role in the methylase reaction. The substitution of dI within the *Eco*RI site of ColE1 DNA also had markedly different effects on the two enzymes (Modrich and Rubin 1977). In contrast to results with the endonuclease, the dI-containing site was an extremely poor substrate for the modification enzyme, with the block being at the level of sequence recognition. Since the degree of substitution in these experiments was extremely limited and since steric effects can be excluded, it seems likely that the 2-amino group of dG, exposed in the minor groove, is used as a contact by the methylase. The distribution of these potential methylase contacts on the B-form structure is presented in Figure 4.

Examination of Table 4 and Figure 4 illustrates the striking differences between the two *Eco*RI enzymes with respect to DNA substituent effects on recognition and/or catalysis. The single exception may be a common requirement for the 6-amino groups of both adenines adjacent to the twofold axis. This lends further support to the conclusion that the two proteins interact with their common recognition sequence in different ways, and, in particular, suggests that the two proteins employ different mechanisms to discriminate a given base pair.

*Hpa*I Restriction and Modification Enzymes

The only other type-II system that has been examined in some detail with respect to DNA substituent effects is *Hpa*I. Using homogeneous preparations of *Hpa*I enzymes, P. Dwyer-Hallquist and K.I. Agarwal (unpubl.) have determined steady-state kinetic parameters governing cleavage and methyl transfer to GGTTAACC and derivatives of this self-complementary octanucleotide containing base-analog substitutions.

The octanucleotide GGT(BrdU)AACC was readily cleaved by *Hpa*I but the K_m and k_{cat} for this substrate were both reduced fivefold relative to the corresponding parameters for GGTTAACC. The parallel decrease in these parameters suggests that although the mechanism of cleavage of the BrdU-containing oligonucleotide may be similar to that of the control, some nonproductive interaction between the enzyme and the substrate analog may occur. In contrast, GGTUAACC and GITTAACC were not substrates for the endonuclease. While failure to cleave the dI-containing oligonucleotide may be at least partially due to reduced stability of the duplex form, the failure to utilize the dU-containing analog implies a requirement for the 5-methyl group of the dT residue adjacent to the site of cleavage for recognition and/or catalysis by this endonuclease.

Berkner and Folk (1979) have also obtained evidence suggesting that the 5-methyl group of thymine is important in the *Hpa*I endonuclease reaction. This work demonstrated that initial rates of double-strand cleavage of PBS2 DNA, which contains dU in place of dT, were 10-fold lower than for λ DNA under comparable conditions. However, the kinetics of double-strand cleavage of the dU-containing substrate were anomalous, being biphasic with respect to time and enzyme concentration. Thus, at moderate enzyme concentrations, products arising as a consequence of double-strand cleavage were only observed after a lag of several hours, but their subsequent rate of appearance *exceeded* that of the λ control. A similar phenomenon was observed with hydroxymethyluracil-containing ϕe DNA. These kinetics were attributed to the introduction of single-strand scissions during early stages of the reaction followed by their conversion to double-strand breaks during the second phase. Unfortunately, this possibility was not directly addressed, rendering it difficult to compare the PBS2 and ϕe results with those obtained in the oligonucleotide experiments mentioned above. It is, nevertheless, not easy to reconcile the results of these two approaches, although it may be pertinent to note that experiments with analog-containing viral DNAs employed preparations of *Hpa*I which were only partially purified, and the location of the sites subject to cleavage were not established.

While analysis of the phosphate contacts of the *Hpa*I endonuclease is not yet complete, it has been found that 5′-OH GGTTAACC is hydrolyzed much more rapidly by the enzyme than 5′-pGGTTAACC. This implicates phosphate residues 5′-external to the recognition sequence as being important in the enzyme-DNA interaction, a finding identical to that cited above for the *Eco*RI endonuclease.

In contrast to the behavior of the *Hpa*I endonuclease with GGTUAACC and GGTBrdUAACC, homogeneous *Hpa*I modification enzyme methylated both analog-containing oligonucleotides with steady-state kinetic parameters that were almost identical to those for the control substrate GGTTAACC (P. Dwyer-Hallquist and K.I. Agarwal, unpubl.). However, like the endonuclease, the methyltransferase failed to use GITTAACC, although, as in the case of the restriction enzyme, this may at least partially reflect the reduced T_m of the dI-containing oligonucleotide. The differential response of the *Hpa*I endonuclease and methylase to dU and BrdU-containing oligonucleotides is reminiscent of analogous results with the *Eco*RI system (above). It would thus appear that in both systems, the restriction endonuclease and the modification methylase interact with their recognition sequence in different ways.

Other Type-II Endonucleases

Only limited information is available on DNA determinants important in recognition or catalysis by other type-II restriction enzymes. Mann and Smith (1977) have used a synthetic duplex oligonucleotide containing overlapping *Hpa*II (CCGG) and *Hae*III (GGCC) sites to examine the effects of abnormal methylation of the 5-position of dC on subsequent cleavage by the restriction endonuclease. For example, a portion of the oligonucleotide contained the sequence: 5′-CCGGCC. Hence, it was possible to assess the effects of *Hpa*II modification (Table 3) on *Hae*III restriction, and vice versa. Although only qualitative, these experiments demonstrated that *Hae*III endonuclease will cleave 5′-GGC$\overset{*}{\mathrm{C}}$ as opposed to the normally modified site 5′-GG$\overset{*}{\mathrm{C}}$C, which is resistant. Similarly, *Hpa*II was found to cleave 5′-CCG$\overset{*}{\text{G}}$ but not 5′-CCGG. These experiments are of interest for two reasons. First, they exclude endonuclease contacts with the 5-position of the dC residue which remains unmodified in each case. Second, they render unlikely the possibility that resistance to restriction is a result of perturbation of DNA secondary structure that might occur as a consequence of modification of this base.

Limited information on a number of systems has also been provided by Marchionni and Roufa (1978) and Berkner and Folk (1979) who have analyzed patterns of cleavage of viral DNAs which naturally contain base analogs or have been prepared in such a way as to contain a BrdU for dT substitution. The results of these experiments, which are summarized in Table 5, are somewhat difficult to interpret for several reasons. Being primarily a survey, these studies did not attribute rate variations with substrate to changes in particular kinetic parameters. Second, with the exception of BrdU-substituted DNAs, the number of recognition sites within an analog-containing viral molecule is unknown in the absence of a positive result. Hence, reduced rates of cleavage relative to the λ standard employed (Berkner and Folk 1979) may reflect a lower density of recognition sites. For example, reduced rates of cleavage of PBS2 DNA (dU in place of dT) by *Hha*I and *Hpa*II, which do not contain dU in their recognition sequences, were attributed to such an effect. However, the possibility that rate reductions were due to the presence of dU in external sequences was not excluded. Indeed, BrdU substitution has been observed to reduce the rates of cleavage by several endonucleases that recognize sequences lacking dT (Marchionni and Roufa 1978; Berkner and Folk 1979), an effect which is clearly due to the presence of the analog in external sequences. Although these studies are preliminary, they do show a differential response of several enzymes to analog substitution (Table 5) and are consistent with the view that discrimination of a given base pair can occur by different mechanisms.

Table 5 Cleavage of base-substituted DNAs by restriction endonucleases

Endonuclease	Recognition sequence	Relative rate of DNA cleavage[a]			
		φe (HmdU)	PBS2 (dU)	T4 (GlcHmdC)	λ, P22 (BrdU)
*Bam*HI	5′ G↓GATCC	poor	good	poor	fair
*Hha*I	5′ GCG↓C	good	fair[b]	inert	—
*Hpa*II	5′ C↓CGG	good	poor[b]	inert	good
*Hin*dII	5′ GTPy↓PuAC	fair	poor	poor	good
*Hin*dIII	5′ A↓AGCTT	fair	poor	poor	fair

Adapted from Berkner and Folk (1979).

[a]Rates are relative to hydrolysis of λ DNA at comparable enzyme and DNA concentrations. Relative rates are good (50–100%), fair (14–33%), and poor (≤10%). The last column summarizes results with BrdU-substituted λ or P22 DNAs.

[b]Reduced rates may at least partially reflect a low density of cleavage sites.

Other Type-II Methylases

Mann and Smith (1979a) have used the random copolymer $d(G,C)_n$ and analogs of this polynucleotide as substrates for partially purified *Hha*I, *Hpa*II, and *Hae*III modification enzymes. While random $d(G,C)_n$ was methylated by all three enzymes, the $d(I,C)_n$ analog supported methylation by *Hha*I and *Hpa*II enzymes only at a low rate. In contrast, the rate of methyl transfer to $d(I,C)_n$ by the *Hae*III enzyme was comparable with that observed with $d(G,C)_n$. Although steady-state kinetic parameters were not evaluated, these findings suggest that, like the *Eco*RI enzyme, *Hha*I and *Hpa*I methylases require the 2-amino of dG while the *Hae*III enzyme evidently does not. The activity of these three methyltransferases was also examined with random $d(N\text{-}AcG,C)_n$ and $d(X,C)_n$ (Mann et al. 1978). While rates of methyl transfer to these copolymers by the *Hha*I enzyme were comparable with those observed with $d(G,C)_n$, these analog polymers were poor substrates for the *Hpa*II and *Hae*III activities. Since the $d(N\text{-}AcG,C)_n$ and $d(X,C)_n$ copolymers employed did not display thermal melting transitions, suggesting the absence of stable secondary structure, it is not clear whether the reduced activity of the latter enzymes on such substrates is due to perturbation of DNA secondary structure or to alteration of a DNA contact.

CONCLUDING REMARKS

Although interest in the type-II restriction enzymes has focused overwhelmingly on their role as reagents, in recent years there has been substantial progress in understanding their enzymatic properties. They clearly provide excellent systems in which to study DNA-protein interactions and may well lead to new insights into the chemical possibilities by which proteins are able to recognize specific sequences. An important theme that has emerged is that multiple mechanisms exist for the discrimination of base pairs. Furthermore specificity can include the recognition of several subsequences that at first sight may not always appear to be closely related. Both principles have implications for other kinds of specific DNA-protein interactions.

As with all sequence-specific proteins, the mechanisms that allow the type-II restriction enzymes to distinguish their canonical sequences from the high background of nonspecific sequences remain unresolved. A particularly startling trend that is apparent among the known recognition sequences is the preponderance of GC-rich sequences. It seems unlikely that this is a statistical sampling problem. It may arise as an evolutionary consequence of the process of sequence recognition, in which case one might surmise that GC base pairs are more easily discriminated in a specific fashion than are AT base pairs.

The successful crystallization of the *Eco*RI endonuclease as well as the existence of co-crystals between it and its recognition sequence promises to yield exciting information about the detailed nature of the contacts involved in recognition. The extensive knowledge already accumulated about the solution properties of this enzyme will be invaluable in correctly interpreting the crystallographic data. Now that a number of other type-II restriction-modification systems have been cloned into *E. coli*, there is hope that a similarly detailed picture will soon be drawn for those systems. With luck, some general principles will emerge that will help to define the processes required for specific DNA-protein interactions.

REFERENCES

Aaij, C. and P. Borst. 1972. The gel electrophoresis of DNA. *Biochim. Biophys. Acta* **269:**192.

Baumstark, B.R., R.J. Roberts, and U.L. RajBhandary. 1979. Use of short synthetic DNA duplexes as substrates for the restriction endonucleases *Hpa*II and *Mno*I. *J. Biol. Chem.* **254:**8943.

Behe, M. and G. Felsenfeld. 1981. Effects of methylation on a synthetic polynucleotide: The B-Z transition in poly(dG-m^{5}dC)-poly(dG-m^{5}dC). *Proc. Natl. Acad. Sci.* **78:**1619.

Berg, O.G., R.B. Winter, and P.H. von Hippel. 1981. Diffusion-driven mechanisms of protein translocation on nucleic acids. I. Models and theory. *Biochemistry* **20:**6929.

Berkner, K.L. and W.R. Folk. 1977. *Eco*RI cleavage and methylation of DNAs containing modified pyrimidines in the recognition sequence. *J. Biol. Chem.* **252:**3185.

Berkner, K.L. and W.R. Folk. 1978. Overmethylation of DNAs by the *Eco*RI methylase. *Nucleic Acids Res.* **5:**435.

———. 1979. The effects of substituted pyrimidines in DNAs on cleavage by sequence-specific endonucleases. *J. Biol. Chem.* **254:**2551.

Bickle, T.A. and K. Ineichen. 1980. The DNA sequence recognized by *Bgl*I. *Gene* **9:**205.

Bickle, T.A., V. Pirotta, and R. Imber. 1980. Purification and properties of the *Bgl*I and *Bgl*II endonucleases. *Methods Enzymol.* (Part I) **65:**132.

Bigger, C.H, K. Murray, and N. Murray. 1973. Recognition sequence of a restriction enzyme. *Nature New Biol.* **244:**7.

Bingham, A.H.A. and A. Atkinson. 1978. Restriction endonucleases and modification methylases in bacteria. *Biochem. Soc. Trans.* **6:**315.

Bingham, A.H.A., T. Atkinson, D. Sciaky, and R.J. Roberts. 1978. A specific endonuclease from *Bacillus caldolyticus*. *Nucleic Acids Res.* **5:**3457.

Bird, A.P., M.H. Taggart, and B.A. Smith. 1979. Methylated and unmethylated DNA compartments in the sea urchin genome. *Cell* **17:**889.

Bishop, J.O. 1979. A DNA sequence cleaved by restriction endonuclease R.*Eco* RI in only one strand. *J. Mol. Biol.* **128:**545.

Bishop, J.O. and J.A. Davies. 1980. Plasmid cloning vectors that can be nicked at a unique site. *Mol. Gen. Genet.* **179:**573.

Blakesley, R.W. and R.D. Wells. 1975. "Single-stranded" DNA from ϕX174 and M13 is cleaved by certain restriction endonucleases. *Nature* **257:**421.

Blakesley, R.W., J.B. Dodgson, I.F. Nes, and R.D. Wells. 1977. Duplex regions in "single-stranded" ϕX174 DNA are cleaved by a restriction endonuclease from *Haemophilus aegyptius*. *J. Biol. Chem.* **252:**7300.

Boyer, H.W., P.J. Greene, R.B. Meagher, M.C. Betlach, D. Russel, and H.M. Goodman. 1975. The methylation of DNA as the biochemical basis of host controlled modification of DNA in bacteria. *FEBS Symp.* **34:**23.

Bron, S. and W. Horz. 1980. Purification and properties of *Bsu*I endonuclease. *Methods Enzymol.* (Part I) **65:**112.

Bron, S. and K. Murray. 1975. Restriction and modification in *B. subtilis*. Nucleotide sequence recognized by restriction endonuclease R.*Bsu* R from strain R. *Mol. Gen. Genet.* **143:**25.

Bron, S., K. Murray, and T.A. Trautner. 1975. Restriction and modification in *B. subtilis*. Purification and general properties of a restriction endonuclease from strain R. *Mol. Gen. Genet.* **143:**13.

Brown, N.L. and M. Smith. 1977. Cleavage specificity of the restriction endonuclease isolated from *Hemophilus gallinarum (Hga*I). *Proc. Natl. Acad. Sci.* **74:**3213.

Brown, N.L., C.A. Hutchison, and M. Smith. 1980. The specific non-symmetrical sequence recognized by restriction endonuclease *Mbo*II. *J. Mol. Biol.* **140:**143.

Catterall, J.F. and N.E. Welker. 1977. Isolation and properties of a thermostable restriction endonuclease (Endo R. *Bst* 1503). *J. Bacteriol.* **129:**1110.

Chou, P.Y. and G.D. Fasman. 1978. Prediction of the secondary structure of proteins from their amino acid sequence. *Adv. Enzymol. Relat. Areas Mol. Biol.* **47:**45.

Clarke, C.M. and B.S. Hartley. 1979. Purification, properties, and specificity of the restriction endonuclease from *Bacillus stearothermophilus*. *Biochem. J.* **177:**49.

Dugaiczyk, A., J.M. Hedgpeth, H.W. Boyer, and H.M. Goodman. 1974. Physical identity of the SV40 deoxyribonucleic acid sequence recognized by the *Eco*RI restriction endonuclease and modification methylase. *Biochemistry* **13:**503.

Endow, S.A. and R.J. Roberts. 1977. Two restriction-like enzymes from *Xanthomonas malvacearum*. *J. Mol. Biol.* **112:**521.

Forsblum, S., R. Rigler, M. Ehrenberg, U. Pettersson, and L. Philipson. 1976. Kinetic studies on the cleavage of adenovirus DNA by restriction endonuclease *Eco*RI. *Nucleic Acids Res.* **3:**3255.

Fuchs, C., E.C. Rosenvold, A. Honigman, and W. Szybalski. 1978. A simple method for identifying the palindromic sequences recognized by restriction endonucleases: The nucleotide sequence of the *Ava*II site. *Gene* **4:**1.

Garfin, D.E. and H.M. Goodman. 1974. Nucleotide sequences at the cleavage sites of two restriction endonucleases from *Haemophilus parainfluenzae*. *Biochem. Biophys. Res. Commun.* **59:**108.

Geier, G.E. and P. Modrich. 1979. Recognition sequence of the *dam* methylase of *Escherichia coli* K12 and mode of cleavage of *Dpn*I endonuclease. *J. Biol. Chem.* **254:**1408.

Gelinas, R.E, P.A. Myers, and R.J. Roberts. 1977. Two sequence-specific endonucleases from *Moraxella bovis*. *J. Mol. Biol.* **114:**169.

George, J., R.W. Blakesley, and J.G. Chirikjian. 1980. Sequence-specific endonuclease *Bam*HI: Effect of hydrophobic reagents on sequence recognition and catalysis. *J. Biol. Chem.* **255:**6521.

Gilbert, W., A. Maxam, and A. Mirzabekov. 1976. *Control of ribosome synthesis*. In *Alfred Benzon Symposium XIII* (ed. N.O. Kjeldgaard and O. Maaloe), pp. 139. Munksgaard, Copenhagen.

Gingeras, T.R., J.P. Milazzo, and R.J. Roberts. 1978. A computer assisted method for the determination of restriction enzyme recognition sites. *Nucleic Acids Res.* **5:**4105.

Godson, G.N. and R.J. Roberts. 1976. A catalogue of cleavages of ϕX174, S13, G4, and ST-1 DNA by 26 different restriction endonucleases. *Virology* **73:**561.

Goppelt, M., A. Pingoud, G. Maass, H. Mayer, H. Koster, and R. Frank. 1980. The interaction of the *Eco*RI endonuclease with its substrate. A physico-chemical study employing natural and synthetic oligonucleotides and polynucleotides. *Eur. J. Biochem.* **104:**101.

Greene, P.J., M. Gupta, H.W. Boyer, W.E. Brown, and J.M. Rosenberg. 1981. Sequence analysis of the DNA encoding *Eco*RI endonuclease and methylase. *J. Biol. Chem.* **256:**2143.

Greene, P.J., M.S. Poonian, A.L. Nussbaum, L. Tobias, D.E. Garfin, H. Boyer, and H.M. Goodman. 1975. Restriction and modification of a self-complementary octanucleotide containing the *Eco*RI site. *J. Mol. Biol.* **99:**237.

Gromkova, R. and S.H. Goodgal. 1972. Action of *Haemophilus* endodeoxyribonuclease on biologically active deoxyribonucleic acid. *J. Bacteriol.* **109:**987.

Gunthert, U., M. Freund, and T.A. Trautner. 1981a. Restriction and modification in *Bacillus subtilis:* Two DNA methyltransferases with *Bsu*RI specificity: Purification and properties. *J. Biol. Chem.* **256:**9340.

Gunthert, U., S. Jentsch, and M. Freund. 1981b. Restriction and modification in *Bacillus subtilis:* Two DNA methyltransferases with *Bsu*RI specificity: Catalytic properties, substrate specificity and mode of action. *J. Biol. Chem.* **256:**9346.

Halford, S.E. and N.P. Johnson. 1980. The *Eco*RI restiction endonuclease with bacteriophage λ DNA. Equilibrium binding studies. *Biochem. J.* **191:**593.

Halford, S.E., N.P. Johnson, and J. Grinstead. 1979. The reactions of *Eco*RI and other restriction endonucleases. *Biochem. J.* **179:**353.

Hattman, S., T. Keisler, and A. Gottehrer. 1978. Sequence specificity of DNA methylases from *Bacillus amyloliquefaciens* and *Bacillus brevis. J. Mol. Biol.* **124:**701.

Hausler, B. 1980. Mapping and sequencing of cloned *Hha*II restriction and modification genes. *Fed. Proc.* **39:**1784.

Hedgpeth, J., H.M. Goodman, and H.W. Boyer. 1972. DNA nucleotide sequence restricted by the RI endonuclease. *Proc. Natl. Acad. Sci.* **69:**3448.

Heininger, K., W. Horz, and H.G. Zachau. 1977. Specificity of cleavage by a restriction nuclease from *Bacillus subtilis. Gene* **1:**291.

Herman, G.E. and P. Modrich. 1981. *Escherichia coli dam* methylase: Physical and catalytic properties of the homogeneous enzyme. *J. Biol. Chem.* **257:**2605.

Hines, J.L. and K.L. Agarwal. 1979. Purification of *Hpa*I and *Hpa*II. *Fed. Proc.* **38:**294.

Hinsch, B. and M.-R. Kula. 1981. Reaction kinetics of some important site-specific endonucleases. *Nucleic Acids Res.* **9:**3159.

Hinsch, B., H. Mayer, and M.-R. Kula. 1980. Binding of nonsubstrate nucleotides to a restriction endonuclease: A model for the interaction of *Bam*HI with its recognition sequence. *Nucleic Acids Res.* **8:**2547.

Horiuchi, K. and N.D. Zinder. 1975. Site-specific cleavage of single-stranded DNA by a *Haemophilus* restriction endonuclease. *Proc. Natl. Acad. Sci.* **72:**2555.

Hsu, M. and P. Berg. 1978. Altering the specificity of restriction endonuclease: Effect of replacing Mg^{2+} with Mn^{2+}. *Biochemistry* **17:**131.

Humphreys, G.O., G.A. Willshaw, H.R. Smith, and E.S. Anderson. 1976. Mutagenesis of plasmid DNA with hydroxylamine: Isolation of mutants of multi-copy plasmids. *Mol. Gen. Genet.* **145:**101.

Jack, W.E., B.J. Terry, and P. Modrich. 1982. Involvement of outside DNA sequences in the major kinetic path by which *Eco*RI endonuclease locates and leaves its recognition sequence. *Proc. Natl. Acad. Sci.* **79:**4010.

Jack, W.E., R.A. Rubin, A. Newman, and P. Modrich. 1980. Structures and mechanisms of *Eco* RI DNA restriction and modification enzymes. In *Gene amplification and analysis,* vol I: *Restriction endonucleases* (ed. J.G. Chirikjian), p. 165. Elsevier-North Holland, New York.

Jovin, T.M. 1976. Recognition mechanisms of DNA-specific enzymes. *Annu. Rev. Biochem.* **45:**889.

Kaplan, D.A. and D.P. Nierlich. 1975. Cleavage of nonglucosylated bacteriophage T4 deoxyribonucleic acid by restriction endonuclease *Eco*RI. *J. Biol. Chem.* **250:**2395.

Kelly, T.J. and H.O. Smith. 1970. A restriction enzyme from *Haemophilus influenzae*. II. Base sequence of the recognition site. *J. Mol. Biol.* **51:**393.

Kiss, A., B. Sain, E. Csordas-Toth, and P. Venetianer. 1977. A new sequence-specific endonuclease (*Bsp*) from *Bacillus sphaericus. Gene* **1:**323.

Kleid, D., Z. Humayun, A. Jeffrey, and M. Ptashne. 1976. Novel properties of a restriction endonuclease isolated from *Haemophilus parahaemolyticus. Proc. Natl. Acad. Sci.* **73:**293.

Koncz, C., A. Kiss, and P. Venetianer. 1978. Biochemical characterization of the restriction-modification system of *Bacillus sphaericus. Eur. J. Biochem.* **89:**523.

Kosykh, V.G., Y.I. Buryanov, and A.A. Bayev. 1980. Molecular cloning of *Eco*RII endonuclease and methylase genes. *Mol. Gen. Genet.* **178:**717.

Lacks, S. and B. Greenberg. 1975. A deoxyribonuclease of *Diplococcus pneumoniae* specific for methylated DNA. *J. Biol. Chem.* **250:**4060.

———. 1977. Complementary specificity of restriction endonucleases of *Diplococcus pneumoniae* with respect to DNA methylation. *J. Mol. Biol.* **114:**153.

Lacks, S.A. 1980. Purification and properties of the complementary endonuclease *Dpn*I and *Dpn*II. *Methods Enzymol.* (Part I) **65:**138.

Langowski, J.M., A. Pingoud, M. Goppelt, and G. Maas. 1980. Inhibition of *Eco*RI by polynucleotides. A characterization of non-specific binding of the enzyme to DNA. *Nucleic Acids Res.* **8:**4727.

Lautenberger, J.A., C.T. White, N.L. Haigwood, M.H. Edgell, and C.A. Hutchison. 1980. The recognition site of type II restriction enzyme *Bgl*I is interrupted. *Gene* **9:**213.

Lee, Y.H. and J.G. Chirikjian. 1979. Sequence-specific endonuclease *Bgl*I: Modification of lysine and arginine residues of the homogeneous enzyme. *J. Biol. Chem.* **254:**6838.

Levy, W.P. and N.E. Welker. 1978. Purification and properties of a modification methylase from *Bacillus stearothermophilus. Fed. Proc.* **37:**1414.

Lu, A.-L., W.E. Jack, and P. Modrich. 1981. DNA determinants important in sequence recognition by *Eco*RI endonuclease. *J. Biol. Chem.* **256:**13200.

Malcolm, A.D.B. 1981. The use of restriction enzymes in genetic engineering. In *Genetic engineering* (ed. R. Williamson), vol. 2, p. 129. Academic Press, London.

Malyguine, E., P. Vannier, and P. Yot. 1980. Alteration of the specificity of restriction endonucleases in the presence of organic solvents. *Gene* **8:**163.

Mann, M.B., R.N. Rao, and H.O. Smith. 1978. Cloning of restriction and modification genes in *E. coli:* The *Hha*II system from *Haemophilus haemolyticus. Gene* **3:**97.

Mann, M.B. and H.O. Smith. 1977. Specificity of *Hpa*II and *Hae*III DNA methylases. *Nucleic Acids Res.* **4:**4211.

Mann, M.B. and H.O. Smith. 1979a. Specificity of DNA methylases from *Haemophilus* sp. In *Transmethylation* (ed. E. Usdin, R. Borchardt, and C. Kreveling), p. 483. Elsevier-North Holland, New York.

———. 1979b. The *Hha*II restriction-modification system. *Fed. Proc.* **38:**293.

Marchionni, M.A. and D.J. Roufa. 1978. Digestion of 5-bromodeoxyuridine-substituted λ-DNA by restriction endonucleases. *J. Biol. Chem.* **253:**9075.

Marinus, M.G. and N.R. Morris. 1973. Isolation of deoxyribonucleic acid methylase mutants of *Escherichia coli* K-12. *J. Bacteriol.* **114:**1143.

Maxam, A.M. and W. Gilbert. 1980. Sequencing end-labeled DNA with base-specific chemical cleavages. *Methods Enzymol.* **65** (Part I):499.

Middleton, J.H., M.H. Edgell, and C.A. Hutchison. 1972. Specific fragments of φX174 deoxyribonucleic acid produced by a restriction enzyme from *Haemophilus aegyptius*, endonuclease Z. *J. Virol.* **10:**42.

Modrich, P. 1979. Structures and mechanisms of DNA restriction and modification enzymes. *Q. Rev. Biophys.* **12:**315.

———. 1982. Studies on sequence recognition by type II restriction and modification enzymes. *CRC Crit. Rev. Biochem.* (in press).

Modrich, P. and R.A. Rubin. 1977. Role of the 2-amino group of deoxyguanosine in sequence recognition by *Eco*RI restriction and modification enzymes. *J. Biol. Chem.* **252:**7273.

Modrich, P. and D. Zabel. 1976. *Eco*RI endonuclease: Physical and catalytic properties of the homogeneous enzyme. *J. Biol. Chem.* **251:**5866.

Molloy, P.L. and R.H. Symons. 1980. Cleavage of DNA·RNA hybrids by type II restriction enzymes. *Nucleic Acids Res.* **8:**2939.

Murray, K. 1973. Nucleotide sequence analysis with polynucleotide kinase and nucleotide "mapping" methods. 5′-Terminal sequences of deoxyribonucleic acid from bacteriophages lambda and 424. *Biochem. J.* **131:**569.

Murray, N.E. and K. Murray. 1974. Manipulation of restriction targets in phage λ to form receptor chromosomes for DNA fragments. *Nature* **251:**476.

Newman, A.K., R.A. Rubin, S.H. Kim, and P. Modrich. 1981. DNA sequences of structural genes for *Eco*RI DNA restriction and modification enzymes. *J. Biol. Chem.* **256:**2131.

Old, R., K. Murray, and G. Roizes. 1975. Recognition sequence of restriction endonuclease III from *Haemophilus influenzae*. *J. Mol. Biol.* **92:**331.

Pirotta, V. 1976. Two restriction endonucleases from *Bacillus globigii*. *Nucleic Acids Res.* **3:**1747.

Polisky, B., P. Greene, D.E. Garfin, B.J. McCarthy, H.M. Goodman, and H.W. Boyer. 1975. Specificity of substrate recognition by the *Eco*RI restriction endonuclease. *Proc. Natl. Acad. Sci.* **72:**3310.

Richter, P.H. and M. Eigen. 1974. Diffusion controlled reaction rates in spheroidal geometry. Application to repressor-operator association and membrane bound enzymes. *Biophys. Chem.* **2:**255.

Roberts, R.J. 1976. Restriction endonucleases. *CRC Crit. Rev. Biochem.* **4:**123.

———. 1982. Restriction and modification enzymes and their recognition sequences. *Nucleic Acids Res.* **10:**r117.

Roberts, R.J., G.A. Wilson, and F.E. Young. 1977. Recognition sequence of specific endonuclease *Bam*HI from *Bacillus amyloliquefaciens* H. *Nature* **265:**82.

Roberts, R.J., P.A. Myers, A. Morrison, and K. Murray. 1976. A specific endonuclease from *Haemophilus haemolyticus*. *J. Mol. Biol.* **103:**199.

Rosenberg, J.M., H.W. Boyer, and P. Greene. 1980. The structure and function of *Eco*RI restriction endonuclease. In *Gene amplification and analysis*, vol. I: *Restriction endonucleases*, (ed. J.G. Chirikjian), p. 131. Elsevier-North Holland, New York.

Rosenberg, J.M., R.E. Dickerson, P.J. Greene, and H.W. Boyer. 1978. Preliminary X-ray diffraction analysis of crystalline *Eco*RI endonuclease. *J. Mol. Biol.* **122:**241.

Roulland-Dussoix, D., R. Yoshimori, P. Greene, M. Betlach, H.M. Goodman, and H. Boyer. 1974. R factor-controlled restriction and modification of deoxyribonucleic acid. In *American Society for Microbiology conference on bacterial plasmids, microbiology*, p. 187. ASM Press,

Roy, P.H. and H.O. Smith. 1973a. DNA methylases of *Haemophilus influenzae* Rd. I. Purification and properties. *J. Mol. Biol.* **81:**427.

———. 1973b. DNA methylases of *Haemophilus influenzae* Rd. II. Partial recognition site base sequences. *J. Mol. Biol.* **81:**445.

Roychoudhury, R. and H. Kossel. 1971. Synthetic polynucleotides. Enzymic synthesis of ribonucleotide terminated oligodeoxynucleotides and their use as primers for the enzymic synthesis of polydeoxynucleotides. *Eur. J. Biochem.* **22:**310.

Roychoudhury, R., E. Jay, and R. Wu. 1976. Terminal labeling and addition of homopolymer tracts to duplex DNA fragments by terminal deoxynucleotidyl transferase. *Nucleic Acids Res.* **3:**863.

Ruben, G., P. Spielman, C.-P.D. Tu, E. Jay, B. Siegel, and R. Wu. 1977. Relaxed circular SV40 DNA as cleavage intermediate of two restriction endonucleases. *Nucleic Acids Res.* **4:**1803.

Rubin R.A., P. Modrich, and T.C. Vanaman. 1981. Partial NH_2- and COOH-terminal sequence analyses of *Eco*RI DNA restriction and modification enzymes. *J. Biol. Chem.* **256:**2140.

———. 1978. Substrate dependence of the mechanism of *Eco*RI endonuclease. *Nucleic Acids Res.* **5:**2991.

———. 1980. Purification and properties of *Eco*RI endonuclease. *Methods Enzymol.* **65** (part I): 96.

Rubin, R.A., P. Modrich, and T.C. Vanaman. 1981. Partial NH_2- and COOH-terminal sequence analyses of *Eco*RI DNA restriction and modification enzymes. *J. Biol. Chem.* **256:**2140.

Sanger, F., A.R. Coulson, T. Friedmann, G.M. Air, B.G. Barrell, N.L. Brown, J.C. Fiddes, C.A. Hutchison, P.M. Slocombe, and M. Smith. 1978. The nucleotide sequence of bacteriophage ϕX174. *J. Mol. Biol.* **125:**255.

Schranner, R. and P.H. Richter. 1978. Rate enhancement by guided diffusion. Chain length dependence of repressor-operator association rates. *Biophys. Chem.* **8:**135.

Sharp, P.A., B. Sugden, and J. Sambrook. 1973. Detection of two restriction endonuclease activities in *Haemophilus parainfluenzae* using analytical agarose-ethidium bromide electrophoresis. *Biochemistry* **12:**3055.

Shen, S., Q. Li, P. Yan, B. Zhou, S. Ye, Y. Lu, and D. Wang. 1980. Restriction endonucleases from three strains of *Haemophilus influenzae. Sci. Sin.* **23:**1435.

Siebenlist, U., R.B. Simpson, and W. Gilbert. 1980. *E. coli* RNA polymerase interacts homologously with two different promoters. *Cell* **20:**269.

Singer, B. 1975. The chemical effects of nucleic acid alkylation and their relation to mutagenesis and carcinogenesis. *Prog. Nucleic Acid Res. Mol. Biol.* **15:**219.

Smith, H.O. and K.W. Wilcox. 1970. A restriction enzyme from *Haemophilus influenzae.* I. Purification and general properties. *J. Mol. Biol.* **51:**379.

Smith, L.A. and J.G. Chirikjian. 1979. Purification and characterization of the sequence-specific endonuclease *Bam*HI. *J. Biol. Chem.* **254:**1003.

Streeck, R.E. 1980. Single-strand and double-strand cleavage at half-modified and fully modified recognition sites for the restriction nucleases *Sau*3A and *Taq*I. *Gene* **12:**267.

Sun, L. and B. Singer. 1975. The specificity of different classes of ethylating agents toward various sites of HeLa cell DNA *in vitro* and *in vivo. Biochemistry* **14:**1795.

Sussenbach, J.S., C.H. Monfoort, R. Schiphof, and E.E. Stobberingh. 1976. A restriction endonuclease from *Staphylococcus aureus. Nucleic Acids Res.* **3:**3193.

Szomolanyi, E., A. Kiss, and P. Venetianer. 1980. Cloning the modification methylase gene of *Bacillus sphaericus* R in *Escherichia coli. Gene* **10:**219.

Takanami, M. 1973. Specific cleavage of coliphage fd DNA by five different restriction endonucleases from *Haemophilus* genus. *FEBS Lett.* **34:**318.

Thomas, M. and R.W. Davis. 1975. Studies on the cleavage of bacteriophage lambda DNA with *Eco*RI restriction endonuclease. *J. Mol. Biol.* **91:**315.

Thomas, M., J.R. Cameron, and R.W. Davis. 1974. Viable molecular hybrids of bacteriophage lambda and eukaryotic DNA. *Proc. Natl. Acad. Sci.* **71:**4579.

Tikchonenko, T.I., E.E. Karamov, B.A. Zavizion, and B.S. Naroditsky. 1978. *Eco*RI* activity: Enzyme modification or activation of accompanying endonuclease. *Gene* **4:**195.

Van Heuverswyn, H. and W. Fiers. 1980. Recognition sequence for the restriction endonuclease *Bgl*I from *Bacillus globigii. Gene* **9:**195.

von Hippel, P.H. 1979. On the molecular bases of the specificity of interaction of transcriptional proteins with genome DNA. *Biological Regulation and Development* **1:**279.

von Hippel, P.H. and J.D. McGhee. 1972. DNA-protein interaction. *Annu. Rev. Biochem.* **41:**231.

Vovis, G.F. and S. Lacks. 1977. Complementary action of restriction enzymes Endo R *Dpn*I and Endo R *Dpn*II on bacteriophage fl DNA. *J. Mol. Biol.* **115:**525.

Vovis, G.E., K. Horiuchi, and N.D. Zinder. 1974. Kinetics of methylation by a restriction endonuclease from *E. coli* B. *Proc. Natl. Acad. Sci.* **71:**3810.

Waalwijk, C. and R.A. Flavell. 1978. DNA methylation at a CCGG sequence in the large intron of the rabbit beta-globin gene: Tissue-specific variations. *Nucleic Acids Res.* **5:**4631.

Walder, R.Y., J.L. Hartley, J.E. Donelson, and J.A. Walder. 1981. Cloning and expression of the *Pst*I restriction-modification system in *E. coli. Proc. Natl. Acad. Sci.* **78:**1503.

Wilson, G.A. and F.E. Young. 1975. Isolation of a sequence-specific endonuclease (*Bam*I) from *Bacillus amyloliquefaciens. J. Mol. Biol.* **97:**123.

Winter, R.B., O.G. Berg, and P.H. von Hippel. 1981. Diffusion-driven mechanisms of protein translocation on nucleic acids. 3. The *Escherichia coli lac* repressor-operator interaction: Kinetic measurements and conclusions. *Biochemistry* **20:**6961.

Woodbury, C.P., R.L. Downey, and P.H. von Hippel. 1980a. DNA site recognition and overmethylation by the *Eco*RI methylase. *J. Biol. Chem.* **255:**11526.

Woodbury, C.P., O. Hagenbuchle, and P.H. von Hippel. 1980b. DNA site recognition and reduced specificity of the *Eco*RI endonuclease. *J. Biol. Chem.* **255:**11534.

Woodhead, J.L. and A.D.B. Malcolm. 1980. Non-specific binding of restriction endonuclease *Eco*RI to DNA. *Nucleic Acids Res.* **8:**389.

Yoo, O.J. and K.L. Agarwal. 1980a. Isolation and characterization of two proteins possessing *Hpa*II methylase activity. *J. Biol. Chem.* **255:**6445.

———. 1980b. Cleavage of single strand oligonucleotides and bacteriophage ϕX174 DNA by *Msp*I endonuclease. *J. Biol. Chem.* **255:**10559.

Young, T.-S., P. Modrich, E. Jay, and S.-H. Kim. 1981. Preliminary X-ray diffraction studies of *Eco*RI restriction endonuclease-DNA complex. *J. Mol. Biol.* **145:**607.

Zabeau, M. and R.J. Roberts. 1979. The role of restriction endonucleases in molecular genetics. In *Molecular Genetics,* Part III (ed. J.H. Taylor), p. 1. Academic Press, New York.

Single-strand-specific Nucleases

Kazuo Shishido and Tadahiko Ando†
Laboratory of Natural Products Chemistry
Tokyo Institute of Technology
Nagatsuta, Yokohama
Kanagawa 227, Japan

†Department of Microbiology
The Institute of Physical and Chemical Research
Wako-shi, Saitama 351, Japan

So far, a number of single-strand-specific nucleases that hydrolyze both DNA and RNA have been isolated from various sources, including fungi, bacteria, yeast, plants, and animals. These enzymes exhibit high selectivity for single-stranded nucleic acids and produce mono- and oligonucleotides terminating in 5′-phosphoryl and 3′-hydroxyl groups. S1 nuclease from *Aspergillus oryzae*, *Neurospora crassa* nuclease, P1 nuclease from *Penicillium citrinum*, *Alteromonas espejiana* BAL 31 nuclease, mung bean nuclease I, and *Ustilago maydis* nuclease are the best characterized. These nucleases degrade single-stranded nucleic acids and single-stranded regions in double-stranded nucleic acids. Information about the biological significance of the enzymes is scant. However, the enzymes have been widely used for purposes such as nucleic acid hybridization, analysis of nucleic acid structure, isolation of specific genes, and gene manipulation. This review summarizes detailed enzyme properties and their applications.

ENZYME PROPERTIES AND FUNCTIONS

S1 Nuclease

An endonuclease designated as nuclease S1 was first isolated by Ando from "Takadiastase," a digestive enzyme preparation obtained from *A. oryzae* (Ando 1966). Nuclease S1 is a single-strand-specific endonuclease and hydrolyzes primarily single-stranded DNA and RNA to yield 5′-nucleoside monophosphates and a very small amount of dinucleotides. The purified enzyme preparation has a molecular weight of 32,000 and is optimally active with denatured DNA at pH 4.5–5.0 in the presence of 10^{-3} M Zn^{2+} (Ando 1966; Vogt 1973). The enzyme is a zinc metalloprotein. At pH 6.4, the enzyme activity is 2.5% and at pH 7.2 it is 0.1% of maximum. Dialysis against 1 mM EDTA completely inactivates the nuclease. Readdition of divalent metals to 1 mM in the reaction mixture, in the presence of 0.1 mM EDTA, can restore at least 70% of the original activity. Zn^{2+} and Co^{2+} are the most effective, whereas Cd^{2+} and Mn^{2+} have marginal stimulatory activity. Mg^{2+}, Cu^{2+}, and

Ca^{2+} have no effect on the enzyme (Vogt 1973). The enzyme is inhibited by low concentrations of phosphate compounds. Half inhibition is caused by 2 mM phosphate or 20 μM pyrophosphate. Nucleotides also inhibit the nuclease. Half inhibition is caused by 85 μM dAMP and 1 μM dATP. However, cAMP has no effect even at 10 mM. Inhibition by the products of digestion might become significant when enough DNA is degraded (Wiegand et al. 1975).

S1 nuclease is quite thermostable. Both deoxyribonuclease and ribonuclease activities are not inactivated by heating at 65°C in the presence of substrate (Ando 1966). The enzyme is relatively insensitive to ionic strength at 37°C. At 0.4 M NaCl, the enzyme degrades DNA at 55% of the maximal rate (Vogt 1973). However, at relatively low temperature (~10°C), an apparent effect of NaCl concentration appears (Shishido and Ando 1972). The sensitivity of the enzyme to several denaturing agents is as follows. Sodium dodecylsulfate, up to 0.6%, has no effect on the rate of DNA hydrolysis. Neither 5% formamide nor 0.8 M urea has any effect on the nuclease activity (Vogt 1973). The enzyme exhibits activity even in the presence of 45–50% formamide, however 10 times the amount of enzyme is necessary (Hofstetter et al. 1976).

At pH 4.6, in 0.05 M NaCl and at 45°C, poly(rU) is degraded as rapidly as denatured DNA, while poly(rC) is hydrolyzed at 5% of this rate. Poly(rA) and poly(rG), on the other hand, are solubilized less than 0.1% as rapidly as DNA (Vogt 1973). Poly(rC), poly(rA), and poly(rG) all have considerable secondary structure under these assay conditions. At pH 6.4, where the nuclease is 5% as active as at pH 4.6, poly(rC) and poly(rA) appear to have a random coil structure. At this pH, the nuclease solubilizes these two polymers approximately 30% and 50% as rapidly as denatured DNA at pH 6.4. The relative rates of digestion of poly(rU) and poly(rG), on the other hand, are unchanged at this pH. Thus, it seems likely that S1 nuclease activity is not affected significantly by the local sequence at the site of enzymatic cleavage.

The purified enzyme exhibits no enzymatic activity on native DNA as measured by acid solubilization. The limit of specificity on native DNA was checked by a more sensitive technique, alkaline sucrose sedimentation (Shishido and Ando 1975a). Native T7 DNA was incubated at pH 5.0 in 60 mM NaCl and at 37°C. Seven units of the enzyme introduced no nicks into the DNA, whereas fourteen units or more introduced one or two nicks. One unit of S1 activity converts 50% of an equivalent amount of heat-denatured T7 DNA to an acid-soluble form and three units are required to render the denatured DNA completely acid-soluble. For stricter single-strand specificity, high salt concentration appears to be important (Vogt 1973).

S1 nuclease can hydrolyze single-stranded regions in duplex DNA and also detect locally altered structures (minor distortions) introduced by physical and chemical procedures. The enzyme can convert superhelical DNA to a nicked circular molecule, and then to a unit-length linear molecule. Shishido (1980) has investigated the number of superhelical turns that are necessary to produce a nonbase-paired site that is sensitive to S1 nuclease cleavage. With the DNAs of colicin E1 and SV40, the enzyme cannot cleave molecules containing only 40% and 30% of the superhelical turns of the naturally occurring DNAs, respectively. By virtue of these specificities, and its facile preparation and stability, S1 nuclease has found widespread application as detailed below.

Neurospora crassa Nuclease

Linn and Lehman (1965a,b) first reported a single-strand-specific endonuclease from the mycelia of *N. crassa*. This enzyme degrades single-stranded DNA, RNA, and synthetic polynucleotides to small oligonucleotides terminated by 5′-phosphomonoester groups. It prefers, but has no absolute specificity for, guanosine or deoxyguanosine residues within a polynucleotide, as judged by the large quantity of guanylic or deoxyguanylic acid found in the mononucleotide fraction after digestion (Linn 1967). The enzyme degrades native DNA, helical RNA, or ordered synthetic polynucleotides at about 2% the rate of their random coil counterparts. By adjustment of the reaction conditions (high ionic strength, high pH, high Mg^{2+} concentration, or addition of thiols), this rate can be reduced to less than 0.1%. It was reported that the enzyme hydrolyzes single-stranded tails of double-stranded DNA produced by treatment with *Escherichia coli* exonuclease III (Linn and Lehman 1965a,b). The purified enzyme is free of any detectable acid or alkaline nucleotidase activity.

The enzyme degrades denatured DNA and RNA at a maximum rate in 0.1–0.2 M Tris-HCl (pH 7.5–8.5), although at least 50% maximum activity is observed at pH 5.8 and 9.3. Greater than half inhibition is observed at salt concentrations below 0.05 M or above 0.25 M. The hydrolysis rate of denatured DNA is stimulated approximately 2.5-fold by the addition of 10 mM Mg^{2+}, Ca^{2+}, or Fe^{2+}. These cations inhibit the hydrolysis of RNA by about 40%, presumably by inducing secondary structure in the RNA substrate. The enzyme appears to contain Co^{2+}, and the presence of 5×10^{-6} to 5×10^{-4} M Co^{2+} produces up to a threefold stimulation in the hydrolysis rate, depending upon the particular enzyme preparation. This stimulation is independent of that found for other divalent cations (Linn 1967). Concentrations of EDTA as low as 10^{-5} M give more than 95% inhibition, even in the presence of 10^{-2} M Mg^{2+}. This inhibition can be completely overcome only by addition of amounts of Co^{2+} equivalent to the EDTA added.

Like S1 nuclease, the enzyme is inhibited by phosphate in the presence or absence of other salts. Less than 20% and 0.5% of the activity remains in 0.035 M and 0.1 M potassium phosphate (pH 7.5), respectively. The enzyme is inhibited by thiols at alkaline pH. At pH 7.5–8.2, 4 mM 2-mercaptoethanol or reduced glutathione inhibits 50% of the activity on denatured DNA. However at pH 5.5, the thiols actually stimulate activity twofold. Oxidized glutathione stimulates activity by around twofold at all pHs. The basis for these effects remains to be determined (Linn 1967).

The hydrolysis rate of denatured DNA and RNA is maximal at temperatures from 47°C to 62°C. Above this range, the enzyme is rapidly inactivated. The protein is globular and has a molecular weight of approximately 55,000.

The *N. crassa* nuclease was found to nick UV-irradiated DNA in direct proportion to the dose. Only single-strand scissions were detected with low concentrations of the enzyme, but double-strand scissions also occurred when the enzyme concentration was 9–15 times higher (Kato and Fraser 1973). The enzyme cleaves superhelical DNA reacted with formaldehyde to generate nicked circular molecules. Unlike S1 nuclease, the enzyme does not produce unit-length linear molecules (Kato et al. 1972). The enzyme is useful for the quantitation of nucleic acid hybridization, detection of sequence heterology, and gene isolation (see below).

Besides the endonuclease stated above, Zain et al. (1972) reported the isolation of an exonuclease specific for single-stranded nucleic acids from the conidia of *N. crassa*. This enzyme has a molecular weight of 72–75,000 and consists of a single polypeptide chain. It degrades single-stranded DNA and RNA from their 5′-termini to give 5′-mononucleotides and oligonucleotides. The exonuclease rapidly destroys the infectivity of single-stranded circular DNA. Enzyme preparations therefore also contain some endonuclease activity. This activity is seen with homoribopolymers in the absence of added Mg^{2+}.

Recently, Fraser and his colleagues have reported systematic studies of the nucleases from *N. crassa* and their genetic background (Fraser et al. 1976; Kwong and Fraser 1978; Chow and Fraser 1979; Fraser 1979; Käfer and Fraser 1979; Fraser et al. 1980). These studies strongly suggest that in sorbose-containing liquid culture medium *Neurospora* produces four major nucleases exhibiting markedly different properties. These enzymes may actually be derived from a single inactive precursor polypeptide (molecular weight of 90K) via different routes of proteolysis. The first is a 75K single-strand-specific, Mg^{2+}-dependent exonuclease, identical with that purified from conidia. It is not found in mycelia. The second is a 65K "endo-exonuclease" having single-strand-specific endonuclease activity and exonuclease activity with double-stranded DNA. The 55K proteolytic product is a single-strand-specific endonuclease and seems to be identical with that originally isolated from the mycelia by Linn and Lehman (1965a,b). Finally, a fourth nuclease, secreted by mycelia, is DNase A, a 65K Ca^{2+}-dependent endonuclease that cleaves both single- or double-stranded DNA but has no RNase activity.

Several repair-deficient mutants of *Neurospora*, *uvs-23*, *uvs-6*, and *nuh-4*, are deficient relative to the wild-type in the release of the single-strand-specific exonuclease and/or the endo-exonuclease. This suggests that the mutants have a lesion either in the protease(s) that controls the level of the nuclease or in some function that regulates the protease(s). These mutants map in the same gene region and may be defective in recombination. They are sensitive to various mutagens and to mitomycin C and show high frequencies of spontaneous, recessive lethal mutations and/or deletions. These results seem to indicate that single-strand-specific exonuclease and/or endo-exonuclease may be involved in DNA repair and/or recombination. Similar implications have been reported with nucleases from *U. maydis* (Holloman and Holliday 1973).

P1 Nuclease

Kuninaka et al. (1961) first reported the presence of a nuclease useful for preparing 5′-mononucleotides from RNA in an aqueous extract of a culture of *P. citrinum*. The enzyme, P1 nuclease, was extensively purified and shown to be single-strand-specific (Fujimoto et al. 1974a,b,c,d; 1975a,b). P1 nuclease attacks RNA and heat-denatured DNA by concerted endo- and exonucleolytic action. The enzyme gives only 5′-mononucleotides. In the early stages of digestion, rapid fragmentation of polynucleotides occurs and both mononucleotides and oligonucleotides are formed simultaneously. The relative amount of the monomer exceeds that of any class of oligomers throughout the process of digestion.

It appears that P1 nuclease preferentially attacks the linkage between the 3′-hydroxyl group of adenosine or deoxyadenosine and the 5′-phosphoryl group of the adjacent nucleotide (pA-N). Additionally, highly purified P1 nuclease exhibits phosphomonoesterase activity toward 3′-nucleotides to yield nucleoside and inorganic phosphate. The phosphodiesterase and phosphomonoesterase activities are not separable and the ratio of their specific activities remains constant through the enzyme purification. This suggests that a single protein is responsible for both functions. Ribonucleoside 3′-monophosphates are hydrolyzed 20 to 50 times faster than the corresponding 3′-deoxyribonucleotides. The base preference of the enzyme for 3′-ribonucleotides is G>A>C>U, whereas that for 3′-deoxyribonucleotides is C≥T>A≥G. The 3′-phosphomonoester bonds in nucleoside 3′, 5′-diphosphates, coenzyme A, and dinucleotides bearing 3′-phosphates are hydrolyzed at a rate similar to that for the corresponding 3′-mononucleotides.

The molecular weight of P1 nuclease is estimated to be 42–50,000 by several methods, including sedimentation velocity, sedimentation equilibrium, gel filtration, and SDS-polyacrylamide gel electrophoresis. The amino acid composition of the enzyme is characterized by the high content of hydrophobic amino acids, especially tyrosine and tryptophan. Like S1 nuclease, the enzyme is found to be a zinc metalloenzyme that contained 3 gram atoms of zinc per mole based upon a molecular weight of 44,000. The enzyme also contains 17.4% carbohydrate, consisting of mannose, galactose, and glucosamine in a molar ratio of 6:2:1. As expected, the enzyme exhibits a high affinity for concanavalin A-Sepharose.

The isoelectric point of the enzyme was estimated to be 4.5 and the optimal temperature of phosphodiesterase and phosphomonoesterase activities of P1 nuclease is 70°C. The optimal pH of the nuclease activities markedly depends on the substrates. The hydrolysis of natural single-stranded DNA and RNA is maximal around pH 5. Poly(A) and poly(C) are rapidly attacked at pH 6, but are highly resistant at pH 4.5, whereas poly(U) and poly(I) are readily cleaved in the range pH 4–5, but very slowly at pH 6. These findings suggest pH-dependent secondary structure changes occur in the synthetic homopolymers.

Phosphomonoesterase activity toward 3′-deoxyribonucleotides is optimal around pH 4.5. The activity on 3′-rCMP and 3′-rUMP is highest at pH 6, whereas the enzyme attacks 3′-rAMP and 3′-rGMP very rapidly at pH 7.2 and 8.5, respectively.

The enzyme is stable in the range pH 5–8 and at temperatures below 60°C. About 50% inactivation occurs at 67°C (pH 6.0) for 15 minutes. The effect of the denaturing reagents urea (8 M, pH 6.0) and guanidine hydrochloride (5 M, pH 6.0) have been investigated. The activities are strongly inhibited, especially by guanidine, in the reaction mixture. When the enzyme was pretreated with these reagents and then diluted, the enzyme activity was only partially lost with guanidine and not lost with urea. Therefore, the denaturation seems to be reversible.

The enzyme is inactivated by dialysis at pH 5.0 and reactivated almost completely by adding 1 mM Zn^{2+}. This dialysis removes one of the three zinc atoms. Upon incubation of the enzyme with 2×10^{-2} M EDTA (pH 5.0), total inactivation occurs by removal of all three zinc atoms. Activity is then only partially restored (24–38%) by adding Zn^{2+}. These results suggest that two zinc atoms are necessary to stabilize the protein conformation. Although the effect of Zn^{2+} can neither be replaced nor influenced by other divalent cations, the addition of Mn^{2+} prior to Zn^{2+} causes a marked reactivation of the nuclease activities.

The molecular conformation of P1 nuclease in aqueous solution has been investigated by measuring its optical rotatory dispersion and circular dichroism (CD). From these analyses, the α-helix content was calculated to be 29–31%. A computer fit of the CD spectrum suggests 6% β-structure and 63% random coil. The helical structure was found to be quite stable to high temperature and denaturing agents. The high stability of P1 nuclease might be due to a high content of hydrophobic amino acid and zinc atoms.

P1 nuclease has proved useful for the preparation of 5′-mononucleotides and for analysis of the cap structures present at the 5′-termini of eukaryotic mRNAs (see below).

Alteromonas BAL 31 Nuclease

Laval (1974) reported a nuclease activity ostensibly associated with phage PM2, which infects the marine bacterium, *A. espejiana*. Subsequently, Gray et al. (1975) found the same activity in the culture fluid of the uninfected host bacterium BAL 31. The organism that produces the nuclease was originally assigned to the genus *Pseudomonas* (Gray et al. 1975), but has now been reassigned to the genus *Alteromonas* species *espejiana* (Gray et al. 1981). *Alteromonas* BAL 31 nuclease is a multifunctional enzyme that contains a highly specific single-strand endodeoxyribonuclease activity and a quasi-processive exonuclease activity that simultaneously degrades both the 3′- and 5′-termini of duplex DNA (Gray et al. 1975; Legerski et al. 1978). It is very likely that ribonuclease activities are expressed by the same enzyme. The products of exhaustive digestion of linear duplex DNA with the *Alteromonas* nuclease are 5′-mononucleotides.

There are apparently two forms (slow [S] and fast [F]) of the *Alteromonas* nuclease, with different kinetic properties. Their presence depends on the initial purification step (Gray et al. 1981). The S form is purified by ion-exchange chromatography on DEAE-cellulose and gel filtration. The F form is obtained by affinity chromotography on 5′-AMP agarose. The purified S form displayed a single band on electrophoresis in heavily loaded nondenaturing polyacrylamide gels, but denatured samples gave rise to two bands in SDS-polyacrylamide gel electrophoresis. Purified F-form preparations yield a very similar pattern in denaturing gels. The relative amounts of material in the two bands revealed by staining with Coomassie Blue varied from preparation to preparation, and small quantitites of the nuclease have been obtained that display only a single band in denaturing gels. Assuming that the more intensely stained band in the denaturing gels represents the nuclease, the *Alteromonas* nuclease (both S and F forms) would be a single polypeptide of 73–83K.

Alteromonas nuclease is very stable on storage in solution at 4°C, both in crude and purified form, and displays no loss of activity after several years. A major advantage of the *Alteromonas* nuclease over other single-strand-specific nucleases, such as S1, P1, and mung bean nucleases, is that its activity is optimal near neutral pH (Gray et al. 1981). The optimal pH for single-stranded DNA is around 8.8, but that for duplex T7 DNA and supercoiled PM2 DNA is close to 8.0. The optimal temperature of the S form for denatured DNA is 60°C. Incubation at 30°C (usual temperature) gives only 13% of the maximal activity (Gray et al. 1981). The activity of the S form for denatured DNA becomes maximal between 0 M and 2 M NaCl and the nuclease retains approximately 40% of its maximal activity in the presence of 4.4 M NaCl. With CsCl, optimal activity is between 0.5–1.8 M. In the presence of 7 M CsCl, the activity is 27% of maximum (Gray et al. 1981). The exonuclease activity for duplex T7 DNA was tested and was found to decrease with increasing CsCl concentration. The activity decreased with ionic strength more sharply than for denatured DNA, although activity was still present at molarities of CsCl as high as 5.5 (Gray et al. 1981).

The *Alteromonas* enzyme is active against both linear duplex and supercoiled DNAs in the presence of 5% SDS and can be preincubated with the detergent without loss of activity provided Ca^{2+} and Mg^{2+} are present at 12.5 mM before the addition of the detergent (Gray et al. 1975, 1981). The effect of urea on the ability of the S form to hydrolyze single-stranded DNA was tested. The enzyme was added to a mixture of all other components immediately before incubation. Over 40% of the maximal activity was still observed in the presence of 6.6 M urea.

0.2 mM Ca^{2+} gives full activity of the S form on single-stranded DNA, and there is no substantial dependence of reaction velocities on Ca^{2+} concentrations up to 50 mM for this substrate. In contrast, the rate of cleavage of PM2 form-I DNA very markedly increases with increasing Ca^{2+} concentration from 1 mM to approximately 10 mM, above which, the reaction rate does not change significantly. Single-stranded DNA is hydrolyzed by the S form in the absence of Mg^{2+} at only 1.5% of the rate at 15 mM Mg^{2+}; PM2 form-I DNA was cleaved in the absence of Mg^{2+} at 13% of the rate measured in the presence of 15 mM Mg^{2+}. For both substrates, the reaction velocities increase markedly with Mg^{2+} concentration between 0 mM and 10 mM and become weakly dependent on the concentration of this cation above approximately 15 mM. The single-stranded circular DNA of phage ϕX174 is readily degraded to dialyzable fragments by the *Alteromonas* nuclease; this indicates that an endonucleolytic activity against single-stranded DNA is present (Gray et al. 1975). Linear duplex DNA is degraded from both 5′- and 3′-ends by the enzyme without the introduction of detectable scissions away from the termini. PM2 form-I DNA is rapidly converted to form-III DNA by the action of the enzyme, and the linear duplex DNA is then degraded from both ends (Lau and Gray 1979). The conversion to form III proceeds by two steps, as with other single-strand-specific nucleases. The *Alteromonas* nuclease also cleaves very highly positively supercoiled DNA prepared from covalently closed, relaxed DNA (form I^0) with ethidium bromide. Initial nicking rates of PM2 form-I DNA mediated by highly purified *Alteromonas* nuclease are readily measurable at superhelical densities as low as -0.02, whereas the initial nicking with nucleases from *N. crassa* and mung bean requires more negative superhelicity. Nicking of positively supercoiled DNA by the *Alteromonas* nuclease becomes detectable at superhelical densities between 0.15 and 0.19.

With highly purified nuclease, linear duplex T7 DNA (molecular weight near 2.5×10^7 daltons) was shortened by 75%. The remaining strands sedimented as a single band in alkali. For removal of several hundred nucleotides or more from the termini of a duplex DNA, the F form should be used. The S form is better if a smaller number of nucleotides are to be removed (Gray et al. 1981). Gray et al. (1975) have reported that linear duplex DNA shortened by the action of the *Alteromonas* nuclease (S or F forms) can contain protruding single-stranded ends long enough to promote joining of the molecules by *E. coli* DNA ligase to give dimeric and oligomeric species. The behavior of the dimeric species obtained on ligase treatment of nearly full-length linear PM2 DNA led to the conclusion that the DNA has joined in a head-to-head fashion, in which joined strands are self-complementary and can form a hairpin structure. This type of joining requires that the single-stranded ends contain complementary nucleotide sequences. These could be formed if the exonuclease activity were slowed when a palindromic sequence long enough to form a hairpin structure is exposed as a protruding single strand.

Talmadge et al. (1980) have shown that a fully base-paired oligonucleotide linker can be ligated with T4 DNA ligase to plasmid DNA that has been shortened by *Alteromonas* nuclease (S form) and that the resulting DNA can be cyclized. However, these authors do not describe the ligation efficiency. Gray et al. (1981) reported that a high percentage of duplex fragments formed by the F-form nuclease undergo T4 DNA ligase-catalyzed joining. However, we find that nibbled linear duplex DNA formed by *Alteromonas* nuclease (purchased from Bethesda Research Laboratories, Inc.) is ligated with 5–10% efficiency with T4 DNA ligase. The ligation efficiency markedly increases to more than 50% after treatment with *E. coli* DNA polymerase I, Klenow fragment, and/or T4 DNA polymerase (K. Shishido et al., unpubl.). This indicates that the majority of the nuclease-treated duplex DNA contains single-stranded portions at the termini. The data also show that, at least, a portion of the nuclease-shortened duplex DNA had fully base-paired termini.

Irradiation of PM2 form-I^0 DNA with UV light gives rise to *Alteromonas* nuclease-sensitive sites. Legerski et al. (1977) found that at approximately seven pyrimidine dimers per molecule, 7% of these lesions served as sites of attack by the nuclease. Binding of base-modifying agents to PM2 form-I^0 DNA also introduces the nuclease-sensitive sites; however, cleavage efficiency with the nuclease is relatively low. Apurinic sites are also a target for the nuclease (Legerski et al. 1977; Gray et al. 1981).

Mung Bean Nuclease I

Sung and Laskowski (1962) originally reported the presence of an endonuclease activity in mung bean sprouts. The purified enzyme specifically hydrolyzed heat-denatured DNA and RNA, and was designated mung bean nuclease I, single-strand-specific (Johnson and Laskowski 1968). The digestion products with the nuclease range from mononucleotides to at least heptanucleotides that are terminated by a 5′-phosphate and 3′-hydroxyl. The nuclease shows a pH optimum at 5.0 and is quite sensitive to ionic strength, 0.025-0.05 M being optimal. The preference for single-stranded DNA, as assayed by the formation of acid-soluble material, depends on the type of DNA and the reaction conditions. At 37°C under the optimal conditions, it is 30,000-fold for T7 DNA, 65-fold for crab poly(dA-dT), and less than 2-fold for synthetic poly(dA-dT). No hydrolysis of poly(dG):poly(dC) occurs under these conditions, even with a 100-fold excess of enzyme. Mung bean nuclease I does not recognize synthetic poly(dA-dT) as a typically double-stranded structure. The enzyme exhibits a preference for Ap↓Nd and T(U)↓pN in single-stranded nucleic acids (Johnson and Laskowski 1970).

The purified enzyme can dephosphorylate 3′-mononucleotides. Also, dinucleotides NpNp are first dephosphorylated to NpN and then the internucleotide linkage is cleaved to form N + pN. This dephosphorylation activity is a 3′-ω-monophosphatase rather than a 3′-nucleotidase (Mikulski and Laskowski 1970). The ω-monophosphatase hydrolyzes ribose mononucleotides 50- to 100-fold faster than the corresponding deoxyribose compounds. It also shows preference for bases in the order A>T(U)>C>G. A similar preference is shown by the nuclease activity, except that poly(dT) or poly(U) were hydrolyzed faster than the corresponding adenine polymers presumably because the former lack an ordered structure.

The nuclease is a glycoprotein consisting of 29% carbohydrate by weight and has a molecular weight of 39,000 as determined by polyacrylamide gel electrophoresis in the presence of SDS (Kowalski et al. 1976). In the absence of the latter, it gives several bands, presumably caused by aggregation. The enzyme contains one sulfhydryl group and three disulfide bonds per molecule. It has a high content (12.6 mole %) of aromatic residues. A portion of the enzyme molecules contains a peptide bond cleavage at a single region in the protein. The two polypeptides, 25K and 15K, are covalently linked by a disulfide bond(s). Both the cleaved and intact forms of the enzyme are equally active in the hydrolysis of phosphate ester linkages in either DNA, RNA, or adenosine 3′-monophosphate.

Like P1 and S1 nucleases, mung bean nuclease I is a zinc metalloenzyme with multiple metal ion-binding sites (Kowalski et al. 1976). The enzymatic activity can be stabilized at pH 5.0 in the presence of 0.1 mM zinc acetate, 1.0 mM cysteine, and 0.001% Triton X-100. After removal of Zn^{2+} and cysteine by dialysis for 24 hours against 0.05 M sodium acetate (pH 5.0) containing 0.001% Triton X-100, the enzyme loses 70–80% of its original activity against denatured DNA. The dialyzed enzyme can be reactivated by addition of Zn^{2+} (0.1 mM) and cysteine (1 mM) and 20–30 minutes incubation at 23°C. Zinc acetate without a sulfhydryl compound is also capable of reactivating the dialyzed enzyme. The rate is very slow, however, since only one-half of the maximal reactivation is achieved in 24 hours with 0.1 mM or 1.0 mM Zn^{2+}. Co^{2+}, Mg^{2+}, Mn^{2+}, Ca^{2+}, Fe^{2+}, and Cu^{2+} under the same conditions were incapable of reactivating the enzyme.

Dialysis of the enzyme against 0.05 M sodium acetate (pH 5.0) containing 1 mM EDTA, followed by further dialysis to remove the EDTA, results in complete loss of the activity against single-stranded DNA. Incubation with 1 mM Zn^{2+} and other divalent cations failed to produce reactivation after 18 hours at 23°C in the presence of 1 mM cysteine and 0.001% Triton X-100. Apparently, zinc ions are totally removed by EDTA dialysis, resulting in irreversible inactivation. This essential metal ion is presumably not totally removed by dialysis in the absence of EDTA, since, under these conditions, both sulfhydryl compounds and Zn^{2+} reactivate the enzyme (compare S1 nuclease above). Also, 0.01% SDS (pH 5.0) completely inactivates the enzyme. Mung bean nuclease I is stable to heat treatment (60–70°C) at pH 5.0 in the presence of both Zn^{2+} and sulfhydryl compounds.

With a high concentration of the enzyme, mung bean nuclease I introduces a limited number of endonucleolytic, double-strand cleavages in linear duplex DNA. Phage λ DNA is cleaved most rapidly in the AT-rich region in the center of the molecule (Johnson and Laskowski 1970) and preferentially at a total of six loci, all of which fall within the three AT-rich blocks revealed by partial denaturation mapping (Kroeker and Kowalski 1978). In addition to these sites, there are other sites that are hydrolyzed more slowly. For phage DNAs of T7, gh-1, and PM2, the digestion products by the nuclease are gene-sized (300 – 2100 bp), suggesting that cleavage occurs at specific sites (Kroeker and Kowalski 1978).

Under conditions that do not favor a tight helical structure and with high concentrations of the enzyme, large duplex polymers such as T7 DNA fragments (300 – 2100 bp) are completely degraded from their termini with a continuous accumulation of mono-, di-, and trinucleotides (Kroeker et al. 1976). The terminally directed activity is an intrinsic property of the enzyme molecule because: (1) it is inactivated and reactivated in parallel with the single-strand activity and (2) the two activities coelectrophorese on analytical gels. The pH optimum for both the hydrolysis of denatured DNA and terminally directed hydrolysis of native DNA becomes more acidic with increasing salt concentration. The relative preference for single-stranded structures increases as the pH becomes more basic.

At near neutral pH, supercoiled DNA is the preferred substrate over the relaxed topoisomer by 30,000-fold. The enzyme converts supercoiled PM2 DNA to the nicked circular form but not to the linear form (Wang 1974; D. Kowalski, J.P. Sanford, and T.J. Foels, pers. comm.). With a high concentration of the enzyme at more acidic pH, the nicked circular DNA is converted to small-sized DNA fragments.

Other Nucleases

Ustilago maydis *Nuclease (DNase I)*

A nuclease highly active on denatured DNA or RNA has been purified by Holloman and Holliday (1973) from the smut fungus *U. maydis*. The nuclease liberates oligonucleotides from heat-denatured DNA with preferential attack at deoxyguanosine residues during the early course of hydrolysis. However, after exhaustive digestion, a large proportion of mononucleotides is found to accumulate. The enzyme requires free ends of DNA molecules for activity, because single-stranded DNA of phage ϕX174 is refractory to hydrolysis. When DNA that is radioactively labeled specifically at either the 3′ or 5′ end is digested with the nuclease, the label is liberated at a faster rate than the total nucleotide (Holloman 1973). The molecular weight of the enzyme is 42,000.

Added divalent cations are not required for activity. EDTA strongly inhibits activity, but simultaneous addition of an excess of Zn^{2+} or Co^{2+} reverses the inhibition. Other inhibitors include potassium phosphate, 2-mercaptoethanol, and ATP (Holloman and Holliday 1973). It is of interest that the nuclease can recognize mismatched base pairs in linear duplex DNA. Single and double-strand nicking at or near the single mismatched site was observed with three out of five pairs of heteroduplexes (Ahmad et al. 1975; Pukkila 1978). Certain *Ustilago* mutant strains that are unable to carry out allelic recombination have reduced amounts of the enzyme. Therefore, this enzyme may play some role in recombination in this organism (Holloman and Holliday 1973).

Physarum polycephalum *Nuclease*

An alkaline nuclease specific for single-stranded nucleic acids was purified by Waterborg and Kuyper (1979) from microplasmodia of *P. polycephalum*. The enzyme has a molecular weight of 32,000 and appears to be a zinc metalloprotein like S1, P1, and mung bean nucleases I. The activity of the nuclease is strongly dependent on ionic strength. Optimal enzyme activity is at pH 8.0. Phosphate ions give a strong competitive inhibition even at very low concentrations. The *Physarum* enzyme seems to have many characteristics in common with S1 nuclease, except pH optimum.

Wheat Seedling Nuclease

A nuclease has been isolated from germinating wheat seedlings by Hanson and Fairley (1969). The enzyme hydrolyzes denatured DNA, RNA, and the phosphomonoester linkage of 3′-AMP at almost the same rate. The optimum pH is around 4.8–5.5. At pH 4.5, Zn^{2+} is essential for stabilizing the three activities. The nuclease has been shown to have endonucleolytic activity toward denatured DNA and primarily exonucleolytic activity toward RNA (Kroeker et al. 1975). These enzyme properties resemble those of mung bean nuclease I. Similar enzymes are distributed among various plants (Wani and Hadi 1979).

Sheep Kidney Nuclease (Nuclease SK)

The enzyme digests single-stranded nucleic acids (Kasai and Grunberg-Manago 1967; Watanabe and Kasai 1978). The optimum pH is 7.5 and the enzyme is most active in the presence of 50 mM Mg^{2+}. The molecular weight is 52–53,000. The nuclease also hydrolyzes 2′–5′ ApA, 3′–5′ ApA, ADP, ATP, 5′-AMP, and cAMP. All 16 deoxydinucleotides are products of nuclease digestion (Watanabe 1978a). Digestion of phage ϕX174 under appropriate conditions results in size-specific fragmentation of the DNA (Watanabe 1978b).

Saccharomyces *Nuclease*

Lee et al. (1968) reported isolation of a nuclease from a hybrid strain of *Saccharomyces*. The pH optimum is 7.6 and Mg^{2+} is required for full activity.

The nuclease attacks denatured DNA and RNA at approximately the same rate. Double- and triple-stranded helical conformations are less susceptible to attack.

Staphylococcus aureus *Nuclease*

The enzyme hydrolyzes both DNA and RNA, but has greater affinity for DNA than for RNA. The activity is greater with denatured DNA than with native DNA. However, the single-strand selectivity of the enzyme is not very high (Cuatrecasas et al. 1967).

Single-strand-specific DNases

E. coli *Exonuclease I*

The enzyme shows a high degree of selectivity for single-stranded DNA (Lehman and Nussbaum 1964; Lehman 1966). It attacks at the 3′-hydroxyl end of a polydeoxyribonucleotide and liberates 5′-mononucleotides in a stepwise manner leaving the 5′-terminal dinucleotide intact. Oligonucleotides are easily hydrolyzed by the enzyme, but it must be used in several hundredfold excess to ensure greater than 90% digestion of denatured DNA. The optimum pH is from 9.2–9.8.

E. coli *Exonuclease VII*

The enzyme is an exonuclease specific for single-stranded DNA or single-stranded termini of duplex DNA (Chase and Richardson 1974a,b). The enzyme does not have an absolute requirement for a divalent cation, but is stimulated 25% by Mg^{2+}. It is active in the presence of EDTA, and optimal activity occurs at pH 7.9 in potassium phosphate buffer. The molecular weight is 88,000. The enzyme can initiate hydrolysis at both 3′ and 5′ termini to yield products that are exclusively oligonucleotides bearing 5′-phosphoryl and 3′-hydroxyl termini. Exonuclease VII is the type of enzyme that is well-suited to serve in both recombination and repair (see Chapter 4).

Phage T4 Endonuclease IV

Sadowski and Hurwitz (1969) isolated an endonuclease specific for single-stranded DNA from phage T4-infected *E. coli*. The enzyme specifically cleaves phosphodiester bonds adjoining deoxycytidine and yields fragments with 5′-phosphoryl termini (Sadowski and Bakyta 1972). The limit product obtained by exhaustive digestion of single-stranded DNA with the enzyme is 25% acid-soluble and consists of oligonucleotides with an average chain length of about 50. The initial structural studies of ϕX174 DNA were greatly facilitated by this enzyme (Ziff et al. 1973).

Phage T7 Gene-3 Product (Endonuclease I)

An endonuclease, the product of gene *3* of phage T7, was isolated from T7-infected *E. coli* (Center and Richardson 1970; Center et al. 1970; Sadowski 1971). The nuclease activity is at least 150 times greater with single-stranded DNA as substrate than with duplex DNA. The limit product obtained by digestion of single-stranded DNA with the enzyme is about 50% acid-soluble and consists of oligonucleotides with a chain length of about 10, bearing 5′-phosphate and 3′-hydroxyl termini. The enzyme also produces single-strand and double-strand breaks in duplex DNA. T7 gene-*3* mutants synthesize only a limited amount of DNA. In addition, they are defective in degrading host DNA, suggesting that the gene-*3* endonuclease is involved in this function. Recently, Panayotatos and Wells (1981) have observed that the nuclease converts superhelical DNA to unit-length linear form under appropriate conditions.

Bacillus subtilis *Deoxyribonuclease*

An endonuclease specific for single-stranded DNA has been isolated from *B. subtilis* (Cobianchi et al. 1978; Matsumoto et al. 1979). The enzyme cleaves single-stranded DNA, producing 5′-phosphate termini. The activity on double-stranded DNA is only about 2% of that on single-stranded DNA. Cobianchi et al. (1978) reported that the enzyme is a heterogeneous aggregate of identical 36K subunits. A mild proteolytic treatment of the aggregate caused the irreversible and nearly quantitative conversion into the free subunit, which also showed catalytic activity. The enzyme activity appears during germination only after the start of DNA synthesis, and is reduced in a recombination-deficient mutant to approximately 13% of the wild-type level (Cobianchi et al. 1980). Matsumoto et al. (1979) reported the isolation of a similar enzyme (named endonuclease MII) from the Marburg strain of *B. subtilis*. The activity requires divalent cations and the molecular weight is about 57,000. The activity on single-stranded DNA relative to that on double-stranded DNA was approximately 5000:1.

Phage SP3 Exonuclease

A single-strand-specific exonuclease was purified from *B. subtilis* infected with phage SP3 by Trilling and Aposhian (1968). The mode of attack on single-stranded DNA is predominantly exonucleolytic with the release of dinucleotides (89–91%) and trinucleotides (10%).

Animal Cell Endonucleases

An endonuclease with a molecular weight of about 70,000 was purified from mouse ascites cells by Otto and Knippers (1976). This enzyme specifically hydrolyzes single-stranded DNA, yielding mainly oligonucleotides with 5–10 residues. Supercoiled DNA is converted to the linear form.

A single-strand-specific endonuclease has also been isolated from human cells by Wang et al. (1978). The enzyme purified from KB cells has a molecular weight of 54,000 and requires Mg^{2+} for activity. The pH optimum is 9.2. The hydrolysis products from denatured DNA are acid-soluble oligonucleotides with 5′-phosphoryl termini. Purified enzyme preparations also hydrolyze poly(U), releasing acid-soluble materials. A similar enzyme activity was found in calf thymus (Wang and Furth 1977).

Ashe et al. (1965) isolated an endonuclease from shark (*Mustelus canis*) liver. The enzyme has 15 times more activity towards thermally denatured DNA than native DNA. The activity is optimal in 0.1 M NaCl and the pH optimum is 8.0. Divalent metals such as Mg^{2+}, Mn^{2+}, and Co^{2+} are required for full activity (Ashe et al. 1965; Jayaraman and Goldberg 1969).

APPLICATIONS

Single-stranded Nucleic Acids

Preparation of Mono- and Dinucleotides

With sufficient enzyme, S1 and P1 nucleases completely degrade RNA to produce 5′-mononucleotides. The enzymes are quite thermostable. Therefore they have been used for the industrial production of 5′-mononucleotides from yeast RNA. Among them, 5′-GMP and 5′-IMP, prepared by deaminating 5′-AMP, are used as flavor enhancers (Kuninaka et al. 1961; Ando 1966).

Sheep kidney nuclease converts single-stranded nucleic acid to 5′-dinucleotides and trinucleotides as final products. Digestion with pancreatic DNase I gives 11 deoxy 5′-dinucleotides, whereas with the sheep kidney nuclease all 16 deoxy 5′-dinucleotides are obtained (Watanabe 1978a).

Estimation of Double-helical Content and Isolation of Rapidly Reannealing Regions

Single-stranded nucleic acids tend to form regions with double-stranded structure. Shishido and Ando (1972) have used S1 nuclease to estimate the double-helical content of various single-stranded nucleic acids. The substrates were digested at 10°C under moderate ionic strength and resistant material was isolated using Sephadex G-50. The order of S1 resistance is double-stranded DNA $\geq$ poly(A):poly(U) $>>$ rRNA, tRNA, Qβ RNA, MS2 RNA, TMV RNA $>$ f1 DNA $>>$ heat-denatured DNA $>>$ poly(U). Poly(U) was almost totally degraded. Using intact and fragmented f1 DNA, the S1 resistance correlates well with the degree of hyperchromicity. Likewise with RNA, S1 resistance agrees with data obtained by optical rotatory dispersion and infrared spectroscopy.

One use of the S1 method described above is to isolate the double-stranded regions of single-stranded nucleic acids (Shishido and Ikeda 1970). Shishido and Ikeda (1971a,b) have succeeded in isolating from phage f1 DNA two sorts of hairpin structures, of 30–50 nucleotides, which are rich in A·T base pairs and G·C base pairs, respectively. The size of looped-out portions in both hairpins seems to be very small since they are resistant to S1. The T_m in standard saline citrate (SSC) of the AT-rich and GC-rich hairpins was 45°C and 70°C, respectively.

Schaller et al. (1969) also have reported the isolation of a GC-rich core fraction from phage fd DNA by successive digestion with *N. crassa* nuclease and *E. coli* exonuclease I. The GC-rich hairpin seems to exert a powerful influence on the overall conformation of the molecule. Wilson and Thomas (1974) have isolated inverted repetitive (fold-back) sequences, 600–2400 nucleotides long, with S1 nuclease from a number of eukaryotic DNAs denatured by alkali. S1 nuclease has also been used to isolate and characterize a double-stranded "stem" sequence from the inversion loop structure formed after denaturation of transposable elements (Ohtsubo and Ohtsubo 1976).

5′ Caps of Eukaryotic mRNAs

Eukaryotic mRNAs contain methylated, blocked 5′-terminal structures: $m^7G(5')ppp(5')N^mpNp$....(cap 1) or, $m^7G(5')ppp(5')$-N^mpN^mpNp....(cap 2). This novel 5′-5′ terminal nucleotide linkage is resistant to P1 nuclease (Furuichi et al. 1975). Digestion with the enzyme leaves intact $m^7G(5')ppp(5')N^m$ while all other sequences are converted to 5′-mononucleotides.

Specific Cleavage of tRNA and Small rRNA

At pH 4.5 in high salt at 20°C, S1 nuclease specifically hydrolyzes tRNAs in the anticodon loop and at the 3′-end (Harada and Dahlberg 1975). *N. crassa* nuclease also exhibits this property with high salt at pH 7.5 and 37°C (Tal 1975). There are multiple cleavage sites in the anticodon, although not all sites are equally susceptible. The dihydro-U loop, the T-Ψ-C-G loop and the variable loop are resistant to both enzymes as expected from their tertiary structures. The specific cleavage in the anticodon loop is very useful for sequence analysis of tRNA.

Khan and Maden (1976) have found that under appropriate conditions S1 nuclease cleaves HeLa-cell 5.8S rRNA preferentially at two internal sites and at the 3′ end of the molecule. One internal site involves the methylated sequence, A-A-U-U-G^m-C-A-Gp and the other site involves a hairpin loop with a GC-rich stem.

Double-stranded Nucleic Acids

Elimination of Short Single-stranded Ends

To construct a recombinant DNA molecule between different restriction endonuclease-derived fragments or an in vitro deletion mutant by exonuclease digestion, it is necessary to remove single-stranded tails before T4 DNA ligase treatment. So far many investigators have used S1 nuclease for this purpose. However, the ligation efficiency of the S1-treated duplex DNA is still not resolved. Shishido and Ando (1981) have recently studied this efficiency. Sticky ends generated by *Eco*RI and *Hpa*II were digested with S1 under conditions where single-stranded DNA was completely converted into an acid-soluble form. *Eco*RI digestion generates a 4-base sticky end and *Hpa*II digestion generates a 2-base sticky end. The ligation efficiency of both ends after S1 treatment was significantly lower than that of a true flush end. However, the efficiency increased up to the same level as the flush end upon treatment with the Klenow fragment of *E. coli* DNA polymerase I. Under the above conditions, S1 did not attack true flush ends. These experimental results indicate that S1 can recognize and hydrolyze even a 2-base extension but that the resulting termini are not necessarily flush. Most probably one base still protrudes at both termini and this is reflected by the low ligation efficiency. In experiments where an excess of S1 was used, *Eco*RI- and *Hpa*II-generated sticky ends were still ligated inefficiently and the ligation efficiency of ligating true flush ends decreases, suggesting a ''nibbling'' reaction of the S1.

Location of Nonbase-paired Regions in Superhelical DNA

Negative superhelicity introduces localized unwinding of helical base pairs. S1 can recognize such regions and cleave both strands of the DNA to generate unit-length linear duplex molecules with no nicks (Beard et al. 1973; Godson 1973; Mechali et al. 1973; Germond et al. 1974; Shishido and Ando 1975a). The reaction proceeds in two steps: (1) Cleavage occurs on either strand and then, (2) the nicked, circular DNA is cleaved on the opposite strand at or near the nick to yield a linear molecule. The nucleases from *N. crassa* (Kato et al. 1972), mung bean sprouts (Wang 1974; D. Kowalski et al., pers. comm.), and *A. espejiana* BAL 31 (Lau and Gray 1979) also cleave supercoiled DNA. *N. crassa* nuclease and mung bean nuclease I convert supercoiled DNA into a nicked circular form, but not a linear form, whereas, *Alteromonas* BAL 31 nuclease generates a linear form.

Shishido (1979) has determined the location of S1 nuclease-cleavage sites on the DNAs of SV40 and polyoma. It was noticed that the specificity was enhanced by using added Mn^{2+} in place of Zn^{2+}. SV40 and polyoma DNA were cleaved at eight and seven specific sites, respectively. The S1 cleavage sites between both viral DNAs were strikingly similar from the point of view of viral genome organization: (1) Both are cleaved at or close to the site at which DNA replication terminates which also corresponds to the 3′-ends of the early and late mRNAs; (2) both DNAs are cleaved in the regions in which the coding sequences for VP2/VP3 terminate while the coding sequence for the *N*-terminus of VP1 lies between two cleavage sites; (3) three other cleavage sites on each DNA are located at around 0.2, 0.4, and 0.8, the distance from the origin of DNA replication in the early region. These S1 cleavage sites coincide with the binding sites of T4 gene-*32* protein and *E. coli* RNA polymerase. Furthermore, it was found that S1 cleavage sites on SV40 exhibit nice agreement with inverted repeated sequences and AT-rich regions.

Shishido and Noguchi have recently observed that the tetracycline-resistance plasmid pNS1 of *B. subtilis* contains eleven S1 cleavage sites and five sites for RNA polymerase binding. Four of the five polymerase binding sites coincide with S1 cleavage sites (K. Shishido and Noguchi, unpubl). Lilley (1980) and Panayotatos and Wells (1981) have reported that plasmid DNAs of pBR322, ColE1, and pVH51 have one or a few strong S1 cleavage site(s), although some DNAs are cleaved at less specific sites by long periods of S1 digestion. They showed that these unique S1 cleavage sites are located at inverted repeated sequences and AT-rich regions, supporting Shishido's previous observation (Shishido 1979). The S1 cleavage site on ColE1 DNA lies between the genes for colicin E1 and immunity. Godson (1973) and Wiegand et al. (1975) have reported that the S1 cleavage sites on replicative form (RF) I DNAs of ϕX174, G4, and G14 are distributed throughout the molecule, although the distribution is not completely random. These experiments were done at 37°C. Lilley (1980) has observed that at 7°C ϕX174 RFI is cleaved by S1 at a single unique site. This site is the inverted repeated sequence recognized, in its single-stranded form, by a DNA replication protein n′ from *E. coli* (Shlomai and Kornberg 1980).

Specific Fragmentation of Double-stranded DNA

In one of its strands, phage T5 DNA contains nicks at defined positions. Each interruption is a scission of only a single phosphodiester bond which can be repaired by DNA ligase. Shishido and Ando (1975b) and Wiegand et al. (1975) have found that under appropriate conditions S1 nuclease can cleave the opposite strand at the preexisting interruptions to generate specific fragments.

Pavlov et al. (1977) have developed a method for selective fragmentation of bacteriophage T7 DNA at AT-rich regions. These regions were melted at a relatively low temperature, fixed with glyoxal, and the DNA subjected to complete digestion with S1. Three specific fragments were produced.

As mentioned earlier, mung bean nuclease I at high concentration introduces double-strand cleavages in linear duplex DNA. Phage λ DNA can be cleaved at six AT-rich loci, the one in the center of the molecule being most rapidly cleaved (Johnson and Laskowski 1970; Kroeker and Kowalski 1978). For bacteriophages T7, gh-1, and PM2, fragments of 300–2100 bp were produced by digestion with this nuclease (Kroeker and Kowalski 1978).

Ordering of Restriction Fragments

As discussed above, *Alteromonas* BAL 31 nuclease can shorten both strands of linear duplex molecules from both termini. By using this activity, Legerski et al. (1978) have developed a convenient method for ordering restriction fragments. A series of progressively shortened samples from a linear duplex

DNA were subjected to subsequent digestion with a restriction endonuclease. Analysis of the resulting fragments by gel electrophoresis then permits the rapid establishment of the order of these fragments throughout the entire genome. This is accomplished by noting from the electropherograms the order in which the various restriction enzyme fragments become noticeably shortened or disappear. They have mapped the five cleavage sites for *Hpa*I in PM2 DNA and 18 of the 24 cleavage sites of *Pst*I in phage λ b_2b_5c DNA. Five of the six remaining *Pst*I cleavage sites have been assigned to specific regions of the genome.

Recovery of Inserts from Hybrid Plasmids Constructed by the Poly(dA):poly(dT) Method

S1 nuclease was successfully used for the specific and efficient excision of the inserted DNA segment from a hybrid plasmid constructed by the poly(dA): poly(dT) tailing method (Hofstetter et al. 1976). The hybrid plasmid, pβG, consisting of plasmid PMB9 DNA with an insert of rabbit globin DNA (596±50 bp) flanked by poly(dA):poly(dT) regions (40–115 bp) was cleaved into two fragments at 50°C under conditions of partial denaturation by 40–50% formamide. Only the smaller fragment (575 bp) contained globin-specific sequences, showing that excision had occurred precisely.

Detection of Locally Altered Structures in Double-stranded DNA

S1, *N. crassa*, and *Alteromonas* BAL 31 nucleases can recognize structural alterations induced by UV-irradiation and introduce single- and double-strand breaks into DNA. The extent of cleavage is dependent upon the dose of UV light, the concentration of the enzyme, and the ionic strength of the reaction. Shishido and Ando (1974) found that several single- or double-strand breaks per 100 pyrimidine dimers containing thymine occurred with S1 used under appropriate conditions. Wiegand et al. (1975), Kato and Fraser (1973), and Legerski et al. (1977) obtained similar results in experiments using S1 nuclease, *N. crassa* nuclease, and *Alteromonas* BAL 31 nuclease, respectively. Heflich et al. (1979) reported that S1 can efficiently remove thymine-containing pyrimidine dimers only when long digestion times and high enzyme concentrations are used.

N-acetoxy-*N*-acetyl-2-aminofluorene reacts with DNA, generating *N*-(deoxyguanosine-8-yl)-*N*-acetyl-2-aminofluorene as a major product. Yamasaki et al. (1977) have found that when *N*-acetyl-2-aminofluorene-modified DNA was incubated with S1 nuclease or *N. crassa* nuclease, there was preferential loss of the guanosine C-8 adduct from the DNA.

Shishido and Ando (1975a) have investigated the action of S1 on T7 DNA subjected to limited depurination in 0.1 M sodium acetate buffer (pH 4.2) at 70°C. The product depended on the degree of depurination; DNA degradation was observed at levels over 0.825% depurination. The size of the products from 3.3% depurinated DNA was approximately one-sixth of intact T7 DNA from neutral sedimentation patterns. This shows efficient double-strand cleavage of the depurinated DNA. Gray et al. (1981) have found similar results with the *Alteromonas* BAL 31 nuclease.

Ben-Hur et al. (1979) have used S1 in a rapid and sensitive method to measure DNA cross-links. Double-stranded DNA was exposed to 4,5′,8-trimethylpsoralen plus 360-nm light and denatured by alkali. Following acid neutralization, cross-linked DNA undergoes spontaneous renaturation and is rendered S1 resistant.

Region-specific Deletion Mutants

Unit-length linear duplex DNA is first prepared from circular DNA by digestion with an appropriate restriction endonuclease and then, the linear DNA is shortened with *Alteromonas* BAL 31 nuclease. Incubation with T4 DNA ligase and transformation in vivo leads to circular DNA molecules that contain deletions in the region centered on the restriction cleavage site. The sizes of the deletions depend on the extent of shortening of the DNA. We have constructed various deletion mutants of drug-resistance plasmid DNA with this procedure (Noguchi et al., unpubl.).

As already mentioned above, a combination of exonuclease III and S1 nuclease can similarly be used to make region-specific deletion mutants.

cDNA Cloning

Reverse transcriptase can be used to synthesize a single-stranded DNA complementary to a mRNA. Reverse transcriptase products, whether partial or complete copies of the mRNA template, often possess short, partially double-stranded hairpins at their 3′-termini. These hairpins can be exploited as primers for *E. coli* DNA polymerase I or the transcriptase. The products of such sequential reverse transcriptase and DNA polymerase reactions still possess a loop at one end. S1 nuclease can cleave the loop to generate a duplex DNA suitable for subsequent insertion into a vector (Efstratiadis et al. 1976; Wickens et al. 1978).

Screening for DNA-binding Substances

A DNA-binding substance, such as ethidium bromide, actinomycin D, mitomycin C, adriamycin, etc., binds to supercoiled DNA, resulting in a decreased number of superhelical turns. This is reflected by increased S1 resistance of the drug-bound, supercoiled DNA (Shishido and Ando 1980). The S1 cleavage patterns can be analyzed by agarose gel electrophoresis after phenol extraction in which the drug is released from the DNA.

Hybrid Molecules

Quantitation of Homology

Single-strand-specific nucleases are widely used for quantitating homology between nucleic acids. The assay relies upon the fact that double-stranded regions formed by hybridization are resistant to a nuclease such as S1 whereas nonbase-paired regions are quantitatively digested to acid-soluble products.

Physical Mapping of Mutations

Heteroduplex DNA molecules formed from the strands of a deletion or insertion mutant and wild-type DNA contain a single-stranded loop at the position of the mutation. Shenk et al. (1975) have found that S1 nuclease can be used to locate the position of this single-stranded loop. The enzyme first digests the loop and subsequently cleaves the complementary strand, yielding fragments whose lengths reflect the position of the loop. Bartok et al. (1974) have reported the use of *N. crassa* nuclease for the same purpose.

Of interest is the finding by Shenk et al. (1975) that S1 can also cleave a heteroduplex formed from the DNAs of a temperature-sensitive mutant and either its revertant or wild-type parent. The cleavage of such heteroduplexes with single base mismatches is slower and more limited in extent than that observed with deletion and insertion heteroduplexes; consequently, the nonspecific background cutting is higher in these cases. With the single-strand-specific *U. maydis* nuclease, Ahmad et al. (1975) and Pukkila (1978) have obtained similar results.

Dodgson and Wells (1977) have investigated, in detail, the action of single-stranded-specific nucleases on base-mismatched heteroduplexes. The sensitivity to S1 and mung bean nucleases I, of model DNAs containing dA·dG and/or dG·dG mismatched regions of defined length was tested by analysis of the distribution of product oligonucleotides. Single base mismatches were extremely resistant to these nucleases, although low levels of cleavage at the heteroduplex nucleotide were observed at high nuclease concentrations. The nuclease sensitivity of dA·dG heteroduplex regions increased gradually as the length of the heteroduplex region increased from one to six nucleotides. The sensitivity of dG·dG heteroduplexes three to five nucleotides long was considerably greater than that of the single dG·dG mismatch.

Construction of Targeted Deletion Mutants via D loops

Closed circular DNA interacts with complementary sequences of single-stranded DNA to form displacement loop (D loop) structures in vitro. The site of D-loop formation can be directed by using single-stranded DNA derived from a selected restriction fragment. Green and Tibbetts (1980) have found that S1 nuclease can linearize circular DNA containing a D loop, removing a limited number of nucleotides from each strand. Incubation with T4 DNA ligase followed by propagation in vivo leads to circular DNA molecules that bear small, single deletions in the region of the single-stranded DNA sequence chosen for the formation of the D loop. The level of mutagenesis obtained by the procedure is quite high.

Mapping RNA Splice Points

Berk and Sharp (1978) have used S1 nuclease for determining the structure of spliced RNAs and mapping splice points. Total cytoplasmic RNA is hybridized to ^{32}P-labeled DNA above the T_m of DNA but below the T_m of the RNA-DNA hybrid. If mRNAs are spliced, RNA-DNA duplex regions form that are flanked by single-stranded DNA and interspersed with loops of nonhybridized single-stranded DNA. These loops occur at the splice points in the mRNA. S1 nuclease hydrolyzes the single-stranded DNA portions, generating a fully duplex structure with discontinuities in the DNA at the splice point. Under the conditions used by Berk and Sharp (1978), the RNA sequence is not cleaved. When the S1-treated hybrids are resolved by electrophoresis through neutral agarose gels, a band is observed migrating as expected for a duplex DNA molecule equal in length to the total mRNA. When these bands are excised and analyzed further by electrophoresis after alkali denaturation, the resulting single-stranded DNA fragments observed are equal in length to each of the sequences spliced together to form the mRNA molecule. If the initial hybridization is performed using restriction fragments, it is possible to map splice points.

Isolation of Specific Genes

Mildly sonicated, denatured DNA is hybridized with specific RNA. During the process of hybridization, reannealing of DNA strands occurs. Short stretches of DNA contiguous with the DNA hybridized to the RNA remain single-stranded, because the RNA is likely to interfere with reannealing of DNA up to the end of the DNA-RNA hybrid. Treatment with a single-strand-specific nuclease destroys unannealed DNA. The RNA-gene hybrids can be separated from reannealed DNA by a combination of column chromatography and density-gradient centrifugation. By using *N. crassa* nuclease, Kohne (1968), Sgaramella et al. (1968), and Marks and Spencer (1970) have succeeded in isolating rRNA-gene hybrids and tRNA-gene hybrids.

Shapiro et al. (1969) have successfully used *N. crassa* nuclease for isolating pure *lac* operon DNA as follows: Two specialized transducing phages that carry the *lac* operon of *E. coli* inserted into their DNA in opposite orientations were used. When the DNAs of these two phages were separated, each into its complementary single strands, the *lac* sense strand was found in the heavy (H) strand of one phage and in the light (L) strand of the other. By mixing separated H strands of the two phages, the *lac* sequences from the two phages annealed to form duplex DNA. The phage sequences, on the other hand, both being H-strand sequences, were noncomplementary and therefore remained single-stranded. The *lac* duplex could be freed of its four single-stranded tails by treatment with the single-strand-specific nuclease to produce pure *lac* operon DNA.

By analogous procedures, Lis and Schleif (1975) have used S1 nuclease to isolate the L-arabinose operon regulatory region of *E. coli*, and Shenk and

Berg (1976) have isolated a segment of the SV40 genome containing the origin of DNA replication.

Other Applications

Specific Cleavage of the Terminal Protein from Adenovirus-5 DNA

It is known that adenovirus-5 DNA has a protein bound to the 5′-terminus of each strand. Ariga et al. (1979) have used S1 nuclease to release the protein from the DNA. This result suggests that a single-stranded structure may be present in the region adjacent to the 5′-termini.

Characterization of a Single-stranded DNA-binding Protein

Meyer et al. (1980) have used S1 nuclease to characterize a single-stranded, DNA-binding protein (SSB) from *E. coli*. Comparison of the wild type with a temperature-sensitive mutant protein showed that the latter was less effective than the wild type in protecting single-stranded DNA from S1 digestion.

Studies of the Structure of DNA in Chromatin

The susceptibility of the DNA in chromatin to single-strand-specific nucleases was examined by Fujimoto et al. (1979) using P1 nuclease and mung bean nuclease I. A stage in the reaction exists where the size range of the solubilized products is similar for each of the two nucleases and is nearly independent of incubation time. During this stage, the chromatin fragments contain duplex DNA ranging from 10^6 to 10^7 daltons. These nucleoprotein fragments are reduced to nucleosomes and their multimers by micrococcal nuclease. Thus, chromatin contains a limited number of DNA sites that are susceptible to single-strand-specific nuclease. These sites occur at intervals of 8−80 nucleosomes and are distributed throughout the chromatin. It is of interest that the minichromosome of SV40 is highly resistant to digestion by P1 nuclease.

E. coli *DNA Membrane Attachment Site*

The chromosome of *E. coli* is attached to the cytoplasmic membrane at approximately 20 sites. On the basis of the finding that exogenous denatured DNA sticks to membranes under conditions where native DNA does not, Abe et al. (1977) have used S1 and mung bean nucleases to study the chemical properties of DNA-membrane interactions. Treatment of gently prepared lysates of *E. coli* with these nucleases resulted in the release of about 90% of the DNA from the membranes in a mode that is consistent with action of the enzymes at or near the membrane attachment sites. The released DNA had an average molecular weight of about 1.2×10^8. DNA near the origin of chromosome replication is not released from the membranes with these enzymes.

Detection of Promoter Melting by RNA Polymerases

Highly purified phage T7 RNA polymerase is frequently contaminated with single-strand-specific endonuclease, probably the T7 gene-*3* product. The endonuclease activity can be used as an enzymatic probe for the regions of double-stranded DNA melted by the polymerase. Strothkamp et al. (1980) have reported that in the absence of nucleoside triphosphates the T7 RNA polymerase melts the 10-bp promoter sequence as detected by the appearance of nuclease cleavages at all phosphodiester bonds of this region in the noncoding strand. In addition, a highly specific, much more efficient cleavage occurs in the center of twofold symmetry in the 6-bp palindrome. No cleavages occurred when the temperature was lowered from 37°C to 20°C. In the presence of NTPs, the cleavages in the initiation region disappear and the phosphodiester bonds most susceptible to nuclease attack are located in the noncoding strand downstream of the promoter in the direction of RNA synthesis.

CONCLUDING REMARKS

The following points emerge from the preceding discussion. The majority of the enzymes are sugar nonspecific and multifunctional. S1 nuclease, one form of *Neurospora* nuclease, P1 nuclease, *Alteromonas* BAL 31 nuclease, and mung bean nuclease I exhibit both endonuclease and exonuclease activities. Single-stranded nucleic acids are degraded into all four 5′-mononucleotides and small amounts of oligonucleotides by the cooperative action of both activities. The activity of the S1, *Neurospora*, P1, and mung bean nucleases are not effective on double-stranded DNA. However, nibbling of the ends does occur with large amounts of enzyme. The activity of the bacterial enzyme *Alteromonas* BAL 31 is remarkable. Linear DNAs are efficiently shortened from their ends.

The single-strand nucleases are quite thermostable enzymes, being active at 60°C or above. Relevant to this stability, P1 and mung bean nucleases contain high percentages of hydrophobic and aromatic amino acids as well as carbohydrates containing mannose, galactose, and glucosamine. Furthermore, these enzymes are zinc metalloproteins with multiple metal ion-binding sites. Removal of all zinc atoms by dialysis against EDTA not only destroys enzymatic activity but results in irreversible denaturation of the protein. S1, *Ustilago, Physarum*, and wheat nucleases are also zinc metalloproteins, whereas *Neurospora* nuclease appears to have Co^{2+} as a prosthetic group and Mg^{2+} and Ca^{2+} are essential for *Alteromonas* nuclease.

Cleavage patterns of supercoiled DNA are different depending on the enzyme. *Neurospora* and mung bean nucleases convert the DNA into a nicked form, but not into a linear form, whereas S1, *Alteromonas* nucleases, and phage T7 endonuclease convert the DNA into a linear form. It seems unlikely that this difference is merely due to the pH of the enzyme reaction.

For the production of "mature" nuclease, there is evidence in the case of *N. crassa* for the involvement of a proteolytic process. The fungus releases several nucleases that may actually be derived from a single inactive precursor polypeptide. It would be of interest to know if this is a more general phenomenon.

Concerning the biological significance of single-strand-specific nucleases, there is some evidence that they play a role in DNA repair and/or recombination. Several repair- and recombination-deficient mutants of *N. crassa* fail to release single-strand exonuclease and/or endo-exonuclease into the culture medium. These observations suggest that the mutants have a lesion in a protease(s) that controls the level of the nuclease or in some function that regulates the protease(s). Certain *Ustilago* mutant strains that are unable to carry out allelic recombination also have reduced amounts of the single-strand-specific endonuclease. *E. coli* exonuclease VII, S1, and *Ustilago* nucleases can recognize base mismatches, making them well suited to serve in a repair capacity (see Chapter 4). S1, *Neurospora*, and *Alteromonas* nucleases can recognize minor distortions in duplex DNA induced by UV light, depurination, and carcinogenic or mutagenic reagents, again implicating an involvement in repair.

By contrast, phage-T7 gene-*3* mutants synthesize only limited amounts of DNA and are defective in carrying out the degradation of host DNA, suggesting an involvement of this endonuclease in this important function.

REFERENCES

Abe, M., C. Brown, W.G. Hendrickson, D.H. Boyd, P. Clifford, R.H. Cote, and M. Schaechter. 1977. Release of *Escherichia coli* DNA from membrane complexes by single-strand endonucleases. *Proc. Natl. Acad. Sci.* **74:**2756.

Ahmad, A., W.K. Holloman, and R. Holliday. 1975. Nuclease that preferentially inactivates DNA containing mismatched bases. *Nature* **258:**54.

Ando, T. 1966. A nuclease specific for heat-denatured DNA isolated from a product of *Aspergillus oryzae*. *Biochim. Biophys. Acta* **114:**158.

Ariga, H., H. Shimojo, S. Hidaka, and K. Miura. 1979. Specific cleavage of the terminal protein from the adenovirus 5 DNA under the condition of single-strand scission by nuclease S1. *FEBS Letters* **107:**355.

Ashe, H., E. Seaman, H. Van Vunakis, and L. Levine. 1965. Characterization of a deoxyribonuclease of *Mustelus canis* liver. *Biochim. Biophys. Acta* **99:**298.

Bartok, K., M.J. Fraser, and G.C. Fareed. 1974. Detection of sequence heterology by use of the *N. crassa* nucleases. *Biochem. Biophys. Res. Commun.* **60:**507.

Beard, P., J.F. Morrow, and P. Berg. 1973. Cleavage of circular, superhelical simian virus 40 DNA to a linear duplex by S1 nuclease. *J. Virol.* **12:**1303.

Ben-Hur, E., A. Prager, and E. Riklis. 1979. Measurement of DNA crosslinks by S1 nuclease: Induction and repair in psoralen-plus-360nm light treated *Escherichia coli*. *Photochem. Photobiol.* **29:**921.

Berk, A.J. and P.A. Sharp. 1978. Spliced early mRNAs of simian virus 40. *Proc. Natl. Acad. Sci.* **75:**1274.

Center, M.S. and C.C. Richardson. 1970. An endonuclease induced after infection of *Escherichia coli* with bacteriophage T7. *J. Biol. Chem.* **245:**6285.

Center, M.S., F.W. Studier, and C.C. Richardson. 1970. The structural gene for T7 endonuclease essential for phage DNA synthesis. *Proc. Natl. Acad. Sci.* **65:**242.

Chase, J.W. and C.C. Richardson. 1974a. Exonuclease VII of *Escherichia coli*. Purification and properties. *J. Biol. Chem.* **249:**4545.

———. 1974b. Exonuclease VII of *Escherichia coli*. Mechanism of action. *J. Biol. Chem.* **249:**4553.

Chow, T.Y.-K. and M.J. Fraser. 1979. The major intracellular alkaline deoxyribonuclease activities expressed in wild type and Rec-like mutants of *Neurospora crassa*. *Can. J. Biochem.* **57:**889.

Cobianchi, F., C. Attolini, A. Falaschi, and G. Ciarrocchi. 1980. *Bacillus subtilis* deoxyribonuclease activity specific for single-stranded deoxyribonucleic acid: Cellular site and variations during germination and sporulation. *J. Bacteriol.* **141:**968.

Cobianchi, F., G. Ciarrocchi, C. Attolini, and A. Falaschi. 1978. Properties of high-molecular weight deoxyribonuclease of *Bacillus subtilis* degrading single-stranded DNA. *Eur. J. Biochem.* **84:**533.

Cuatrecasas, P., S. Fuchs, and C.B. Anfinsen. 1967. Catalytic properties and specificity of the extracellular nuclease of *Staphylococcus aureus*. *J. Biol. Chem.* **242:**1541.

Dodgson, J.B. and R.D. Wells. 1977. Action of single-strand specific nucleases on model DNA heteroduplexes of defined size and sequence. *Biochemistry* **16:**2374.

Efstratiadis, A., F.C. Kafatos, A.M. Maxam, and T. Maniatis. 1976. Enzymatic *in vitro* synthesis of globin genes. *Cell* **7:**279.

Fraser, M.J. 1979. Alkaline deoxyribonucleases released from *Neurospora crassa* mycelia: Two activities not released by mutants with multiple sensitivities to mutagens. *Nucleic Acids Res.* **6:**231.

Fraser, M.J., T.Y.-K. Chow, and E. Käfer. 1980. Nucleases and their control in wild type and *nuh* mutants of *Neurospora*. In *DNA repair and mutagenesis in eukaryotes* (ed. W.M. Generoso, M.D. Shelby, and F.J. deSerres), chapter 5, p. 63. Plenum Publishing Corp., New York.

Fraser, M.J., R. Tjeerde, and K. Matsumoto. 1976. A second form of the single-strand specific endonuclease of *Neurospora crassa* which is associated with a double-strand exonuclease. *Can. J. Biochem.* **54:**971.

Fujimoto, M., K. Fujiyama, A. Kuninaka, and H. Yoshino. 1974d. Mode of action of nuclease P1 on nucleic acids and its specificity for synthetic phosphodiesters. *Agric. Biol. Chem.* **38:**2141.

Fujimoto, M., A. Kalinski, A.E. Pritchard, D. Kowalski, and M. Laskowski, Sr. 1979. Accessibility of some regions of DNA in chromatin (chicken erythrocytes) to single-strand specific nucleases. *J. Biol. Chem.* **254:**7405.

Fujimoto, M., A. Kuninaka, and H. Yoshino. 1974a. Purification of a nuclease from *Penicillium citrinum*. *Agric. Biol. Chem.* **38:**777.

———. 1974b. Identity of phosphodiesterase and phosphomonoesterase activities with nuclease P1 (a nuclease from *Penicillium citrinum*). *Agric. Biol. Chem.* **38:**785.

———. 1974c. Substrate specificity of nuclease P1. *Agric. Biol. Chem.* **38:**1555.

———. 1975a. Some physical and chemical properties of nuclease P1. *Agric. Biol. Chem.* **39:**1991.

———. 1975b. Secondary structure of nuclease P1. *Agric. Biol. Chem.* **39:**2145.

Furuichi, Y., M. Morgan, S. Muthukrishnan, and A.J. Shatkin. 1975. Reovirus messenger RNA contains a methylated, blocked 5′-terminal structure: $m^7G(5')ppp(5')G^mpCp$-. *Proc. Natl. Acad. Sci.* **72:**362.

Germond, J.-E., V.M. Vogt, and B. Hirt. 1974. Characterization of the single-strand specific nuclease S1 activity on double-stranded supercoiled polyoma DNA. *Eur. J. Biochem.* **43:**591.

Godson, G.N. 1973. Action of the single-stranded DNA specific nuclease S1 on double-stranded DNA. *Biochim. Biophys. Acta* **308:**59.

Gray, Jr., H.B., D.A. Ostrander, J.L. Hodnett, R.J. Legerski, and D.L. Robberson. 1975. Extracellular nucleases of *Pseudomonas* BAL 31. I. Characterization of the single-strand-specific deoxyriboendonuclease and double-strand deoxyriboexonuclease. *Nucleic Acids Res.* **2:**1459.

Gray, Jr., H.B., T.P. Winston, J.L. Hodnett, R.J. Legerski, D.W. Nees, C.-F. Wel, and D.L. Robberson. 1981. The extracellular nuclease from *Alteromonas espejiana*: An enzyme highly specific nonduplex structure in nominally duplex DNAs. In *Gene amplification and analysis, vol. 2. Structural analysis of nucleic acid*, p. 169. Elsevier North-Holland, New York.

Green, C. and C. Tibbetts. 1980. Targeted deletions of sequences from closed circular DNA. *Proc. Natl. Acad. Sci.* **77:**2455.

Hanson, D.M. and J.L. Fairley. 1969. Enzymes of nucleic acid metabolism from wheat seedlings. I. Purification and general properties of associated deoxyribonuclease, ribonuclease and 3′-nucleotidase activities. *J. Biol. Chem.* **244:**2440.

Harada, F. and J.E. Dahlberg. 1975. Specific cleavage of tRNA by nuclease S1. *Nucleic Acids Res.* **2:**865.

Heflich, R.H., E. Mahoney-Leo, V.M. Maher, and J.J. McCormick. 1979. Removal of thymine-containing pyrimidine dimers from UV light-irradiated DNA by S1 endonuclease. *Photochem. Photobiol.* **30:**247.

Hofstetter, H., A. Schamböck, J. Van den Berg, and C. Weissmann. 1976. Specific excision of the inserted DNA segment from hybrid plasmids constructed by the poly(dA)·poly(dT) method. *Biochim. Biophys. Acta* **454:**587.

Holloman, W.K. 1973. Studies on a nuclease from *Ustilago maydis*. II. Substrate specificity and mode of action of the enzyme. *J. Biol. Chem.* **248:**8114.

Holloman, W.K. and R. Holliday. 1973. Studies on a nuclease from *Ustilago maydis*. I. Purification, properties and implication in recombination of the enzyme. *J. Biol. Chem.* **248:**8107.

Jayaraman, R. and E.B. Goldberg. 1969. A genetic assay for mRNAs of phage T4. *Proc. Natl. Acad. Sci.* **64:**198.

Johnson, P.H. and M. Laskowski, Sr. 1968. Sugar-unspecific mung bean nuclease I. *J. Biol. Chem.* **243:**3421.

———. 1970. Mung bean nuclease I. II. Resistance of double-stranded deoxyribonucleic acid and susceptibility of regions rich in adenosine and thymidine to enzymatic hydrolysis. *J. Biol. Chem.* **245:**891.

Käfer, E. and M. Fraser. 1979. Isolation and genetic analysis of nuclease halo (*nuh*) mutants of *Neurospora crassa*. *Molec. Gen. Genet.* **169:**117.

Kasai, K. and M. Grunberg-Manago. 1967. Sheep kidney nuclease. Hydrolysis of tRNA. *Eur. J. Biochem.* **1:**152.

Kato, A.C. and M.J. Fraser. 1973. Action of single-strand specific *Neurospora crassa* endonuclease on ultraviolet light-irradiated native DNA. *Biochim. Biophys. Acta* **312:**645.

Kato, A.C., K. Bartok, M.J. Fraser, and D.T. Denhardt. 1972. Sensitivity of superhelical DNA to a single-strand specific endonuclease. *Biochim. Biophys. Acta* **308:**68.

Khan, M.S.N. and B.E.H. Maden. 1976. Conformation of mammalian 5.8S ribosomal RNA. S1 nuclease as a probe. *FEBS Letters* **72:**105.

Kohne, D.E. 1968. Isolation and characterization of bacterial ribosomal RNA cistron. *Biophys. J.* **8:**1104.

Kroeker, W. and D. Kowalski. 1978. Gene-sized pieces produced by digestion of linear duplex DNA with mung bean nuclease. *Biochemistry* **17:**3236.

Kroeker, W.D., D.M. Hanson, and J.L. Fairley. 1975. Activity of wheat seedlings nuclease toward single-stranded nucleic acids. *J. Biol. Chem.* **250:**3767.

Kroeker, W., D. Kowalski, and M. Laskowski, Sr. 1976. Mung bean nuclease I. Terminally directed hydrolysis of native DNA. *Biochemistry* **15:**4463.

Kowalski, D., W.D. Kroeker, and M. Laskowski, Sr. 1976. Mung bean nuclease I. Physical, chemical and catalytic properties. *Biochemistry* **15:**4457.

Kuninaka, A., M. Kibi, H. Yoshino, and K. Sakaguchi. 1961. Studies on 5′-phosphodiesterase in microorganisms. Part II. Properties and application of *Penicillium citrinum* 5′-phosphodiesterase. *Agric. Biol. Chem.* **25:**693.

Kwong, S. and M.J. Fraser. 1978. *Neurospora* endoexonuclease and its inactive (precursor?) form. *Can. J. Biochem.* **56:**370.

Lau, P.P. and H.B. Gray, Jr. 1979. Extracellular nucleases of *Alteromonas espejiana* BAL 31. IV. The single-strand-specific deoxyribonuclease activity as a probe for regions of altered secondary structure in negatively and positively supercoiled closed circular DNA. *Nucleic Acids Res.* **6:**331.

Laval, F. 1974. Endonuclease activity associated with purified PM2 bacteriophage. *Proc. Natl. Acad. Sci.* **71:**4965.

Lee, S.Y., Y. Nakao, and R.M. Bock. 1968. The nucleases of yeast. II. Purification, properties and specificity of an endonuclease from yeast. *Biochim. Biophys. Acta* **151:**126.

Legerski, R.J., H.B. Gray, Jr., and D.L. Robberson. 1977. A sensitive endonuclease probe for lesions in deoxyribonucleic acid helix structure produced by carcinogenic or mutagenic agents. *J. Biol. Chem.* **252:**8740.

Legerski, R.J., J.L. Hodnett, and H.B. Gray, Jr. 1978. Extracellular nucleases of *Pseudomonas* BAL 31. III. Use of the double-strand deoxyribonuclease activity as the basis of a convenient method for the mapping of fragments of DNA produced by cleavage with restriction enzymes. *Nucleic Acids Res.* **5:**1445.

Lehman, I.R. 1966. Exonuclease I (phosphodiesterase) from *Escherichia coli*. In *Procedures in nucleic acid research* (ed. G.L. Cantoni, D.R. Davies), vol. 1, p. 203. Harper and Row, New York.

Lehman, I.R. and A.L. Nussbaum. 1964. The deoxyribonucleases of *Escherichia coli*. V. On the specificity of exonuclease I (phosphodiesterase). *J. Biol. Chem.* **239:**2628.

Lilley, D.M.J. 1980. The inverted repeat as a recognizable structural feature in supercoiled DNA molecules. *Proc. Natl. Acad. Sci.* **77:**6468.

Linn, S. 1967. An endonuclease from *Neurospora crassa* specific for polynucleotides lacking an ordered structure. *Methods Enzymol.* **12A:**247.

Linn, S. and I.R. Lehman. 1965a. An endonuclease from *Neurospora crassa* specific for polynucleotides lacking an ordered structure. I. Purification and properties of the enzyme. *J. Biol. Chem.* **240:**1287.

———. 1965b. An endonuclease from *Neurospora crassa* specific for polynucleotides lacking an ordered structure. II. Studies of enzyme specificity. *J. Biol. Chem.* **240:**1294.

Lis, J.T. and R. Schleif. 1975. The regulatory region of the L-arabinose operon: Its isolation on a 1000 base-paired fragment from DNA heteroduplexes. *J. Mol. Biol.* **95:**409.

Marks, A. and J.H. Spencer. 1970. Isolation of *Escherichia coli* transfer RNA-gene hybrids. *J. Mol. Biol.* **51:**115.

Matsumoto, K., T. Ando, H. Saito, and Y. Ikeda. 1979. Purification and characterization of an endonuclease specific for single-stranded DNA from *Bacillus subtilis* Marburg. *J. Biochem.* **86:**627.

Mechali, M., A.-M. DeRecondo, and M. Girard. 1973. Action of the S1 nuclease from *Aspergillus oryzae* on simian virus 40 supercoiled component I DNA. *Biochem. Biophys. Res. Comm.* **54:**1306.

Meyer, R.R., J. Grassberg, J.V. Scott, and A. Kornberg. 1980. A temperature-sensitive single-stranded DNA-binding protein from *Escherichia coli. J. Biol. Chem.* **255:**2897.

Mikulski, A.J. and M. Laskowski, Sr. 1970. Mung bean nuclease I. III. Purification procedure and (3′)-ω-monophosphatase activity. *J. Biol. Chem.* **245:**5026.

Ohtsubo, H. and E. Ohtsubo. 1976. Isolation of inverted repeated sequences, including IS1, IS2, and IS3, in *Escherichia coli* plasmids. *Proc. Natl. Acad. Sci.* **73:**2316.

Otto, B. and R. Knippers. 1976. An endonuclease from mouse cells specific for single-stranded DNA. *Eur. J. Biochem.* **71:**617.

Panayotatos, N. and R.D. Wells. 1981. Cruciform structures in supercoiled DNA. *Nature* **289:**466.

Pavlov, V.M., Yu. L. Lyubchenko, S.A. Borovik, and Yu. S. Lazurkin. 1977. Specific fragmentation of T7 phage DNA at low-melting sites. *Nucleic Acids Res.* **4:**4053.

Pukkila, P.J. 1978. The recognition of mismatched base pairs in DNA by DNase I from *Ustilago maydis. Molec. Gen. Genet.* **161:**245.

Sadowski, P.D. 1971. Bacteriophage T7 endonuclease I. Properties of the enzyme purified from T7 phage-infected *Escherichia coli* B. *J. Biol. Chem.* **246:**209.

Sadowski, P.D. and I. Bakyta. 1972. T4 endonuclease IV. Improved purification procedure and resolution from T4 endonuclease III. *J. Biol. Chem.* **247:**405.

Sadowski, P.D. and J. Hurwitz. 1969. Enzymatic breakage of deoxyribonucleic acid. II. Purification and properties of endonuclease IV from T4 phage-infected *Escherichia coli. J. Biol. Chem.* **244:**6192.

Schaller, H., H. Voss, and S. Gucker. 1969. Structure of the DNA of bacteriophage fd. II. Isolation and characterization of a DNA fraction with double-strand-like properties. *J. Mol. Biol.* **44:**445.

Sgaramella, V., S. Spadari, and A. Falaschi. 1968. Isolation of the hybrid between ribosomal RNA and DNA of *Bacillus subtilis. Cold Spring Harbor Symp. Quant. Biol.* **33:**839.

Shapiro, J., L. Machattie, L. Eron, G. Ihler, K. Ippen, and J. Beckwith. 1969. Isolation of pure *lac* operon DNA. *Nature* **224:**768.

Shenk, T.E. and P. Berg. 1976. Isolation and propagation of a segment of the simian virus 40 genome containing the origin of DNA replication. *Proc. Natl. Acad. Sci.* **73:**1513.

Shenk, T.E., C. Rhodes, P.W.J. Rigby, and P. Berg. 1975. Biochemical method for mapping mutational alterations in DNA with S1 nuclease: The location of deletions and temperature-sensitive mutations in simian virus 40. *Proc. Natl. Acad. Sci.* **72:**989.

Shishido, K. 1979. Location of S1 nuclease-cleavage sites on circular, superhelical DNAs between polyoma virus and simian virus 40. *Agric. Biol. Chem.* **43:**1093.

———. 1980. Relationship between S1 endonuclease-sensitivity and number of superhelical turns in a negatively-twisted DNA. *FEBS Letters* **111:**333.

Shishido, K. and T. Ando. 1972. Estimation of the double-helical content in various single-stranded nucleic acids by treatment with a single-strand specific nuclease. *Biochim. Biophys. Acta* **287:**477.

———. 1974. Cleavage of ultraviolet light-irradiated DNA by single-strand specific S1 endonuclease. *Biochem. Biophys. Res. Commun.* **59:**1380.

———. 1975a. Action of single-strand specific S1 endonuclease on locally altered structures in double-stranded DNA. *Agric. Biol. Chem.* **39:**673.

———. 1975b. Site-specific fragmentation of bacteriophage T5 DNA by single-strand specific S1 endonuclease. *Biochim. Biophys. Acta* **390:**125.

———. 1980. Enzymatic studies on binding of mutagenic principles in tryptophan pyrolysate to DNA. *Agric. Biol. Chem.* **44:**1609.

———. 1981. Efficiency of T4 DNA ligase-catalyzed end joining after S1 endonuclease treatment on duplex DNA containing single-stranded portions. *Biochim. Biophys. Acta.* **656:**123.

Shishido, K. and Y. Ikeda. 1970. Some properties of the polynucleotide segments isolated from heat-denatured DNA by digestion with a nuclease specific for single-stranded DNA. *J. Biochem.* **67:**759.

———. 1971a. Isolation of double-helical regions rich in adenine·thymine base pairing from bacteriophage f1 DNA. *J. Mol. Biol.* **55:**287.

———. 1971b. Isolation of double-helical regions rich in guanine·cytosine base pairing from bacteriophage f1 DNA. *Biochem. Biophys. Res. Comm.* **42:**482.

Shlomai, J. and A. Kornberg. 1980. An *Escherichia coli* replication protein that recognizes a unique sequence within a hairpin region in ϕX174 DNA. *Proc. Natl. Acad. Sci.* **77:**799.

Strothkamp, R.E., J.L. Oakley, and J.E. Coleman. 1980. Promoter melting by T7 ribonucleic acid polymerase as detected by single-strand endonuclease digestion. *Biochemistry* **19:**1074.

Sung, S.-C. and M. Laskowski, Sr. 1962. A nuclease from mung bean sprouts. *J. Biol. Chem.* **237:**506.

Tal, J. 1975. The cleavage of transfer RNA by a single-strand specific endonuclease from *Neurospora crassa. Nucleic Acids Res.* **2:**1073.

Talmadge, K., S. Stahl, and W. Gilbert. 1980. Eukaryotic signal sequence transports insulin antigen in *Escherichia coli. Proc. Natl. Acad. Sci.* **77:**3369.

Trilling, D.M. and H.V. Aposhian. 1968. Sequential cleavage of dinucleotides from DNA by phage SP3 DNase. *Proc. Natl. Acad. Sci.* **60:**214.

Vogt, V.M. 1973. Purification and further properties of single-strand specific nuclease from *Aspergillus oryzae. Eur. J. Biochem.* **33:**192.

Wang, J.C. 1974. Interactions between twisted DNAs and enzymes: The effects of superhelical turns. *J. Mol. Biol.* **87:**797.

Wang, E.C. and J.J. Furth. 1977. Mammalian endonuclease, DNase V. Purification and properties of enzyme of calf thymus. *J. Biol. Chem.* **252:**116.

Wang, E.C., J.J. Furth, and J.A. Rose. 1978. Purification and characterization of a DNA single-strand specific endonuclease from human cells. *Biochemistry* **17:**544.

Wani, A.A. and S.M. Hadi. 1979. Partial purification and properties of an endonuclease from germinating pea seeds specific for single-stranded DNA. *Arch. Biochem. Biophys.* **196:**138.

Watanabe, T. 1978a. Application of sheep kidney nuclease to nucleic acid research: Enzymatic preparation of deoxy-dinucleotides. *Anal. Biochem.* **88:**62.

———. 1978b. Studies on sheep kidney nuclease. II. Limited digestion of ϕX174 single-strand DNA. *Biochim. Biophys. Acta* **520:**61.

Watanabe, T. and K. Kasai. 1978. Studies on sheep kidney nuclease. I. An improved purification method and some properties. *Biochim. Biophys. Acta* **520:**52.

Waterborg, J. and C.M.A. Kuyper. 1979. Purification of an alkaline nuclease from *Physarum polycephalum. Biochim. Biophys. Acta* **571:**359.

Wickens, M.P., G.N. Buell, and R.T. Schimke. 1978. Synthesis of double-stranded DNA complementary to lysozyme, ovomucoid, and ovalbumin mRNAs. *J. Biol. Chem.* **253:**2483.

Wiegand, R.C., G.N. Godson, and C.M. Radding. 1975. Specificity of the S1 nuclease from *Aspergillus oryzae. J. Biol. Chem.* **250:**8848.

Wilson, D.A. and C.A. Thomas, Jr. 1974. Palindromes in chromosomes. *J. Mol. Biol.* **84:**115.

Yamasaki, H., P. Pulkrabek, D. Grunberger, and I.B. Weinstein. 1977. Differential excision from DNA of the C-8 and N^2 guanosine adducts of *N*-acetyl-2-aminofluorene by single-strand specific endonucleases. *Cancer Res.* **37:**3756.

Zain, E.Z., H. Tenenhouse, and M.J. Fraser. 1972. An exonuclease of *Neurospora crassa* specific for single-stranded nucleic acids. *Biochim. Biophys. Acta* **259:**50.

Ziff, E.B., J.W. Sedat, and F. Galibert. 1973. Determination of the nucleotide sequence of a fragment of bacteriophage ϕX174 DNA. *Nature New Biol.* **241:**34.

Nuclease Activities Involved in DNA Replication

David R. Brown, Jerard Hurwitz, Danny Reinberg,* and Stephen L. Zipursky†
Department of Developmental Biology and Cancer
Albert Einstein College of Medicine
Bronx, New York 10461

SITE-SPECIFIC ENDONUCLEASES REQUIRED FOR DNA REPLICATION

The genomes of both the icosahedral and filamentous single-stranded DNA bacteriophages code for site-specific endonucleases essential for duplex phage DNA replication in vivo. Two prototypic and extensively studied endonucleases are the ϕX174 gene-*A* protein and the fd gene-*2* protein. The proteins function at equivalent points in the closely related life cycles of these two phage types. They initiate duplex DNA synthesis by introducing a single specific nick in the viral (+) strand of supercoiled replicative form (RFI)

Present address: *Department of Molecular Biology, University of California, Berkeley, Berkeley, California 94702; †Division of Biology, California Institute of Technology, Pasadena, California 90025.

DNA, and terminate viral strand synthesis such that unit-length circular progeny molecules are produced. Early attempts to understand the properties and functions of these endonucleases involved the analysis of effects of mutations in the ϕX gene *A* and the fd gene *2* on specific aspects of the phage life cycles. More recently, emphasis has been placed on the characterization of activities associated with the purified gene products. The roles played by the gene-*A* protein and the gene-*2* protein in the life cycles of icosahedral and filamentous phages, respectively, will be described below, followed by a review of the currently characterized enzymatic activities. In addition, the ϕX174 *A** protein, a related but poorly understood endonuclease, will be discussed. Table 1 summarizes the properties of the ϕ X gene-*A* and -*A** proteins and the fd gene-2 protein, and serves as an outline of the information to follow.

Role of Site-specific Endonucleases in Phage Life Cycles

The viral DNA of single-stranded circular (SS[C]) DNA phages replicates by way of a duplex intermediate, the replicative form (RF) (Sinsheimer et al. 1962). Synthesis of the parental RFI DNA from infecting SS(C) DNA molecules (SS→RF), semiconservative replication of RF DNA (RF→RF), and production of progeny SS(C) molecules from RF templates (RF→SS) define the three basic stages of the life cycles of icosahedral and filamentous phages (Sinsheimer 1968). Only host functions are required for SS→RF replication of both ϕX174 (Denhardt 1975) and fd (Geider and Kornberg 1974) DNA. However, the phage-encoded endonucleases are required for both RF→RF and RF→SS DNA replication.

RF→RF Replication

Mutations in the *A* gene of the icosahedral phage S13 block RF replication in vivo (Tessman 1966). The *A* gene product was shown to be a *cis*-acting endonuclease (Tessman 1966; Sinsheimer 1968) which converted RFI DNA to the relaxed RFII derivative (Franke and Ray 1971, 1972). The protein introduced a single nick specifically into the viral strand of RFI DNA (Franke and Ray 1972) within the *A* cistron at the origin of viral-strand replication (Baas and Jansz 1972a,b; Eisenberg and Denhardt 1974; Johnson and Sinsheimer 1974; Baas et al. 1976). In vitro studies with the purified ϕX174 gene-*A* protein (Henry and Knippers 1974; Eisenberg et al. 1976a; Ikeda et al. 1976) allowed the determination of the precise cleavage site (Langeveld et al. 1978).

The filamentous phage M13 gene-*2* product was also implicated as an endonuclease when it was shown to be required in vivo for conversion of RFI DNA to the RFII derivative (Fidanian and Ray 1972; Lin and Pratt 1972). Studies with the purified fd gene-*2* protein showed that it specifically cleaved the viral strand of fd RFI DNA (Meyer and Geider 1979b) at the origin of fd viral strand replication (Meyer and Geider 1978; Meyer et al. 1979; Schaller 1979).

The association of both the gene-*A* and gene-*2* endonucleases with the initiation of DNA replication fit well with Gilbert and Dressler's (1969) rolling circle model of DNA replication. The model involved the elongation of a 3′-hydroxyl DNA terminus at a nick within a circular duplex DNA template.

RF→SS Replication

In vivo studies suggested that both fd gene-*2* protein (Lin and Pratt 1972; Mazur and Model 1973) and ϕX174 gene-*A* protein (Fujisawa and Hayashi 1976) are involved in SS (viral) DNA synthesis. The gene-*A* and gene-*2* proteins are also essential for synthesis in vitro of SS(C) DNA from RFI templates (Eisenberg et al. 1977; Harth et al. 1981; Meyer et al. 1981).

ϕX174 Gene-*A* Protein

The bacteriophage ϕX174 gene-*A* protein is a 59-kilodalton polypeptide. The protein is required early in the phage life cycle and accumulates to high levels in the host cell soon after infection.

Purification Strategies

The gene-*A* protein was required for ϕX DNA replication in vitro (Eisenberg et al. 1976b; Ikeda et al. 1976; Sumida-Yasumoto et al. 1976; Scott et al. 1977; Sumida-Yasumoto and Hurwitz 1977). Based on this requirement, two assays for the protein were developed, a complementation assay (Ikeda et al. 1976) using extracts from infected and uninfected cells, and a reconstitution assay (Eisenberg and Kornberg 1979) with purified enzymes. The *A* protein was purified to near homogeneity as an activity essential for specific ϕX DNA synthesis in vitro.

The endonuclease activity of the ϕX *A* protein is highly specific for ϕX RFI DNA (Henry and Knippers 1974; Ikeda et al. 1976, 1979; Langeveld et al. 1978; van Mansfeld et al. 1979). This specificity was exploited by Langeveld et al. (1980) in an attempt to obtain an *A* protein preparation free of the partially copurifying *A** protein. The *A* and *A** activities were separated based on the distinct products of ϕX RFI DNA cleavage formed in the presence of Mg^{2+} or Mn^{2+} cations.

DNA Binding Properties in the Absence of Cleavage

Ikeda et al. (1979) obtained evidence in support of specific and reversible binding of the ϕX *A* protein to ϕX RFI DNA. Using a Millipore filter binding assay, 95% of the protein preparation bound ϕX RFI DNA, but not ϕX RFI′ (relaxed, covalently closed, circular duplex), fd RFI, Colicin EI form I, or PM2 DNAs. In the presence of excess DNA, approximately one

Table 1 Properties of the ϕX gene-*A* protein, the ϕX *A** protein, and the fd gene-*2* protein

	ϕX *A* protein	ϕX *A** protein	fd gene *2* protein
Mass (daltons)	59,000	35,000	46,000
Purification assays	ϕX DNA synthesis (complementation and reconstitution)	filter binding	fd DNA synthesis (complementation)
		inhibition of ϕX RF→SS DNA synthesis	
	specific ϕX RFI DNA cleavage	ϕX RFI DNA cleavage with Mn^{2+}, not Mg^{2+}	
DNA binding in the absence of cleavage			
Binds duplex DNA	yes	yes	yes
Mg^{2+}	strong (K_d 5×10^{-8}) reversible, polymers	nonspecific	protects complementary strand from BAL 31
No cation	none	nonspecific, noncooperative	not studied
Binds single-stranded DNA	yes	yes	not studied
Mg^{2+}	not studied	not studied	—
No cation	nonspecific, noncooperative	nonspecific, noncooperative, stronger interaction with ϕX SS than ϕX DS	—
Endonuclease activity on RFI DNA	yes	yes	yes
Sequence specificity			
Mg^{2+}	specific	no activity	specific
Mn^{2+}	not	moderately specific	less specific

Covalent DNA-protein complex	yes	not studied	no
Blocked 5′ end at nick	yes	not studied	no
Free 3′ end at nick	yes	not studied	yes
Products			
(Mg^{2+})	RFII · A, RFI′ · A	none	RFII, RFI′
(Mn^{2+})	RFII · A, RFI′ · A (?)	RFII · A, RFIII · A, RFI′	RFIII
Inhibition by SS DNA	yes	not studied	no
Endonuclease activity on SS DNA	yes	yes	yes; only at replication fork
Sequence specificity			
Mg^{2+}	specific	moderately specific	specific
Mn^{2+}	not studied	not studied	not specific
Covalent DNA-protein complex	yes	yes	no
Blocked 5′ end of product	yes	yes	no
Free 3′ end of product	yes	yes	yes
Role in DNA replication			
Specific cleavage of viral strand in RFI	yes	no	yes
Interaction with *E. coli* *rep* protein	yes	not studied	yes
Circularize displaced viral strands	yes	unclear	yes
Inhibit DNA replication in vitro	no	yes	no

protein molecule was bound per DNA-protein complex, with an apparent K_d of 5×10^{-8} M. In the absence of excess DNA, between 6 and 12 protein molecules were bound per DNA-protein complex. The incubation of ϕX RFI DNA with excess *A* protein under conditions which inhibited nonspecific aggregation, followed by cross-linking with dimethyl suberimidate, yielded protein complexes containing up to 10 protomers. Complex formation required ϕX RFI DNA. If excess ϕX RFI DNA was added just prior to the cross-linking reaction, no oligomeric RFI DNA-dependent complexes were detected. These results suggested that the ϕX *A* protein reversibly formed polymeric complexes specifically with ϕX RFI DNA.

The binding of the ϕX *A* protein to single-stranded (SS) and double-stranded (DS) DNA has been recently studied with electron microscopy (A. van der Ende et al., in prep.). Results suggested that in the absence of divalent cations, the *A* protein bound ϕX SS DNA nonspecifically and probably noncooperatively. Under similar conditions, no binding of ϕX *A* protein to ϕX RFI or ϕX RFII DNA was observed. Incubation in the presence of Mg^{2+} yielded only endonucleolytic cleavage products. The findings are difficult to interpret, as electron microscopic studies do not allow the determination of binding constants and the physiologic significance of binding in the absence of Mg^{2+} is unclear.

Endonuclease Activity on Duplex DNA

The ϕX *A* protein specifically nicked ϕX RFI DNA at a single site in the viral (+) strand (Franke and Ray 1972; Henry and Knippers 1974; Ikeda et al. 1976, 1979; Langeveld et al. 1978), but was inactive against ϕX RFI′, Colicin EI form I, fd RFI, or PM2 DNAs (Ikeda et al. 1976, 1979; Marians et al. 1977). The cleavage product was a stable RFII DNA-*A* protein complex (RFII·*A*) containing a free 3′-hydroxyl DNA terminus and the *A* protein covalently bound to the 5′ end of the DNA by an alkali-stable, presumably phosphodiester linkage (Eisenberg and Kornberg 1979; Ikeda et al. 1979). When purified by velocity sedimentation through sucrose gradients, the RFII·*A* complex functioned as a highly active template for ϕX DNA synthesis in vitro, in the absence of added free *A* protein (Ikeda et al. 1976). Yields in excess of 5 SS(C) product molecules per input RFII·*A* complex were obtained (Eisenberg and Kornberg 1979). Careful stoichiometric studies showed that the *A* protein was required in a 10–20-fold molar excess over ϕXRFI DNA for quantitative formation of the RFII·*A* cleavage product (Ikeda et al. 1979). Although 95% of the *A* protein preparation was active in filter binding assays, experiments could not easily be controlled for a preferential loss of nuclease activity. However, an analysis of both the binding data and cleavage obtained from products of DNA titration experiments suggested that no loss of nuclease activity had occurred. Following sedimentation of the RFII·*A* complex through sucrose gradients in the presence of 0.2 M NaCl, the ϕX RFII DNA and *A* protein were recovered in a molar ratio of approximately 1 to 1 (Eisenberg and Kornberg 1979). This purified RFII·*A* complex was active in supporting net synthesis in vitro of ϕX SS(C) DNA. These experiments suggest that following the initial cleavage event, a molar excess of *A* protein was not required for elongation and termination of ϕX DNA replication.

Under conditions of DNA excess, an RFI′·*A* protein covalent complex was also isolated. The covalently closed, relaxed duplex circles were converted to RFII structures with proteinase K or phenol treatment whereas detergents had no effect (Ikeda et al. 1976, 1979). This RFI′·*A* complex may have represented an intermediate of the so-called termination or ligation reaction in which displaced unit-length SS DNA is circularized to form the SS(C) product. These results suggested that the *A* protein was covalently bound to both 3′ and 5′ termini of the DNA discontinuity, possibly at each of two active sites (first proposed by Eisenberg et al. 1977). RFI′ DNA was also formed in reaction mixtures containing ϕX RFI DNA, the *A* or *A** proteins, and Mn^{2+} (Langeveld et al. 1980). The sensitivity to proteinase K of this RFI′ product was not tested. This product may have represented a covalently closed DNA-protein complex, or may have reflected a nicking-closing activity of the *A* and *A** proteins in the presence of Mn^{2+}.

The sequence specificity for the endonuclease activity of the ϕX *A* protein on RFI DNA has been studied extensively. RFI DNAs of several related icosahedral phages including ϕX174, G4, St-1, α3, U3, G14, and ϕX were all specifically nicked by the ϕX *A* protein (Duguet et al. 1979; van Mansfeld et al. 1979; Heidekamp et al. 1980; F. Heidekamp et al. 1982). The nucleotide sequences surrounding the nick sites have been determined and compared (Fiddes et al. 1978; Godson et al. 1978; Langeveld et al. 1978; Sanger et al. 1978; van Mansfeld et al. 1979; Heidekamp et al. 1980, 1982). In each case studied, the nick was introduced at the origin of replication between the seventh and eighth residues of an AT-rich, 30-nucleotide, conserved sequence (Table 2). The origin sequences of the phages St-1 and α3 were the only exceptions in that they shared two identical base substitutions within the conserved sequence (Heidekamp et al. 1980, 1982). This 70% AT, 30-nucleotide, conserved sequence, which is flanked by GC-rich regions, was clearly implicated as a target for the specific endonuclease activity of the ϕX *A* protein.

The results of several workers suggest that the duplex DNA target sequence of the ϕX *A* protein must be present in a partially denatured or single-stranded conformation to participate in the cleavage reaction. AT-rich regions in supercoiled DNA are known to exhibit partially single-stranded character (Jacob et al. 1974). ϕX174 duplex DNA is only nicked by the ϕX *A* protein if supercoiled (Marians et al. 1977). Synthetic oligonucleotides representing portions of the 30-nucleotide conserved sequence have been constructed (van Mansfeld et al. 1980) and cloned into the plasmid pACYC177 (Heidekamp et al. 1981). Although the single-stranded decamer, homologous to the first 10

Table 2 Origin of ϕX viral strand DNA replication

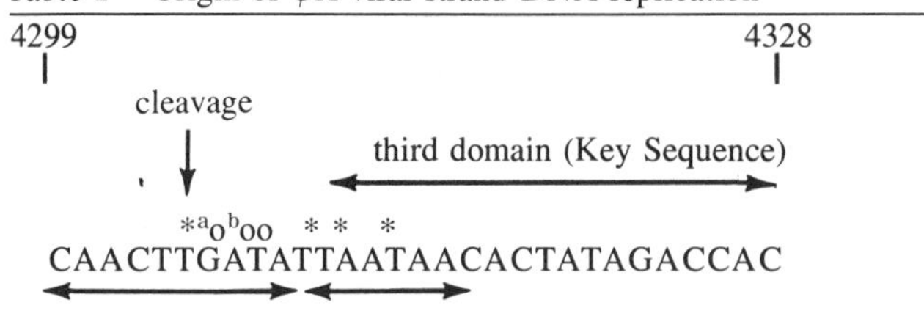

Essential features of the origin of ϕX RF DNA replication.

[a] * Refers to nucleotides that can be changed without effect on ϕX gene-*A* protein nicking or DNA replication ability: at position 4304, T → C; at position 4309, T → A; at position 4310, T → C; at position 4312, A → G.

[b] o Refers to nucleotides that cannot be changed without effect on ϕX gene-*A* protein nicking or DNA replication ability (lethal mutations): at position 4305, G → A; at position 4306, A → G or T; at position 4307, T → G.

Data from Baas et al. (1981) and Heidekamp et al. (1980, 1982).

residues of the 30-nucleotide conserved sequence, was efficiently cleaved between the seventh and eighth positions by the ϕX *A* protein (van Mansfeld et al. 1980), the duplex vigesimer, comprising the first 20 residues, cloned into the RFI plasmid, could not be cut (Heidekamp et al. 1981). Similarly, the single-stranded hexadecamer, homologous to the first 16 residues of the conserved sequence, was accurately cut by the ϕX *A* protein, but an RFI plasmid carrying this hexadecamer sequence within an AT-rich region of 33 nucleotides was not nicked by the *A* protein.

Recent studies in vivo (Baas et al. 1981) corroborated the above in vitro findings by characterizing certain residues within the 30-nucleotide conserved region as nonessential or possibly essential in the determination of a biologically active ϕX174 replication origin (Table 2). *E. coli* spheroplasts were transfected with ϕX174 SS(C) complementary-strand DNA containing preselected single nucleotide changes in the origin region. Transfecting with DNA containing base changes at the sixth or twelfth positions of the 30-nucleotide conserved sequence gave rise to progeny phage. As both mutations were third-position codon changes in the *A* cistron, neither affected the *A*-protein amino acid sequence. No progeny phage were isolated following transfection with DNA containing single nucleotide changes at the seventh, eighth, or ninth residues of the 30-nucleotide sequence (the *A* protein cleaves between the seventh and eighth residues). The four mutations tested all altered the *A* cistron such that single amino acid changes in the *A* protein would ensue. The failure to detect progeny phage may have resulted from altered DNA structure, altered *A* protein structure, or both.

From the above studies, a model has been developed in which the 30-nucleotide conserved sequence has been divided into three domains as summarized in Table 2 (adapted from Heidekamp et al. 1982). The first domain spanned residues 1–10 of the 30-nucleotide conserved sequence, contained the site nicked in ϕXRFI DNA by the ϕX *A* protein, and, when present as a single-stranded decanucleotide, was cleaved in vitro by the *A* protein. A T→C transition within the first domain, which did not preclude origin activity in vivo, has been identified (Baas et al. 1981), though as noted above changes at positions 7, 8, and 9 did prevent activity. The second domain was defined as a short, variable, AT-rich sequence of unspecified length, immediately 3′ to the first domain. Silent mutations within this second domain at positions 11, 12, and 14 of the 30-nucleotide conserved sequence have been found (Heidekamp et al. 1980; Baas et al. 1981). Heidekamp et al. (1981, 1982) have invoked the requirement for a third domain since recombinant RFI plasmids containing inserts homologous to sequences of the first and second domains, within AT-rich regions, were not nicked by the *A* protein. This sequence-specific third domain, which has not been studied directly, was called the "key sequence," and was said to be essential for recognition by the *A* protein of the first domain in duplex supercoiled DNA (Heidekamp et al. 1981, 1982). Following a site-specific interaction with the third domain, the *A* protein was thought to induce local unwinding or "melting" of the DNA helix, facilitated by the AT-richness of the second domain, exposing the first domain in single-stranded form and allowing cleavage to ensue. The exact sequence requirements for this third domain are currently under study.

Recombinant plasmids containing inserts derived from the viral-strand replication origin regions of icosahedral phage DNAs have been constructed which, as RFI molecules, are reactive with the ϕX *A* protein. Plasmids containing the ϕX174 origin region were specifically nicked in vivo (Weisbeek et al. 1981) and in vitro by the ϕX *A* protein and functioned in vitro as replication templates in DNA-synthesizing systems which required the ϕX *A* protein (Zipursky et al. 1980). Recombinant plasmids carrying the ϕX174 *Hae*III-6b fragment, an insert 281 bp in length that contained the 30-nucleotide conserved sequence, behaved exactly as did ϕX174 DNA in these in vitro systems (Zipursky et al. 1980; Reinberg et al. 1981). Furthermore, when recombinant plasmids containing two such inserts were incubated in vitro in the DNA replication reaction mixtures, initiation and termination events were shown to occur at either *A*-protein recognition sequence (Reinberg et al. 1982). A series of recombinant plasmids (pIPori series, *i*cosahedral *p*hage *ori*gin) have recently been constructed that contain inserts of 29–49 bp in length that share homology with the 30-nucleotide conserved sequence (Brown et al. 1982a,b). While all RFI plasmids of the series were efficiently nicked in vitro by the ϕX *A* protein, they were utilized with varying efficiencies as templates in ϕX-specific in vitro DNA replication systems. Homologies to positions 1, 29, or 30 of the 30-nucleotide conserved sequence were not required for cleavage by the ϕX *A* protein, however homology to position 1 but not 29 or 30 was required for efficient template utilization (Brown et al. 1982b).

In summary, supercoiled recombinant plasmids containing inserts homologous to the first 20 residues of the 30-nucleotide conserved sequence were unreactive in vitro with the ϕX *A* protein (Heidekamp et al. 1981). Plasmids containing inserts homologous to residues 2–30 or 1–28 of the 30-nucleotide conserved sequence (Brown et al. 1982b) were nicked. The smallest insert studied which conferred complete *A*-protein-specific activity in vitro was 30 nucleotides in length and homologous to residues 1–28 of the 30-nucleotide conserved sequence (Brown et al. 1982b).

Endonuclease Activity on Single-Stranded DNA

The ϕX *A* protein cleaves ϕX viral SS(C) DNA only once, and at precisely the same site where ϕX RFI DNA is nicked (Langeveld et al. 1979). The product is a linear unit-length SS DNA-protein complex (SS[L]·*A*) with a free 3′-hydroxyl terminus and an alkali-stable covalent linkage between the *A* protein and the 5′ DNA terminus (Eisenberg 1980). With a large molar excess of *A* protein over ϕX SS(C) DNA, Eisenberg (1980) reported the formation of smaller than unit-length linear DNA fragments, a result possibly due to contaminating *A** protein (see below) in the *A*-protein preparation (Langeveld et al. 1979, 1981). Studies with synthetic DNA oligonucleotides described in the previous section, showed that the smallest SS DNA molecule efficiently cleaved by the ϕX *A* protein was the decamer homologous to the first 10 residues of the 30-nucleotide conserved region (van Mansfeld et al. 1980). The octamer (homologous to residues 3—10) was cleaved at a very low efficiency, and again may have reflected a contaminating *A** activity (the *A** protein cleaved the octamer efficiently).

ϕX174 *A** Protein

The ϕX *A** protein, a 35-kilodalton polypeptide of unknown function in vivo, is derived from the ϕX174 gene *A*. Translation of the *A** protein is initiated from an internal start codon, read in the same frame as the ϕX *A* protein, and shares the same stop codon (Linney and Hayashi 1973, 1974; Sanger et al. 1977).

Purification Strategies

The *A** protein has several properties that are distinct from the *A* protein and have been exploited in developing assays for *A** protein purification. Ikeda et al. (1979) found that the *A** protein avidly bound all superhelical and relaxed DS DNAs tested, whereas the *A* protein bound only ϕX RFI DNA. In relatively pure preparations, this nonspecific DS DNA binding activity provided an assay for the *A** protein and has been exploited for its purification (Ikeda et al. 1979). The distinct endonuclease activities of the *A* and *A** proteins, directed against ϕX RFI DNA, in the presence of Mg^{2+} or Mn^{2+}

were useful in distinguishing these two proteins during purification (Langeveld et al. 1980). In the presence of Mg^{2+}, the *A* protein cleaved ϕX RFI DNA to the RFII derivative, whereas the *A** protein was inactive. When Mn^{2+} was used as the divalent cation, the *A* protein cleavage products were RFII and RFI′ DNA; under similar conditions, the *A** protein gave rise to RFII, RFIII, and RFI′ DNA. Eisenberg and Ascarelli (1981) reported that ϕX SS(C) DNA synthesis in vitro was inhibited by high levels of the *A** protein. This inhibition of ϕX DNA synthesis in vitro was used as an assay system to measure *A** activity in crude extracts, and to purify the *A** protein.

DNA Binding Properties

The binding of *A** protein to DNA has been studied using electron microscopy and velocity sedimentation analysis (A. van der Ende et al., in prep.) and using filter binding assays (Ikeda et al. 1979; Eisenberg and Ascarelli 1981). The *A** protein binds to circular or linear duplex DNA nonspecifically in the presence (Ikeda et al. 1979) or absence (Eisenberg and Ascarelli 1981) of Mg^{2+}. A. van der Ende et al. (in prep.) observed nonspecific and noncooperative binding of the *A** protein to linearized ϕX RFIII DNA, and to ϕX SS(C) DNA in the absence of divalent cations. In competitive binding and sedimentation experiments, the *A** protein had a higher affinity for the ϕX SS DNA than for the linearized ϕX RFIII DNA.

The SS DNA-specific endonuclease and the duplex DNA binding activities of the purified *A** protein have suggested several possible in vivo functions. Martin and Godson (1975) proposed that the protein shuts down host DNA replication. The *A** protein could inhibit host DNA synthesis by cleaving SS DNA at replication forks (Langeveld et al. 1981) or by binding to DS DNA (Eisenberg and Ascarelli 1981). As high levels of *A** protein occur only late in infection, functional roles in RF→SS(C) replication or SS(C) DNA packaging into phage heads have been proposed (Fujisawa and Hayashi 1976). Functions in the maturation or processing of SS(C) viral DNA, or in regulation of the shift from RF→RF to RF→SS(C) modes of DNA synthesis have also been suggested (Langeveld et al. 1979, 1981).

Endonuclease Activity On Duplex DNA

In the presence of Mg^{2+}, no DS DNA-directed endonucleolytic activities could be attributed to the ϕX *A** protein (Langeveld et al. 1980), nor could the *A** protein substitute for the *A* protein in DNA replication systems in vitro (Ikeda et al. 1979). But when incubated with ϕX RFI DNA and Mn^{2+}, RFII, RFIII, and RFI′ products were formed (Langeveld et al. 1980). Upon further study (van der Ende et al. 1981), the product distribution was found to depend on the input ϕX RFI DNA to *A** protein ratio, with the yield of RFIII increasing in roughly linear proportion to the *A** protein level. The RFI′ product was shown to remain covalently closed following proteinase-K treatment, and therefore did not resemble the RFI′·*A* complex previously described by Ikeda et al. (1976, 1979). Both the RFII and RFIII products were visualized as DNA-protein complexes (RFII·*A**, RFIII·*A**) with electron microscopy. The *A** protein apparently recognized secondary cleavage sites in ϕX RFI DNA in the presence of Mn^{2+}, as only 80% of the RFII·*A** complexes contained origin-specific nicks. In contrast, all of the RFIII·*A** molecules examined were cleaved at the origin. One terminus of the RFIII product was covalently bound to the *A** protein, and the two DNA strands at the opposite terminus were covalently joined in a proteinase-K-resistant linkage. The formation of these proteinase-K-resistant RFI′ and terminally cross-linked RFIII·*A** products in the presence of Mn^{2+} may have reflected an aberrant expression of a ϕX *A* protein-related ligation activity of the *A** protein.

Endonuclease Activity On Single-Stranded DNA

Langeveld et al. (1979) have shown that in the presence of Mg^{2+}, the *A** protein was a SS DNA-specific endonuclease. The cleavage product was a stable linear DNA-protein complex (SS[L]·*A**) with a free 3′-hydroxyl terminus and covalent linkage between the *A** protein and the 5′-terminus (Eisenberg and Finer 1980). The *A** protein preferentially cut ϕX SS(C) DNA at the replication origin (the site at which the *A* protein nicked ϕX RFI DNA), although secondary sites were cut as well (Langeveld et al. 1979). Thirty-four of these secondary sites have been mapped in detail (Langeveld et al. 1981) and the flanking nucleotide sequences were found to possess variable and sometimes quite limited homology to the first 12 residues of the 30-nucleotide conserved sequence (5′-CAACTTG↓ATATT-3′; *A*-protein cleavage site shown by arrow). Some flanking sequences were homologous only to the left portion (5′ of the cleavage site), and others only to the right portion (3′ of the cleavage site) of the above sequence. From these observations, Langeveld et al. (1981) proposed that the ϕX *A** protein, and perhaps the *A* protein as well, contained two domains that were both involved in nicking and ligation functions. The authors suggested that one domain interacted specifically with the sequence 5′-CAACTTG-3′, and the other domain with the sequence 5′-GATATT-3′. Interactions between these domains were discussed in terms of the multiple functions attributed to the ϕX *A* protein by Eisenberg et al. (1977).

Recent studies by Eisenberg and Finer (1980) suggest that the *A** protein may retain a ligation activity in the presence of Mg^{2+}. SS(L)·*A** complexes purified by gel filtration were incubated in the presence of Mg^{2+}, and were subsequently analyzed by velocity sedimentation, by banding in CsCl density gradients and with electron microscopy. Results suggested that SS(L)·*A** complexes gave rise to SS(C) DNA plus free *A** protein in a Mg^{2+}-dependent reaction. However, the proteinase-K sensitivity of putative SS(C) DNA products was not tested.

As mentioned above, the synthetic single-stranded octamer (homologous to residues 3–10 of the 30-nucleotide conserved sequence) was efficiently cleaved by the *A** protein and the decamer (homologous with residues 1–10) was similarly cleaved by the *A* protein, but the *A** protein failed to cut ϕX RFI DNA, (Langeveld et al. 1980). Based on these observations, Langeveld et al. (1979, 1981) have speculated that the amino acid sequence shared by the *A* and *A** proteins contains the endonucleolytic active site. They further suggest that the *N*-terminal domain of the *A* protein that is absent in the *A** protein confers the ability to nick ϕX RFI DNA specifically. This *N*-terminal domain may interact with the "key sequence" (Heidekamp et al. 1981) within the 30-nucleotide conserved region, and induce the DNA helix destabilization or localized melting apparently required for expression of the SS DNA-directed endonuclease activity.

fd Gene-2 Protein

Purification Strategy

The gene-*2* product is required for filamentous phage RF DNA replication in vivo (Fidanian and Ray 1972; Lin and Pratt 1972). In addition, specific fd DNA synthesis in vitro was observed only in the presence of extracts derived from fd-infected cells. As with the ϕX *A* protein (Ikeda et al. 1976, 1979), the fd gene-*2* product could therefore be purified as an activity required for specific synthesis of fd DNA in vitro using a complementation assay (Meyer and Geider 1979a).

Endonucleolytic Activities

Meyer and Geider (1979b) have characterized the endonucleolytic properties of the purified fd gene-*2* protein. Like the ϕX *A* protein, the enzyme has an absolute requirement for divalent cations, and in the presence of Mg^{2+}, introduces a single specific nick in the viral strand of fd RFI DNA. The nick has been precisely mapped to the origin of viral strand replication, and was closely related to two regions of dyad symmetry (Meyer et al. 1979; Schaller 1979). The cleavage products were singly nicked RFII DNA with free 3′-hydroxyl and 5′-phosphate termini, and RFI′ DNA which was insensitive to proteinase K, phenol, salt, detergents, and alkali. The ratio of RFII to RFI′ products was influenced by the Mg^{2+} concentration. The endonuclease activity has an absolute requirement for substrate superhelicity and shows maximal activity with superhelical densities in the physiological range. Only the RFI DNAs of fd and the related filamentous phages f1 and M13 are nicked. Surprisingly, fd SS(C) viral DNA is not cleaved. Detailed stoichiometric studies have not been done, but preliminary studies suggested that a molar excess of the gene-*2* protein over fd RFI DNA was required for quantitative conversion to the RFII derivative.

The endonucleolytic specificity of the gene-*2* protein was altered by non-physiological cations. In the presence of Ca^{2+} or Ba^{2+}, RFII, but not RFI′ DNA products were formed. Substituting Mn^{2+} for Mg^{2+} resulted in complete conversion of fd RFI to fd RFIII DNA, partial conversion of ϕX RFI to ϕX RFII DNA, and cleavage of SS DNA into small fragments.

An important distinction between the fd gene-*2* protein and the ϕX gene-*A* protein is the failure of the gene-*2* protein to cleave fd SS(C) DNA. The specificity of gene-*2* protein for superhelical substrates cannot be explained simply as the requirement for a target sequence with "single-stranded" character. The protein must interact with some specific higher-order structure, or helix distortion, present only in RFI or replicating DNA.

Behavior of fd Gene-2 Protein In DNA Replication Systems In Vitro

The reconstituted system of purified *E. coli* enzymes which supported RF→SS(C) ϕX174 DNA replication in vitro was adapted to the in vitro replication of fd DNA (Geider and Meyer 1979; Geider et al. 1982). The *E. coli* DNA polymerase III elongation system and *E. coli rep* and SSB proteins catalyzed the synthesis of unit-length fd SS(C) and SS(L) DNA from fd RFI DNA templates in fd gene-*2* protein-dependent reactions (Geider et al. 1982). Purified fd $RFII_0$ DNA (fd RFI DNA specifically nicked by the gene-*2* protein) was also utilized as template in this system, provided that the fd gene-*2* protein was added to reaction mixtures. fd RFI or $RFII_0$ DNAs were fully unwound in the presence of the *E. coli rep* and SSB proteins, the fd gene-*2* protein, and ATP, but randomly nicked fd RFII DNA was unreactive. The complementary (−) strand of fd $RFII_0$ DNA opposite from the specific nick could be cleaved by the single-strand-specific endonuclease activity of BAL 31 in the absence, but not in the presence, of the gene-*2* protein. The gene-*2* protein did not protect randomly nicked fd RFII DNA from linearization to the RFIII derivative by BAL 31. These observations suggested that following cleavage of fd RFI DNA, the gene-*2* protein remained noncovalently associated with the DNA at the origin of leading strand synthesis in a manner that provided an entry site for the *rep* protein helicase activity, and protected the (−) strand from endonucleolytic attack. This noncovalent complex between fd RFII DNA and the gene-*2* protein could be reconstituted from separately isolated $RFII_0$ DNA and gene-*2* protein. Removal of the 5′-terminal phosphate from $RFII_0$ DNA did not impede the unwinding reaction, however when dephosphorylated $RFII_0$ DNA was used as template in RF SS DNA replication reactions in vitro, unit-length fd SS(L) DNA molecules were the major products. Therefore, the terminal phosphate of fd $RFII_0$ DNA is not required for expression of the gene-*2* endonuclease activity, but may be required for the circularization reaction.

Replication of fd DNA in vitro has also been obtained using reconstituted systems of purified phage T4 (Meyer et al. 1981) and phage T7 (Harth et al. 1981) enzymes. In the absence of fd gene-*2* protein, nicked fd RFII templates were replicated by the T4 system, and rolling circles with single-stranded tails of greater than unit length were formed. If gene-*2* protein was added to these reaction mixtures postreplicatively, the single-stranded tails were not cleaved, consistent with the inactivity of gene-*2* protein against fd SS(C) DNA. However, if gene-*2* protein was present during replication, unit length SS(L) and SS(C) products were formed. The presence of fd gene-*2* protein in a roughly threefold molar excess over template DNA reduced the rate of deoxyribonucleotide incorporation by 50%. A molar excess of approximately 25-fold was required for maximal processing of nascent DNA strands to SS(L) and SS(C) products and produced no further decrease in total DNA synthesis. The inclusion of fd gene-*2* protein also resulted in an enrichment of rolling circle intermediates with exactly unit-length tails as seen with electron microscopy. The T7 enzyme system also supported specific fd SS DNA synthesis in reactions that required the fd gene-*2* protein. Most of the SS DNA formed was circularized to the unit-length SS(C) derivative (Harth et al. 1981). However, the majority of the newly incorporated nucleotide was present as RFII DNA.

Meyer et al. (1981) have assembled the above observations into a model of multifunctional fd gene-*2* protein action during DNA replication. They suggested that a sequence-specific helix distortion, recognized by fd gene-*2* protein, was present both in RFI fd DNA and in relaxed fd DNA at the site of helix unwinding, immediately ahead of the replication fork. Having initiated DNA synthesis, the gene-*2* protein was envisioned to bind noncovalently to the template strand at the origin of replication and to stall the replication fork at the origin after a round of synthesis was complete. An interaction between the bound and the free gene-*2* protein was thought to cause processing of the unit-length nascent DNA strands to SS(C) products.

Many aspects of the fd gene-2 protein have yet to be studied in detail, including the DNA:protein stoichiometry associated with the binding, cleavage, and ligation reactions and the DNA target sequence specificity. As with ϕX *A* protein, the mechanism by which displaced nascent DNA is processed to covalently closed unit-length SS(C) DNA, the so-called termination event, has yet to be resolved. The specificity expressed for fd RFI DNA over fd SS(C) DNA has important implications regarding higher-order DNA structures associated with both superhelicity and helix unwinding. A complete characterization of the gene-*2* protein must address these complex aspects of DNA structure.

Concluding Remarks

The ϕX gene-*A* protein and the fd gene-*2* protein are highly complex enzymes, essential for multiple stages of phage DNA replication. Although 4 years have passed since a multifunctional role was proposed for the ϕX *A* protein (Eisenberg et al. 1977), much still remains unknown, particularly the events associated with specific termination of replication. The *A** protein remains a puzzle, as a role for the protein in vivo has yet to be determined.

The fd gene-*2* protein is also a complex problem but methodologies previously developed for the ϕX system may facilitate our understanding of gene-*2*-mediated DNA replication.

It is probable that proteins of this type will be discovered and characterized in other systems. The site-specific breakage and rejoining activities are well tailored for the processing and maturation of nucleic acid structure. In addition to the conservation of phosphate-bond energy (Wang 1971; Champoux and Dulbecco 1972; Champoux 1976), the covalently bound ϕX *A* protein of the RFII·*A* complex is ideally suited to protect the displaced DNA strand from exonucleases and to direct specifically the fate of the DNA terminus to which it is bound. The ϕX *A* protein or fd gene-*2* protein may represent prototypic activities required for recombination and transposition (Galas and Chandler 1981; Harshey and Bukhari 1981).

NUCLEASE ACTIVITIES ASSOCIATED WITH DNA POLYMERASES

3′→5′ Exonuclease Activity

All three prokaryotic DNA polymerases (I, II, and III) possess a 3′→5′ exonuclease activity (Kornberg 1980) (see Table 3). It is generally believed that this activity plays an important role in the proofreading of incorporated nucleotides which are dictated by the template DNA strand. In this proofreading function, errors in incorporation are sensed by the exonuclease and are then removed (Goulian et al. 1968; Englund 1971; Brutlag and Kornberg 1972). Evidence supporting this function for the 3′→5′ exonuclease activity has come from studies on the DNA polymerase induced by bacteriophage T4 (Speyer et al. 1966; Drake et al. 1969). Conditional lethal point mutations which alter the mutation frequencies have been found in *E. coli* and in the bacteriophage T4. Some of the mutations in the latter case were shown to be in gene *43*, the gene coding for T4 DNA polymerase (Muzyczka et al. 1972). In a series of important studies, it was shown that mutations in gene *43* could result in either increased or decreased spontaneous mutations. This variation is related to the level of 3′→5′ exonuclease activity associated with T4 DNA polymerase. Mutants manifesting a high spontaneous mutation rate were found to have a low ratio of 3′→5′ exonuclease to DNA polymerase activity; conversely, mutants showing a low mutation rate yielded T4 DNA polymerase possessing a high ratio of 3′→5′ exonuclease to polymerase activity (Lo and Bessman, 1976a,b; Bessman and Reha-Krantz 1977; Reha-Krantz and Bessman 1977, 1981).

There are a number of examples of mutant prokaryotic DNA polymerases with altered exonuclease activities, and in some instances this has been correlated with an altered DNA polymerase activity (Hershfield 1973; Gillin and Nossal 1976; Engler and Bessman 1978).

Table 3 Activities associated with DNA polymerases

Enzyme	Exonuclease	Endonuclease
DNA polymerase I (*E. coli*)	$3' \rightarrow 5'$ SS $5' \rightarrow 3'$ DS	yes[a]
DNA polymerase II (*E. coli*)	$3' \rightarrow 5'$ SS	none
DNA polymerase III (*E. coli*)	$3' \rightarrow 5'$ SS $5' \rightarrow 3'$	yes[a]
T4 DNA polymerase	$3' \rightarrow 5'$	none
DNA polymerase α, β, and γ higher eukaryotes	none	none
Reverse transcriptase	RNase H	DNase

[a]Nicks D-loop substrates.

5′→3′ Exonuclease Activity

5′→3′ exonuclease activity has been found associated with DNA polymerases I and III (Kornberg 1980). The 5′→3′ exonuclease activity of DNA polymerase I is the best characterized. This activity results in the generation of mononucleotides as well as oligonucleotides from a 5′ terminus. It can act on mismatched as well as matched base pairs at the 5′ end of a duplex. It is nonspecific in that the end can be a 5′-hydroxyl, mono-, di-, or triphosphate; DNA or RNA are cleaved. The action of the 5′→3′ exonuclease is increased with concurrent polymerization. This property has been exploited as a method by which DNA chains can be labeled (Rigby et al. 1977).

In general, the 5′→3′ exonuclease appears to act at sites that contain alterations of DNA structures as one would expect near replication forks or in mismatched regions (Kelly et al. 1969). Under certain conditions, the polymerase can make an endonucleolytic scission in the absence of a 5′ terminus. When superhelical circular DNA containing a D-loop structure (formed by the interpolation of a single-stranded DNA or RNA fragment which serves as a primer) is incubated with DNA polymerase I, DNA synthesis initiates at the primer fragment and the template DNA is nicked (Champoux and McConaughy 1975; Liu and Wang 1975).

In the case of the DNA polymerase III, Livingston and Richardson (1975) have shown that in contrast to DNA polymerase I, its 5′→3′ exonuclease activity requires a single-stranded 5′ extension; the enzyme will hydrolyze the single-stranded extension and then migrate into the duplex area. In the absence of a single-stranded 5′ region, this activity does not function. Like DNA polymerase I, the 5′→3′ exonuclease activity of DNA polymerase III can form oligonucleotides. The full significance of this activity is unclear.

The higher eukaryotic DNA polymerases α, β, and γ do not contain endogeneous 5′→3′ or 3′→5′ exonuclease activities (Kornberg 1980). Two DNA polymerases have been isolated from yeast and one of these enzymes, DNA polymerase II, has been reported to contain a 3′→5′ exonuclease (Chang 1977; Plevani and Chang 1977; Wintersberger 1978).

Nuclease Activities Of Reverse Transcriptase

Reverse transcriptase isolated from avian retrovirus displays a number of resident activities. These include an RNA-directed DNA polymerase required for the eventual synthesis of double-stranded DNA, a bidirectional exonucleolytic RNaseH activity (see Chapter 9), and a third activity which acts as a DNA endonuclease.

The enzyme is encoded by the viral *pol* gene and contains two subunits, α (62K) and β (92K) (Grandgenett et al. 1973; Lai and Verma 1978). It has been shown that the α subunit is a polypeptide derived from the β subunit (Gibson and Verma 1974; Rho et al. 1975). The smaller 32K product formed by the cleavage of the β subunit, has been identified in virions of Rous sarcoma virus (RSV) and avian myeloblastosis virus. This smaller product was identified as a nucleic acid-binding protein with DNA endonuclease activity and has been designated pp32 (Collett et al. 1978; Grandgenett et al. 1978; Schiff and Grandgenett 1978).

It has been shown that the DNA polymerase and RNaseH activities reside solely in the α subunit (Molling et al. 1971; Grandgenett et al. 1973). Interestingly, the holoenzyme ($\alpha\beta$) contains a DNA endonuclease activity, which is not displayed by the isolated α subunit and which prefers covalently closed circular DNA as substrate. The activity is detected in the presence of Mn^{2+} but not Mg^{2+}. The endonuclease activity resides within the β subunit (α + the pp32) of the holoenzyme and is clearly different from the activity shown by the pp32 protein alone. The latter possesses a higher specific activity and can be assayed with either Mn^{2+} or Mg^{2+}.

The endonuclease activity of the holoenzyme is linked with the *pol* gene, since several well-characterized *pol ts* mutants from RSV are thermolabile in both polymerase and endonuclease activity (Golomb et al. 1981). Other *pol* gene mutants show thermolability in DNA polymerase and RNaseH activities but are not thermolabile in the DNA endonuclease activity. The specificity of the endonuclease activity has not been characterized.

Though circular DNA is one of the products leading to provirus formation, the ligase responsible for this covalent circular DNA synthesis has not been identified. While highly purified reverse transcriptase is free of DNA ligase activity, it is possible that the enzyme interacts with a cellular DNA ligase to form a complex. If this were the case, the reverse transcriptase would have some properties in common with those of the bacteriophage ϕX174 *A* protein and the bacteriophage fd gene-*2* protein (superhelical endonuclease- and ligase-type activities).

REFERENCES

Baas, P.D. and H.S. Jansz. 1972a. Asymmetric information transfer during φX174 DNA replication. *J. Mol. Biol.* **63:**557.

Baas, P.D. and H.S. Jansz. 1972b. φX174 replicative form DNA replication, origin and direction. *J. Mol. Biol.* **63:**569.

Baas, P.D., H.S. Jansz, and R.L. Sinsheimer. 1976. Bacteriophage φX174 DNA synthesis in a replication-deficient host: Determination of the origin of φX DNA replication. *J. Mol. Biol.* **102:**633.

Baas, P.D., W.R. Teertstra, A.D.M. van Mansfeld, H.S. Jansz, G.A. van der Marel, G.H. Veeneman, and J.H. van Boom. 1981. Construction of viable and lethal mutations in the origin of bacteriophage φX174 using synthetic oligodeoxyribonucleotides. *J. Mol. Biol.* **152:**615.

Bessman, M.J. and J.J. Reha-Krantz. 1977. Studies on the biochemical basis of spontaneous mutation. V. Effect of temperature on mutation frequency. *J. Mol. Biol.* **116:**115.

Brown, D.R., D. Reinberg, T. Schmidt-Glenewinkel, S.L. Zipursky, and J. Hurwitz. 1982. Analysis of the φX174 gene A protein using *in vitro* DNA replication systems. *Methods Enzymol.* (in press).

Brown, D.R., D. Reinberg, T. Schmidt-Glenewindel, M. Roth, S.L. Zipursky, and J. Hurwitz. 1982. DNA structures required for φX174 A protein-directed initiation and termination of DNA replication. *Cold Spring Harbor Symp. Quant. Biol.*, **47** (in press).

Brutlag, D. and A. Kornberg. 1972. Enzymatic synthesis of deoxyribonucleic acid. XXXVI. A proofreading function for the 3′→5′ exonuclease activity in deoxynucleic acid polymerases. *J. Biol. Chem.* **247:**241.

Champoux, J.J. 1976. Evidence for an intermediate with a single-strand break in the reaction catalyzed by the DNA untwisting enzyme. *Proc. Natl. Acad. Sci.* **73:**3488.

Champoux, J.J. and R. Dulbecco. 1972. An activity from mammalian cells that untwists superhelical DNA—A possible swivel for DNA replication. *Proc. Natl. Acad. Sci.* **69:**143.

Champoux, J.J. and B.L. McConaughy. 1975. Priming of superhelical SV40 DNA by *Escherichia coli* RNA polymerase for *in vitro* DNA synthesis. *Biochemistry* **14:**307.

Chang, L.M.S. 1977. DNA polymerases from baker's yeast. *J. Biol. Chem.* **252:**1873.

Collett, M.S., J.P. Leis, M.S. Smith, and A.J. Faras. 1978. Unwinding-like activity associated with avian retrovirus RNA-directed DNA polymerase. *J. Virol.* **26:**498.

Denhardt, D.T. 1975. The single-stranded DNA phages. *CRC Crit. Rev. Microbiol.* **4:**161.

Drake, J.W., E.F. Allen, S.A. Forsberg, R.-M. Preparata, and E.O. Greening. 1969. Spontaneous mutation. *Nature* **221:**1128.

Duguet, M., G. Yarranton, and M. Gefter. 1979. The *rep* protein of *Escherichia coli*: Interaction with DNA and other proteins. *Cold Spring Harbor Symp. Quant. Biol.* **43:**335.

Eisenberg, S. 1980. Cleavage of φX174 single-stranded DNA by gene *A* protein and formation of a tight protein-DNA complex. *J. Virol.* **35:**409.

Eisenberg, S. and R. Ascarelli. 1981. The A* protein of φX174 is an inhibitor of DNA replication. *Nucleic Acids Res.* **9:**1991.

Eisenberg, S. and D.T. Denhardt. 1974. Structure of nascent φX174 replicative form: Evidence for discontinuous DNA replication. *Proc. Natl. Acad. Sci.* **71:**984.

Eisenberg, S. and M. Finer. 1980. Cleavage and circularization of single-stranded DNA: A novel enzymatic activity of φX174 A* protein. *Nucleic Acids Res.* **8:**5305.

Eisenberg, S. and A. Kornberg. 1979. Purification and characterization of φX174 gene A protein. A multifunctional enzyme of duplex DNA replication. *J. Biol. Chem.* **254:**5328.

Eisenberg S., J. Griffith, and A. Kornberg. 1977. φX174 cistron *A* protein is a multifunctional enzyme in DNA replication. *Proc. Natl. Acad. Sci.* **74:**3198.

Eisenberg, S., J.F. Scott, and A. Kornberg. 1976a. An enzyme system for replication of duplex circular DNA: The replicative form of phage ϕX174. *Proc. Natl. Acad. Sci.* **73:**1594.

——————. 1976b. Enzymatic replication of viral and complementary strands of duplex DNA of phage ϕX174 proceeds by separate mechanisms. *Proc. Natl. Acad. Sci.* **73:**3151.

Engler, M.J. and M.J. Bessman. 1978. Characterization of a mutator DNA polymerase I from *Salmonella typhimurium. Cold Spring Harbor Symp. Quant. Biol.* **43:**929.

Englund, P.T. 1971. The initial step of *in vitro* synthesis of deoxyribonucleic acid by the T4 deoxyribonucleic acid polymerase. *J. Biol. Chem.* **246:**5684.

Fidanian, H.M. and D.S. Ray. 1972. Replication of bacteriophage M13. VII. Requirement of the gene *2* protein for the accumulation of a specific RFII species. *J. Mol. Biol.* **72:**51.

Fiddes, J.C., B.G. Barrell, and G.N. Godson. 1978. Nucleotide sequences of the separate origins of synthesis of bacteriophage G4 viral and complementary strands. *Proc. Natl. Acad. Sci.* **75:**1081.

Franke, B. and D.S. Ray. 1971. Formation of the parental replicative form DNA of bacteriophage ϕX174 and initial events in its replication. *J. Mol. Biol.* **61:**565.

——————. 1972. *Cis*-limited action of the gene-A product of bacteriophage ϕX174 and the essential bacterial site. *Proc. Natl. Acad. Sci.* **69:**475.

Fujisawa, H. and M. Hayashi. 1976. Gene A product of ϕX174 is required for site-specific endonucleolytic cleavage during single-stranded DNA synthesis *in vivo. J. Virology* **19:**416.

Galas, D.J. and M. Chandler. 1981. On the molecular mechanisms of transposition. *Proc. Natl. Acad. Sci.* **78:**4858.

Geider, K. and A. Kornberg. 1974. Conversion of the M13 viral single strand to the double-stranded replicative forms by purified proteins. *J. Biol. Chem.* **249:**3999.

Geider, K. and T.F. Meyer. 1979. Gene-II protein of bacteriophage fd in enzymatic replication of viral duplex DNA. *Cold Spring Harbor Symp. Quant. Biol.* **43:**59.

Geider, K., I. Baumel, and T.F. Meyer. 1982. Intermediate stages in enzymatic replication of bacteriophage fd duplex DNA. *J. Biol. Chem.* **257:**6488.

Gibson, W. and I.M. Verma. 1974. Studies on the reverse transcriptase of RNA tumor viruses. Structural relatedness of two subunits of avian RNA tumor viruses. *Proc. Natl. Acad. Sci.* **71:**4991.

Gilbert, W. and D. Dressler. 1969. DNA replication: The rolling circle model. *Cold Spring Harbor Symp. Quant. Biol.* **33:**473.

Gillin, F.D. and N.G. Nossal. 1976. Control of mutation frequency by bacteriophage T4 DNA polymerase. II. Accuracy of nucleotide selection by the L88 mutator, CB120 antimutator, and wild type phage T4 DNA polymerases. *J. Biol. Chem.* **251:**5225.

Godson, G.N., B.G. Barrell, R. Staden, and J.C. Fiddes. 1978. Nucleotide sequence of bacteriophage G4 DNA. *Nature* **276:**236.

Golomb, M., D.P. Grandgenett, and W. Mason. 1981. Virus-coded DNA endonuclease from avian retrovirus. *J. Virol.* **38:**548.

Goulian, M., Z.J. Lucas, and A. Kornberg. 1968. Enzymatic synthesis of deoxyribonucleic acid. XXV. Purification and properties of deoxyribonucleic acid polymerase induced by infection with phage T4. *J. Biol. Chem.* **243:**627.

Grandgenett, D.P., G.F. Gerard, and M. Green. 1973. A single subunit from avian myeloblastosis virus with both RNA-directed DNA polymerase and ribonuclease H activity. *Proc. Natl. Acad. Sci.* **70:**230.

Grandgenett, D.P., A.C. Vora, and R.D. Schiff. 1978. A 32,000-dalton nucleic acid-binding protein from avian retrovirus cores possesses DNA endonuclease activity. *Virolology* **89:**119.

Harshey, R.M. and A.I. Bukhari. 1981. A mechanism of DNA transposition. *Proc. Natl. Acad. Sci.* **78:**1090.

Harth, G., I. Baumel, T.F. Meyer, and K. Geider. 1981. Bacteriophage fd gene-2 protein. Processing of phage fd viral strands replicated by phage T7 enzymes. *Eur. J. Biochem.* **119:**663.

Heidekamp, F., P.D. Baas, and H.S. Jansz. 1982. Nucleotide sequences at the ϕX gene A protein cleavage site in replicative form I DNAs of bacteriophages U3, G14, and 3. *J. Virol.* **42:**91.

Heidekamp, F., P.D. Baas, J.H. van Boom, G.H. Veeneman, S.L. Zipursky, and H.S. Jansz. 1981. Construction and characterization of recombinant plasmid DNAs containing sequences of the origin of bacteriophage ϕX174 DNA replication. *Nucleic Acids Res.* **9:**3335.

Heidekamp, F., S.A. Langeveld, P.D. Baas, and H.S. Jansz. 1980. Studies of the recognition sequence of ϕX174 gene A protein. Cleavage site of ϕX gene A protein in St-1 RFI DNA. *Nucleic Acids Res.* **8:**2009.

Henry, T.J. and R. Knippers. 1974. Isolation and function of the gene *A* initiator of bacteriophage ϕX174, a highly specific DNA endonuclease. *Proc. Natl. Acad. Sci.* **71:**1549.

Hershfield, M.S. 1973. On the role of deoxyribonucleic acid polymerase in determining mutation rates. Characterization of the defect in the T4 deoxyribonucleic acid polymerase caused by the *TS* L88 mutation. *J. Biol. Chem.* **248:**1417.

Ikeda, J.-E., A. Yudelevich, and J. Hurwitz. 1976. Isolation and characterization of the protein coded by gene *A* of bacteriophage ϕX174 DNA. *Proc. Natl. Acad. Sci.* **73:**2669.

Ikeda, J.E., A. Yudelevich, N. Shimamoto, and J. Hurwitz. 1979. Role of polymeric forms of the bacteriophage ϕX174 coded gene A protein in ϕX RFI DNA cleavage. *J. Biol. Chem.* **254:**9416.

Jacob, R.J., J. Lebowitz, and A.K. Kleinschmidt. 1974. Locating interrupted hydrogen bonding in the secondary structure of PM2 circular DNA by comparative denaturation mapping. *J. Virol.* **13:**1176.

Johnson, P.H. and R.L. Sinsheimer. 1974. Structure of an intermediate in the replication of bacteriophage ϕX174 deoxyribonucleic acid: The initiation site for DNA replication. *J. Mol. Biol.* **83:**47.

Kelly, R.B., M.R. Atkinson, J.A. Huberman, and A. Kornberg. 1969. Excision of thymine dimers and other mismatched sequences by DNA polymerase of *Escherichia coli. Nature* **224:**495.

Kornberg, A. 1980. *DNA replication.* W.H. Freeman and Co., San Francisco.

Lai, M.-H.T. and I.M. Verma. 1978. Reverse transcriptase of RNA tumor viruses. V. *In vitro* proteolysis of reverse transcriptase from avian myeloblastosis virus and isolation of a polypeptide manifesting only RNase H activity. *J. Virol.* **25:**652.

Langeveld, S.A., G.A. van Arkel, and P.J. Weisbeek. 1980. Improved method for the isolation of A and A* proteins of bacteriophage ϕX174. *FEBS Lett.* **114:**269.

Langeveld, S.A., A.D.M. van Mansfeld, P.D. Baas, H.S. Jansz, G.A. van Arkel, and P.J. Weisbeek. 1978. Nucleotide sequence of origin of replication in bacteriophage ϕX174 RF DNA. *Nature* **271:**417.

Langeveld, S.A., A.D.M. van Mansfeld, J.M. de Winter, and P.J. Weisbeck. 1979. Cleavage of single-stranded DNA by the A and A* proteins of bacteriophage ϕX174. *Nucleic Acids Res.* **7:**2177.

Langeveld, S.A., A.D.M. van Mansfeld, A. van der Ende, J.H. van de Pol, G.A. van Arkel, and P.J.Weisbeek. 1981. The nuclease specificity of the bacteriophage ϕX174 A* protein. *Nucleic Acids Res.* **9:**545.

Lin, N.S.C. and D. Pratt. 1972. Role of bacteriophage M13 gene-*2* in viral DNA replication. *J. Mol. Biol.* **72:**37.

Linney, E. and M. Hayashi. 1973. Two proteins of gene A of ϕX174. *Nature New Biol.* **245:**6.

Linney, E. and M. Hayashi. 1974. Intragenic regulation of the synthesis of ϕX174 gene *A* proteins. *Nature* **249:**345.

Liu, L.F. and J.C. Wang. 1975. In *DNA synthesis and its regulation* (ed. M. Goulian and P. Hanawalt), p. 30. Benjamin, Menlo Park.

Livingston, D.M. and C.C. Richardson. 1975. Deoxyribonucleic acid polymerase III of *Escherichia coli*. Characterization of associated exonuclease activities. *J. Biol. Chem.* **250:**470.

Lo, K.-Y. and M.J. Bessman. 1976a. An antimutator deoxyribonucleic acid polymerase. I. Purification and properties of the enzyme. *J. Biol. Chem.* **251**:2475.

———. 1976b. An antimutator deoxyribonucleic acid polymerase. II. *In vitro* and *in vivo* studies of its temperature sensitivity. *J. Biol. Chem.* **251**:2480.

Marians, K.J., J.-E. Ikeda, S. Schlagman, and J. Hurwitz. 1977. Role of DNA gyrase in ϕX replication in vitro. *Proc. Natl. Acad. Sci.* **74**:1965.

Martin, D.F. and G.N. Godson. 1975. Identification of a ϕX174 coded protein involved in the shut-off of host DNA replication. *Biochem. Biophys. Res. Commun.* **65**:323.

Mazur, B. and P. Model. 1973. Regulation of coliphage f1 single-stranded DNA synthesis by a DNA-binding protein. *J. Mol. Biol.* **78**:285.

Meyer, T.F. and K. Geider. 1978. Isolation and partial characterization of gene II protein from phage-fd-infected cells. In *The single-stranded DNA phages* (ed. D.T. Denhardt et al.), pp. 389-392. Cold Spring Harbor Laboratory, Cold Spring Harbor, New York.

———. 1979a. Bacteriophage fd gene II-protein. I. Purification, involvement in RF replication, and the expression of gene II. *J. Biol. Chem.* **254**:12636.

———. 1979b. Bacteriophage fd gene II-protein. II. Specific cleavage and relaxation of supercoiled RF from filamentous phages. *J. Biol. Chem.* **254**:12642.

Meyer, T.F., I. Baumel, K. Geider, and P. Bedinger. 1981. Replication of phage fd RF with fd gene 2 proteins and phage T4 enzymes. *J. Biol. Chem.* **256**:5810.

Meyer, T.F., K. Geider, C. Kurz, and H. Schaller. 1979. Cleavage site of bacteriophage fd gene II-protein in the origin of viral strand replication. *Nature* **278**:365.

Molling, K., D.P. Bolognesi, H. Bauer, W. Bausen, H.W. Plassmann, and P. Hausen. 1971. Association of viral reverse transcriptase with an enzyme degrading the RNA moiety of RNA-DNA hybrids. *Nature New Biol.* **234**:240.

Muzyczka, N., R.L. Poland, and M.J. Bessman. 1972. Studies on the biochemical basis of spontaneous mutation. I. A comparison of the deoxyribonucleic acid polymerases of mutator, antimutator and wild type strains of bacteriophage T4. *J. Biol. Chem.* **247**:7116.

Plevani, P. and L.M.S. Chang. 1977. Enzymatic initiation of DNA synthesis by yeast DNA polymerases. *Proc. Natl. Acad. Sci.* **74**:1937.

Reha-Kranz, L.J. and M.J Bessman. 1977. Studies on the biochemical basis of mutation. IV. Effect of amino acid substitution on the enzymatic and biological properties of bacteriophage T4 DNA polymerase. *J. Mol. Biol.* **116**:99.

Reha-Krantz, L.J. and M.J. Bessman. 1981. Studies on the biochemical basis of mutation. VI. Selection and characterization of a new bacteriophage T4 mutator DNA polymerase. *J. Mol. Biol.* **145**:677.

Reinberg, D., S.L. Zipursky, and J. Hurwitz. 1981. Separate requirements for leading and lagging strand DNA synthesis during ϕX A protein-dependent RF→RF DNA replication *in vitro*. *J. Biol. Chem.* **256**:13143.

Reinberg, D., S.L. Zipursky, P. Weisbeek, D. Brown, and J. Hurwitz. 1982. Studies on the ϕX A protein-mediated termination of leading strand DNA synthesis. *J. Biol. Chem.* (in press).

Rho, H.M., D.P. Grandgenett, and M. Green. 1975. Sequence relatedness between the subunits of avian myeloblastosis virus reverse transcriptase. *J. Biol. Chem.* **250**:5278.

Rigby, P.W.J., M. Dieckmann, C. Rhoades, and P. Berg. 1977. Labeling deoxyribonucleic acid to high specific activity *in vitro* by nick translation with DNA polymerase I. *J. Mol. Biol.* **113**:237.

Sanger, F., G.M. Air, B.G. Barrell, N.L. Brown, A.R. Coulson, J.C. Fiddes, C.A. Hutchison III, P.M. Slocombe, and M. Smith. 1977. Nucleotide sequence of bacteriophage ϕX174 DNA. *Nature* **265**:687.

Schaller, H. 1979. The intergenic region and the origins for filamentous phage DNA replication. *Cold Spring Harbor Symp. Quant. Biol.* **43**:401.

Schiff, R.D. and D.P. Grandgenett. 1978. Virus-coded origin of a 32,000-dalton protein from avian retrovirus cores: Structural relatedness of p32 and the polypeptide of the avian retrovirus DNA polymerase. *J. Virol.* **28**:279.

Scott, J.F., S. Eisenberg, L.L. Bertsch, and A. Kornberg. 1977. A mechanism of duplex DNA replication revealed by enzymatic studies of phage φX174: Catalytic strand separation in advance of replication. *Proc. Natl. Acad. Sci.* **74:**193.

Sinsheimer, R.L. 1968. Bacteriophage φX174 and related viruses. *Prog. Nucleic Acid Res. Mol. Biol.* **8:**115.

Sinsheimer, R.L., B. Starman, C. Nagler, and S. Guthrie. 1962. The process of infection with bacteriophage φX174. I. Evidence for a "replicative form." *J. Mol. Biol.* **4:**142.

Speyer, J.F., J.D. Karam, and A.B. Lenny. 1966. On the role of DNA polymerase in base selection. *Cold Spring Harbor Symp. Quant. Biol.* **31:**693.

Sumida-Yasumoto, C. and J. Hurwitz. 1977. Synthesis of φX174 viral DNA *in vitro* depends on φX replicative form DNA. *Proc. Natl. Acad. Sci.* **74:**4195.

Sumida-Yasumoto, C., A. Yudelevich, and J. Hurwitz. 1976. DNA synthesis *in vitro* dependent upon φX174 replicative form I DNA. *Proc. Natl. Acad. Sci.* **73:**1887.

Tessman, E.S. 1966. Mutants of bacteriophage S13 blocked in infectious DNA synthesis. *J. Mol. Biol.* **17:**218.

van der Ende, A., S.A. Langeveld, R. Teertstra, G.A. van Arkel, and P.J. Weisbeek. 1981. Enzymatic properties of the bacteriophage φX174 A* protein on superhelical φX174 DNA: A model for the termination of the rolling circle DNA replication. *Nucleic Acids Res.* **9:**2037.

van Mansfeld, A.D.M., S.A. Langeveld, P.D. Baas, H.S. Jansz, G.A. van der Marel, G.H. Veeneman, and J.H. van Boom. 1980. Recognition sequence of bacteriophage φX174 gene *A* protein—an initiator of DNA replication. *Nature* **288:**561.

van Mansfeld, A.D.M., S.A. Langeveld, P.J. Weisbeek, P.D. Baas, G.A. van Arkel, and H.S. Jansz. 1979. Cleavage site of φX174 gene-*A* protein in φX and G4 RFI DNA. *Cold Spring Harbor Symp. Quant. Biol.* **43:**331.

Wang, J.C. 1971. Interaction between DNA and an *Escherichia coli* protein ω. *J. Mol. Biol.* **55:**523.

Weisbeek, P., A. van der Ende, H. van der Avoort, R. Teertstra, F. van Mansfeld, and S. Langeveld. 1981. Viral DNA sequences and proteins important in the φX174 DNA synthesis. In *The initiation of DNA replication* (ed. D.S. Ray), p. 211. Academic Press, New York.

Wintersberger, E. 1978. Yeast DNA polymerases: Antigenic relationship, use of RNA primer and associated exonuclease activity. *Eur. J. Biochem.* **84:**167.

Zipursky, S.L., D. Reinberg, and J. Hurwitz. 1980. *In vitro* DNA replication of recombinant plasmid DNAs containing the origin of progeny replicative form DNA synthesis of phage φX174. *Proc. Natl. Acad. Sci.* **77:**5182.

Ribonucleases H

Robert J. Crouch and Marie-Luise Dirksen
Laboratory of Molecular Genetics
National Institute of Child Health and Human Development
National Institutes of Health
Bethesda, Maryland 20205

A ribonuclease specific for the RNA strand of a DNA-RNA hybrid was first recognized by Stein and Hausen (1969) and named RNaseH (Hausen and Stein 1970). Other authors have used the term hybridase. Unfortunately, enzymes that degrade the RNA or DNA of hybrids could also be considered hybridases. In this review, discussion is limited to RNaseH as defined by Hausen and Stein.

Many cellular RNasesH have been described but the majority of papers is devoted to the RNaseH activities associated with the viral reverse transcriptases. Despite much study, the metabolic role of these enzymes has remained elusive. A continuing problem is the lack of a uniform nomenclature and a systematic comparison of enzymes characterized in different laboratories.

STRUCTURAL ASPECTS OF DNA-RNA HYBRIDS RELEVANT TO RNaseH ACTIVITY

The characteristic conformation of DNA-RNA hybrids, as exemplified for poly(rA)·poly(dT), is the A form (Fig. 1A). In the A form, the major and minor grooves are similar in size and the bases are relatively inaccessible. The sugar-phosphate backbone, of particular importance for RNaseH substrate specificity, is the most prominent part of the molecule.

DNA-RNA hybrids, unlike DNA-DNA and RNA-RNA duplexes, are polar, with one strand having ribose and the other deoxyribose. This polarity in substrate provides a possible means for the enzyme to move along the substrate in an oriented manner.

The minimum hybrid region necessary to determine susceptibility to an RNaseH is not known precisely. Colicin E1 (ColE1) plasmid DNA, which

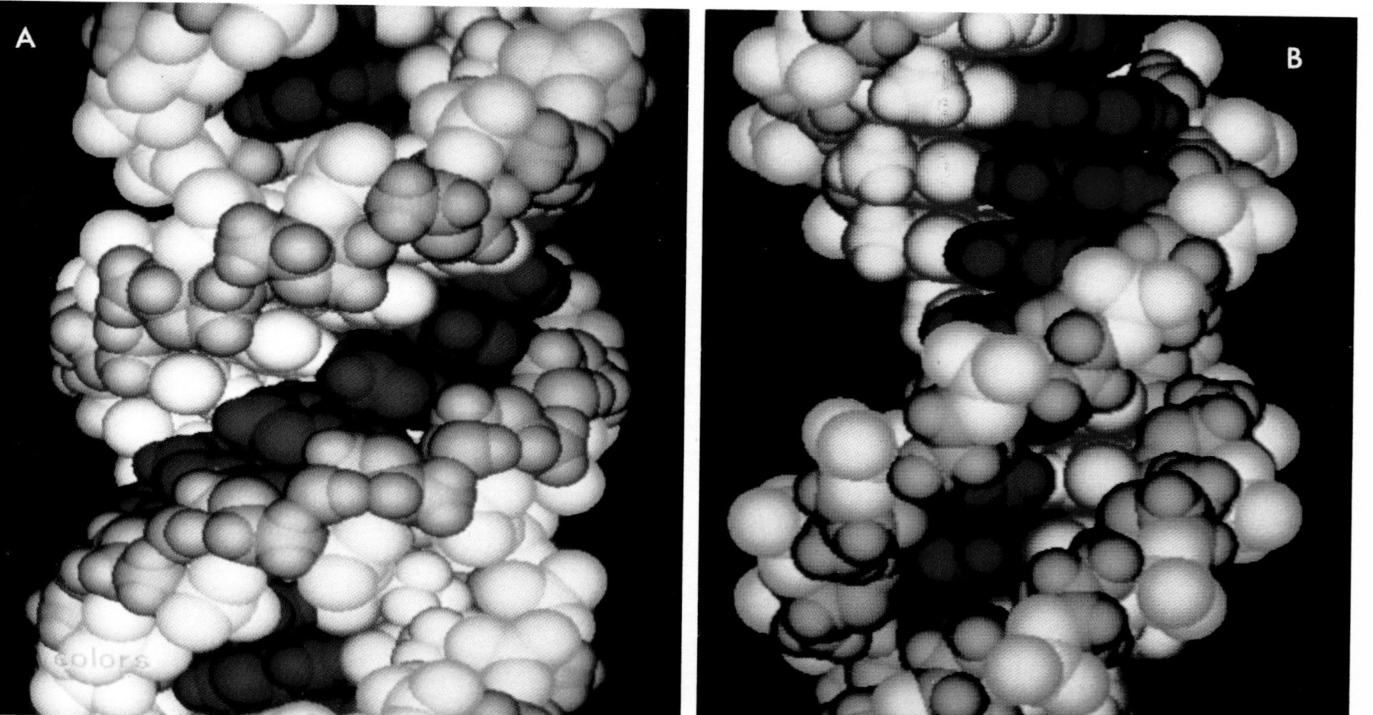

Figure 1 (*A*) Poly(rA)·poly(dT) structure in the A configuration. (*B*) Poly(dA)·poly(dT) structure in the B configuration. Red=adenine, pink=thymine, blue=sugar, white=phosphate. (Computer graphics are by Richard Feldman, NIH Computer Center.)

contains a small region with ribonucleotide substitutions (Blair et al. 1972), is a substrate for endonucleolytic RNasesH (Keller and Crouch 1972). Mitochondrial DNAs from mouse and HeLa cells, which also contain small amounts of ribonucleotides, are susceptible to chick embryo RNaseH (Grossman et al. 1973). Oligomeric deoxyribonucleotides as short as tetramers have been annealed to RNAs and serve as substrates for *Escherichia coli* or calf thymus RNaseH (Donis-Keller 1979).

CELLULAR RNasesH

RNasesH have been described from organisms and cells as diverse as *Tetrahymena* (Tashiro et al. 1976), plant cells (Sawai et al. 1979b), tumor cells (O'Cuinn et al. 1973; Sarngadharan et al. 1975), frog oocytes (Wasserman et al. 1974), HeLa cells (Ferrari et al. 1977), Herpes virus-infected cells (Kleinicke et al. 1974), and chick brain mitochondria (Soriano et al. 1974). To avoid duplication this section will focus on calf thymus, yeast, and *E. coli* RNasesH as representative of enzymes from higher eukaryotes, lower eukaryotes, and prokaryotes, respectively. When appropriate, additional and supporting information concerning RNasesH from other organisms will be discussed.

Higher Eukaryotic RNaseH

Two size classes of RNaseH are present in calf thymus (Büsen and Hausen 1975). The larger of the two has a native molecular weight of $\lesssim$70K (Haberkern and Cantoni 1973; Stavrianopoulos and Chargaff 1973; Büsen and Hausen 1975; Büsen 1980a), whereas the smaller has a molecular weight of $\lesssim$30K (Büsen and Hausen 1975; Büsen 1980b). The larger species has a pI around pH 5 (Stavrianopoulos and Chargaff 1973; Büsen, 1980a) and seems to be composed of two polypeptides, one of $\lesssim$32K and one of $\lesssim$25K (Stavrianopoulos and Chargaff 1978; Büsen 1982). Isoelectric focusing of the high-molecular-weight RNaseH yields two forms with different divalent metal ion requirements but which appear to have the same subunit structure (Stavrianopoulos and Chargaff 1978; Büsen 1980a, 1981, 1982). The relationship between these two forms is unknown but their relative amounts are related to the age of the calf from which the enzyme is isolated and to the metabolic stage of the cell.

Antibody directed against calf thymus RNaseH (high molecular weight) purified by Büsen (1980b) reacts with commercial preparations of RNaseH purified by the method of Stavrianopoulos and Chargaff (1978) as well as RNaseH from HeLa cells and *Xenopus laevis* oocyte and egg but does not cross-react with the smaller-molecular-weight species from calf thymus. Using the antibody, Büsen (1980b, 1981) has established the large RNaseH to be a nuclear enzyme. The earlier finding of calf thymus RNaseH in the cytoplasm

(Haberkern and Cantoni 1973) may be an artefact of nuclear leakage (see also Sawai et al. 1981).

Large- and small-molecular-weight RNasesH have been found in bovine lymphocytes (Büsen et al. 1977), in rat liver (Sawai et al. 1978; Tashiro and Ueno 1978a,b; Sawai et al. 1979a; M. Todorava and A. Avramova, pers. comm.), rat brain (Sawai et al. 1980a), Krebs-II ascites cells (Cathala et al. 1979), Ehrlich ascites cells (Natori et al. 1973; Avramova and Galcheva-Gargova 1980; Galcheva-Gargova et al. 1980), and KB cells (W. Keller and R.J. Crouch, unpubl.). The small-molecular-weight RNaseH from Krebs-II ascites cells can be run on SDS-PAGE and detected by the renaturation assay (see Section V) and thus is probably active as a single polypeptide (Cathala et al. 1979). High-molecular-weight RNaseH from Krebs-II ascites cells was not renatured, a result consistent with a multisubunit enzyme (compare calf thymus RNaseH).

Lower Eukaryotic RNaseH

Three yeast enzymes have been described that degrade DNA-RNA hybrids (Wyers et al. 1973, 1976a,b; Huet et al. 1976, 1977, 1979; Dezelee et al. 1977; Iborra et al. 1979). One is a 40K protein that degrades the RNA of a DNA-RNA hybrid when assayed by the SDS-PAGE renaturation technique (Huet et al. 1978; Iborra et al. 1979). It remains to be established whether it is an RNaseH.

A second, RNaseH_2, is a typical RNaseH both in specificity and products and has a molecular weight of 21K and a pI of $\lesssim 10$. RNaseH_2 requires a divalent metal ion and a reducing agent and is optimally active at alkaline pH.

The third, yeast RNaseH_1, is a 49K protein that has been found in association with RNA polymerase A. The original isolate of RNaseH_1 was contaminated with another protein of 49K that proved to be a DNA-binding protein (Iborra et al. 1979; F. Iborra, unpubl.). This preparation of RNaseH_1 showed no requirement for a divalent metal ion, although it was stimulated about twofold by Mg^{2+}. RNaseH, when associated with RNA polymerase A, has an absolute requirement for a divalent metal ion (Huet et al. 1977). The most highly purified fractions of RNaseH_1, which have been separated from the RNA polymerase A, have not been examined for their ionic requirements. One intriguing feature of RNaseH_1, in association with RNA polymerase A, is its inhibition by ATP if poly(rA)·poly(dT) is the substrate or by GTP if poly(rG)·poly(dC) is the substrate.

Prokaryotic RNaseH

E. coli RNaseH is similar to eukaryotic RNaseH in its requirement for a divalent metal ion and a sulfhydryl reagent, and it gives similar products (Berkower et al. 1973; Henry et al. 1973; Miller et al. 1973). Different laboratories have reported molecular weights ranging from 19,000 to 40,000 (Miller et al. 1973; Darlix 1975; Arrendes et al. 1982; R.J. Crouch, unpubl.). Results from SDS-PAGE renaturation studies are consistent with a 15 – 19K polypeptide (Huet et al. 1978; Carl et al. 1980), possibly forming dimers to generate the active enzymes (R.J. Crouch, unpubl.). Like yeast RNaseH$_2$, *E. coli* RNaseH has a very high pI ($\lesssim$9 – 9.6) (Arrendes et al. 1982; R.J. Crouch; unpubl.). Arrendes et al. (1982) have found that wild-type RNaseH exhibits two peaks of activity upon isoelectric focusing at a pI of 9.0 and of 9.6, while a mutant with altered levels of RNaseH has a single peak of activity with a pI of 9.4.

Enzymes in *E. coli*, other than RNaseH, have been shown to degrade the RNA of DNA-RNA hybrids. DNA polymerase I (Keller and Crouch 1972; Berkower et al. 1973) and exonuclease III (Keller and Crouch 1972; Weiss et al. 1978) both degrade RNA in DNA-RNA hybrids but also degrade duplex DNA. Early reports that RNase III degrades RNA in DNA-RNA hybrids (Robertson and Zinder 1968; Robertson et al. 1968; Robertson 1971) were not confirmed, since further purification revealed that contamination with RNaseH was responsible (Crouch 1974; Robertson and Dunn 1975; Darlix 1975) (see also Weatherford et al. 1972).

Products of Digestion

All of the RNasesH described to date produce a 5′-phosphate and 3′-hydroxyl termini. Cellular RNasesH can be divided into two categories based on the product distribution: (1) *E. coli* RNaseH (Berkower et al. 1973; Henry et al. 1973; Miller et al. 1973), yeast RNaseH$_2$ (Wyers et al. 1976a), KB-cell RNaseH (Keller and Crouch 1972), calf thymus RNaseH (Haberkern and Cantoni 1973), rat liver RNaseH (Roewenkamp and Sekeris 1974; Sawai et al. 1979a), and Krebs ascites cell RNaseH$_2$ (Cathala et al. 1979) produce oligonucleotides 2–9 in length. (2) Yeast RNaseH$_1$ (Wyers 1976a), *Ustilago maydis* RNaseH (Banks 1974), and Krebs ascites cell RNaseH$_1$ produce mono- and dinucleotides.

A striking similarity of products has been observed between RNaseH from KB cells (Keller and Crouch 1972) and *E. coli* (R.J. Crouch, unpubl.). To examine the formation of such products, a series of reactions with *E. coli* RNaseH was performed in which the temperature, monovalent cation concentration, and divalent cation concentration were varied. These changes, which alter the T_m of the hybrid substrate, did not affect the relative amounts of each oligoribonucleotide, suggesting that the product distribution is an inherent property of the enzyme. One possibility is that the enzyme acts as a dimer of two identical subunits that has two catalytically active sites, and the oligomeric products result from simultaneous cleavages (R.J. Crouch, unpubl.).

RNaseH$_1$ from Krebs ascites cells may be contaminated with an exonuclease (Hall and Crouch 1977) that converts the products of RNaseH to mono- and dinucleotides. A similar contaminating exonuclease activity has been separated from preparations of chick embryo RNaseH (R.J. Crouch and W. Keller, unpubl.). The fungal enzymes produce mono- and dinucleotides and may contain additional exonucleolytic activity. However, Banks (1974) has shown that the *Ustilago maydis* RNaseH, although it produces mono- and dinucleotides, is an endonuclease.

Base Specificity of RNasesH

Under usual assay conditions, *E. coli* RNaseH will degrade both homopolymeric and heteropolymeric substrates (Berkower et al. 1973; Leis et al. 1973; Robertson and Dunn 1975). In a hybrid between viral f1 and its complementary RNA, a slight preference for cleavage adjacent to pyrimidines was observed (Robertson and Dunn 1975).

Changes in divalent metal ions influence the degradation of homopolymeric RNA in hybrids by calf thymus RNaseH (high-molecular-weight form) (Stavrianopoulos et al. 1976). Mn^{2+} is commonly used in the assay for this enzyme, and in its presence all homopolymeric hybrids are degraded. Substitution of Mg^{2+} for Mn^{2+} permits cleavage of only poly(rG)·poly(dC) and poly(rA)·poly(dT). Only the latter is degraded when Co^{2+} is used. Roewekamp and Sekeris (1974) have observed preferential degradation of poly(rA)·poly(dT) and poly(rG)·poly(dC) by rat liver RNaseH in the presence of Mg^{2+}.

Viral DNA hybridized to complementary RNA is degraded regardless of the divalent metal ion used in the reaction (Stavrianopoulos et al. 1976). All four bases are found at the 3′-terminus of the cleavage products (Stavrianopoulos and Chargaff 1978), although the relative amounts vary slightly depending upon the divalent metal ion used.

E. coli RNaseH is inhibited, in a selective manner, by Dextran (Dirksen and Crouch 1981). ϕXDNA-RNA hybrid degradation is unaltered but poly(rA)·poly(dT) becomes completely resistant. A similar inhibition of Krebs ascites RNaseH by Dextran has been reported by Cathala et al. (1979).

Inhibitors

Many cellular RNasesH have been reported to be inhibited by *N*-ethylmaleimide (NEM). In general, reverse transcriptase RNasesH are relatively less sensitive to sulfhydryl reagents than cellular enzymes (Leis et al. 1973; Moelling 1974b, 1976; Gorecki and Panet 1978). Yeast $RNaseH_1$ unassociated with RNA polymerase A is insensitive to NEM (Wyers et al. 1976b) whereas the same activity when associated with polymerase is NEM sensitive (Huet et al. 1977). RNasesH of the low-molecular-weight class of higher eukaryotes are always sensitive to parachloromercuribenzoate (PCMB)(Sawai et al. 1980a) or NEM (Büsen and Hausen 1975). With Mg^{2+} as activator, all high-molecular-weight RNasesH are likewise sensitive, but resistance is observed with Mn^{2+} as the divalent metal ion (Büsen and Hausen 1975; Cathala et al. 1979; Sawai et al. 1979a; Avramova and Galcheva-Gargova 1980; Galcheva-Gargova et al. 1980).

Other inhibitors of RNasesH include some rifampicin derivatives such as AF/103 (Moelling et al. 1971; Sekeris and Roewekamp 1972; Haberkern and Cantoni 1973) and mycotoxins such as patulin (Tashiro et al. 1979). Various nucleic acids are sometimes inhibitory (Haberkern and Cantoni 1973; Cathala et al. 1979), and one report of pyrophosphate inhibition of an RNaseH can be found (Sawai et al. 1980a). *S*-adenosylmethionine (SAM) inhibits calf thymus (high-molecular-weight) RNaseH (Stavrianopoulos et al. 1976) and baby hamster kidney (BHK) RNasesH (Cooper et al. 1974). Inhibition of calf thymus RNaseH by SAM is irreversible at 37°C but reversible at 0°C.

RETROVIRAL REVERSE TRANSCRIPTASE RNaseH

Distribution

Moelling et al. (1971) first showed that avian myeloblastosis virus (AMV) particles contained RNaseH activity and that RNaseH and reverse transcriptase activities did not separate through several protein purification steps. Subsequently, Keller and Crouch (1972), Baltimore and Smoler (1972), Leis et al. (1972, 1973), and Watson et al. (1973) showed that homogeneous reverse transcriptase, indeed, has RNaseH activity. An extensive survey of RNaseH in the virions of other RNA tumor viruses was made by Grandgenett et al. (1972) in which they were able to detect RNaseH activity in Rous sarcoma virus (RSV), murine sarcoma virus-murine leukemia virus (MSV-MLV), murine leukemia virus (MLV), feline sarcoma leukemia virus, feline leukemia virus, RD-feline leukemia virus, and Mason-Pfizer monkey virus. Both Moloney (Mo) MSV-MLV (Gerard and Grandgenett 1975; Verma 1975b; Rucheton et al. 1979) and Rauscher (Ra) MSV-MLV (Lai et al. 1978; Sarngadharan et al. 1978) also contain RNaseH. In addition, RNaseH has been found associated with reverse transcriptase of Friend murine tumor virus (Moelling 1974a,b, 1976; Weimann et al. 1974), Friend leukemia virus (Rho and Gallo 1980), murine mammary tumor virus (Dion et al. 1977), avian reticuloendotheliosis virus (Moelling et al. 1975), and an endogenous virus of mouse MOPC315 (Hizi and Yaniv 1980).

RNaseH activity has been reported absent from Kirsten MSV-MLV (Wang and Duesberg 1973; Wu et al. 1974), Mo-MSV-MLV (Wang and Duesberg 1973), Ra-MLV (Wu et al. 1974), and hamster leukemia virus (Verma et al. 1974). These early reports are in disagreement with those mentioned in the previous paragraph.

RNaseH and Reverse Transcriptase Are Present in a Single Polypeptide

Three lines of evidence have been used to show the presence of RNaseH and reverse transcriptase activities within a single polypeptide. First, biochemical purification yields a single polypeptide chain that is homogeneous on SDS-PAGE and contains both activities (Grandgenett et al. 1973; Green et al. 1974; Gerard and Grandgenett 1975; Verma 1975b; Rucheton et al. 1979). In one instance, RNaseH activity has been demonstrated following renaturation of the enzyme after SDS-PAGE (Rucheton et al. 1979). Second, viral mutants that exhibit temperature-sensitive growth properties have been isolated. Among these are mutants that have temperature-sensitive polymerase. Purified reverse transcriptase from such mutants also have temperature-sensitive RNasesH (Panet et al. 1974; Verma 1975a; Lai et al. 1978; Moelling and Friis 1979). Third, antibodies directed against reverse transcriptase cross-react with RNaseH (Papas et al. 1977; Rho and Gallo 1980).

RNaseH and Reverse Transcriptase Are Differentially Inhibited

Although RNaseH and reverse transcriptase reside on a single polypeptide chain, they can be differentially inhibited. NaF preferentially inhibits the RNaseH activity (Flügel et al. 1972; Brewer and Wells 1974). It can also be irreversibly inactivated by incubation at pH 4.3 without affecting polymerase activity (Gorecki and Panet 1978). DNA (Modak and Marcus 1977) and RNA (Marcus et al. 1978; Sarngadharan et al. 1978), both of which can serve as templates for polymerization, inhibit RNaseH. Pyridoxal phosphate (Modak 1976; Srivastava and Modak 1980), NEM (Leis et al. 1972; Gorecki and Panet 1978), *o*-phenanthroline (Srivastava and Modak 1979, 1980), 5,5′-dithiobis(2-nitrobenzoic acid) (Gorecki and Panet 1978), *p*-hydroxymercuribenzoate (Gorecki and Panet 1978), *N*-methyl isatin β-thiosemicarbazone-copper complex (Wang and Levinson 1978), and sodium pyrophosphate (Srivastava and Modak 1979) preferentially inhibit the polymerase activity. Viral protein p12 inhibits RNaseH but stimulates polymerase (Sykora and Moelling 1981).

Antibodies raised by injecting avian myeloblastosis virus (AMV) $\alpha\beta$ form of reverse transcriptase (see below for a discussion of α and $\alpha\beta$ forms of AMV reverse transcriptase) inhibit $\alpha\beta$ and α DNA polymerase but only the RNaseH of the $\alpha\beta$ form. Antibodies raised against α inhibit α and $\alpha\beta$ polymerase but only α RNaseH (Papas et al. 1977). Additionally, proteolytic cleavage of reverse transcriptase can lead to an RNaseH of viral origin that lacks polymerase activity (Gerard 1978; Lai and Verma 1978).

Limited Proteolysis of Reverse Transcriptase

Mild proteolysis of reverse transcriptase can lead to the formation of a smaller polypeptide that retains both polymerase and RNaseH activity (Moelling 1974a, 1976; Gerard 1978; Lai and Verma 1978; Rho and Gallo 1980). Moelling (1974a, 1976) observed changes in the size of polypeptides following prolonged incubation of the enzyme on a DEAE column at 4°C or aging of reverse transcriptase at −20°C in 50% glycerol for 6 months. Conversion of

AMV reverse transcriptase from a structure of $\alpha\beta$ to α upon limited digestion with trypsin was observed by Moelling and suggested that α could be a proteolytic product of β.

Subunit Structure of AMV and RSV Reverse Transcriptases

Kacian et al. (1971) purified AMV reverse transcriptase to homogeneity and on SDS-PAGE found two polypeptides which they designated as $\alpha(\lesssim 64K)$ and β ($\lesssim$90–110K). Grandgenett et al. (1972) showed that α, which could be isolated separately from the $\alpha\beta$ complex, possesses both polymerase and RNaseH activities. The suggestion of Moelling (1974a, 1976) that α is related to β led to a direct demonstration by Gibson and Verma (1974) and by Rho et al. (1975) that tyrosine-containing tryptic peptides of α could be found among the β tryptic peptides. Recently, 17 amino acids from the amino terminus of both α and β have been shown to be identical (Copeland et al. 1980).

Lai et al. (1978) have studied further the products of partial proteolysis of AMV reverse transcriptase by following the RNaseH activity. A polypeptide indistinguishable from α by SDS-PAGE is generated together with a smaller polypeptide whose tyrosine-containing tryptic peptides resemble those formed from β but which are not found in α. Further cleavage produces a polypeptide (A) that has RNaseH activity but no polymerase activity. Two of the tyrosine-containing tryptic peptides in α seem to be the same as those found in A, suggesting that A is derived from α.

The sequence of RSV RNA has recently been determined (D. Schwartz et al., pers. comm.) and the polymerase gene has been located in the sequence using the data of Copeland et al. (1980). AMV and RSV polymerases are very similar by the following criteria: (1) RSV polymerase reacts with antiserum against AMV polymerase (Nowinski et al. 1972); (2) the tyrosine-containing tryptic peptide pattern is the same for AMV and RSV (Gibson and Verma 1974); and (3) the amino terminal portions of α and $\alpha\beta$ polypeptides of AMV are identical with those predicted for the same protein from the RSV sequence. We have constructed Figure 2 to suggest that the RNaseH activity is located on the amino terminus of the reverse transcriptase polypeptide. This placement seems to fit best with tryptic peptide analyses.

$\beta\beta$, $\alpha\beta$, and α forms of reverse transcriptase have been isolated from avian sarcoma virus (ASV) grown on duck embryo fibroblasts (Hizi and Joklik 1977; Hizi et al. 1977).

Mode of Action

AMV reverse transcriptase RNaseH has been shown to be an exonuclease (Keller and Crouch 1972; Leis et al. 1972, 1973). DNA-RNA hybrids in which the RNA has no free ends (either ColE1 DNA containing ribonucleotides [Keller and Crouch 1972] or circles of poly[A] annealed to poly[dT] [Leis et al. 1972, 1973]) are not substrates. By constructing a variety of

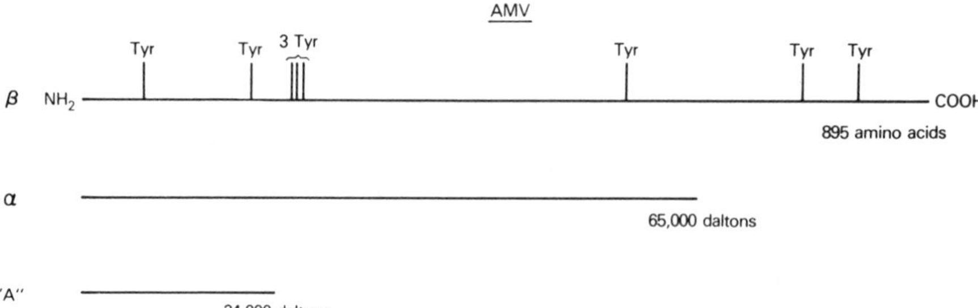

Figure 2 Structural relatedness of retroviral RNasesH. Using information described in the text, we have located the tyrosines on the β polypeptide and aligned the amino terminus of α and β. The A polypeptide has RNaseH activity and its location at the amino terminus of the β polypeptide is consistent with the data from peptide maps.

DNA-RNA hybrids whose RNA termini were blocked either at the 3′ or 5′ end, Leis et al. (1972, 1973, 1976) were able to show that the RNaseH was able to degrade RNA from either end. The $\alpha\beta$ form of AMV reverse transcriptase RNaseH is a processive exoribonuclease, however, the α form is a dispersive exoribonuclease (i.e., after each, or a few, cleavages the enzyme dissociates from its substrate) (Grandgenett and Green 1974; Green et al. 1974; Grandgenett 1975; Grandgenett and Rho 1975).

Mo-MLV RNaseH is also a processive exoribonuclease (either 5′→3′ or 3′→5′) that generates oligonucleotides terminated with 5′-phosphate and 3′-hydroxyl moieties (Verma 1975b; Gerard 1981b). Gerard (1978, 1981a) has described a smaller polypeptide ($\lesssim$30K) from Mo-MLV that seems to be derived from Mo-MLV reverse transcriptase by proteolysis. This RNaseH is no longer processive but is a random exoribonuclease whose products are relatively long, especially for an exonuclease.

Digestion Products of AMV RNaseH

Several investigators have shown that AMV RNaseH produces oligoribonucleotides between 2 and 10 nucleotides in length that have 5′-phosphate and 3′-hydroxyl termini (Baltimore et al. 1972; Baltimore and Smoler 1972; Flügel et al. 1972; Keller and Crouch 1972; Leis et al. 1972, 1973). Others have found products ranging up to 30 nucleotides in length (Grandgenett and Green 1974).

AMV RNaseH cleaves homopolymeric RNAs with the exception of poly(rU)·poly(dA) (Baltimore and Smoler 1972; Leis et al. 1973). Failure to cleave this substrate may reflect the fact that hybrids between poly(rU) and poly(dA) frequently form poly$(rU)_2$·poly(dA).

CELLULAR CHANGES AND RNaseH

RNaseH activity has been followed in cells during changes of metabolic state. Several such studies (summarized in Fig. 3) are discussed below.

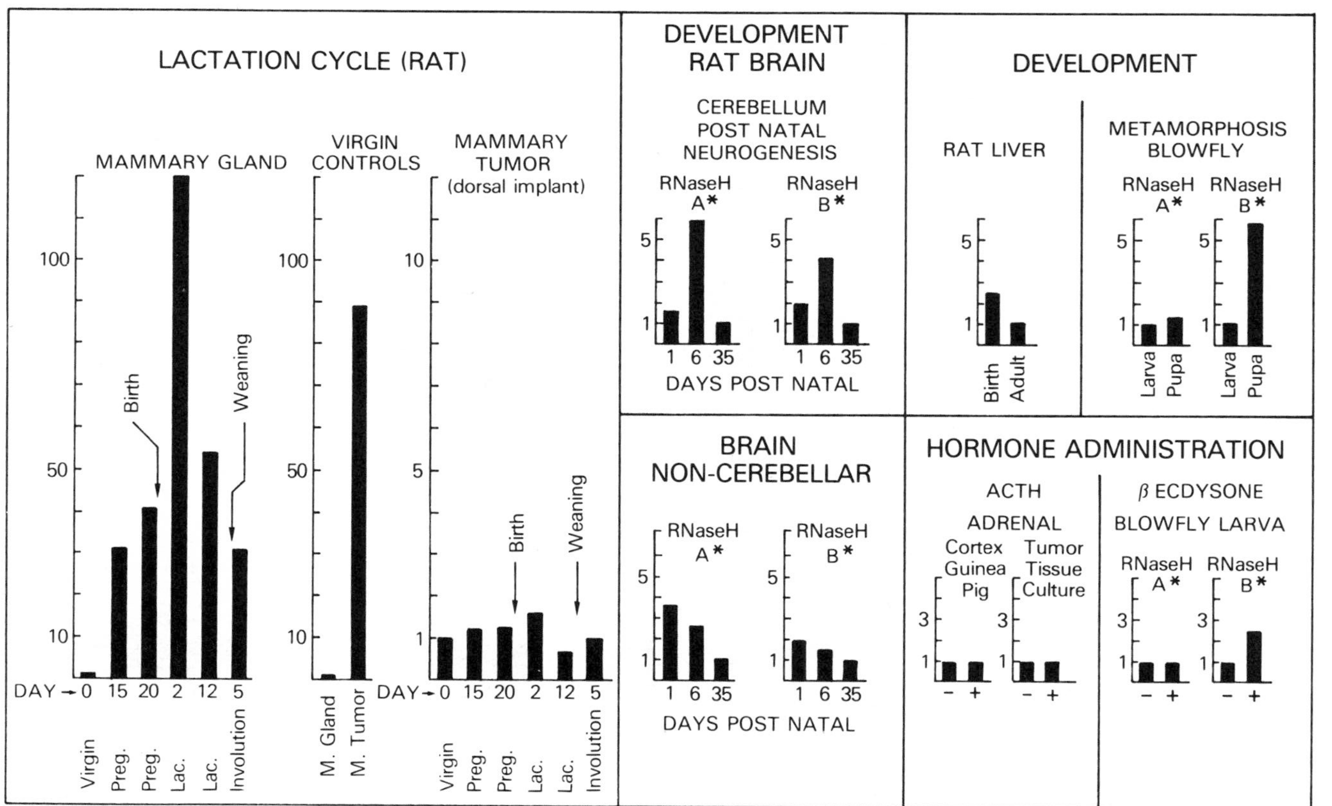

Figure 3. See page 225 for legend.

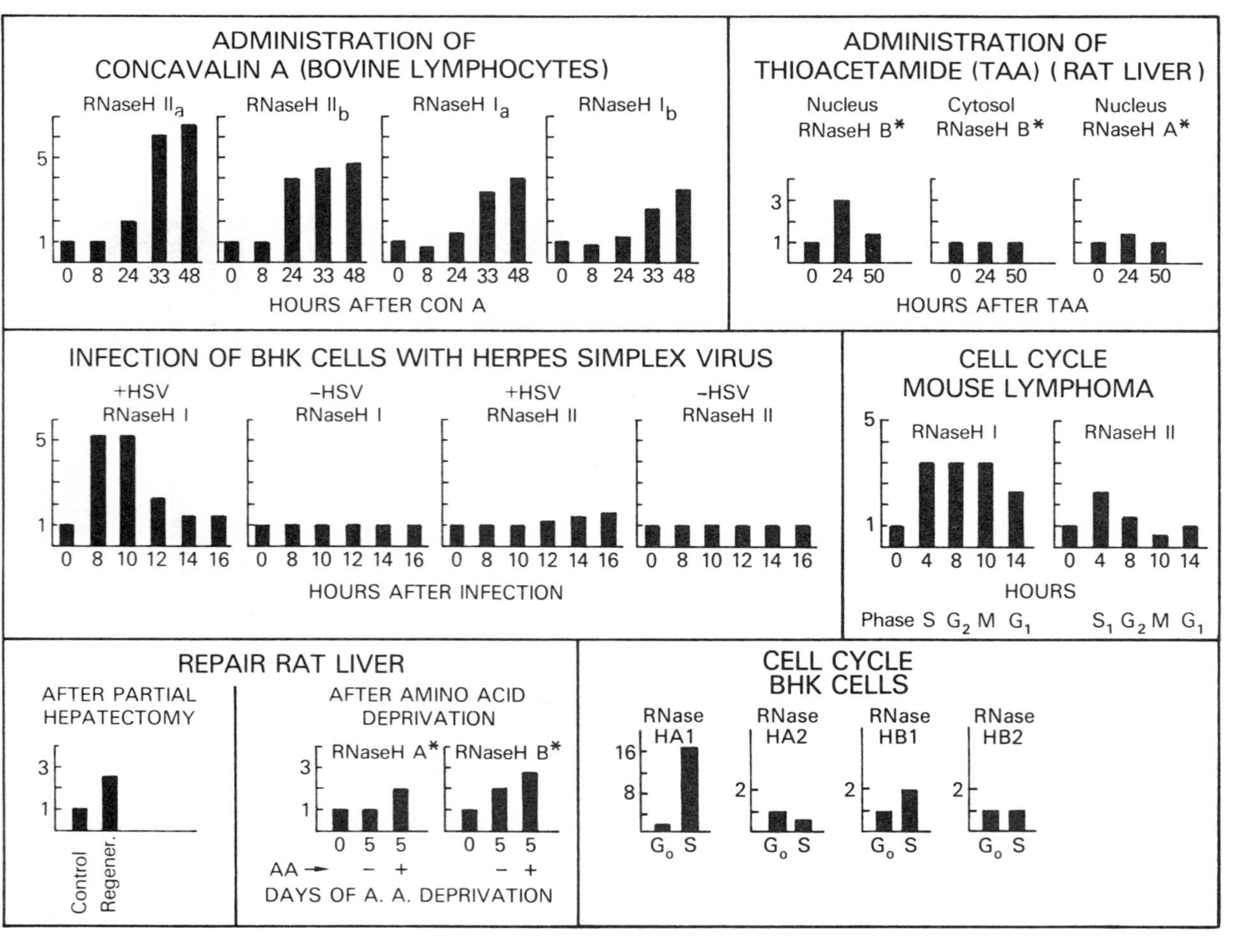

RELATIVE RNaseH ACTIVITY
ADMINISTRATION OF
CONCAVALIN A (BOVINE LYMPHOCYTES)
RNaseH IIa
RNaseH IIb
RNaseH Ia
RNaseH Ib
HOURS AFTER CON A
ADMINISTRATION OF
THIOACETAMIDE (TAA) (RAT LIVER)
Nucleus RNaseH B*
Cytosol RNaseH B*
Nucleus RNaseH A*
HOURS AFTER TAA
INFECTION OF BHK CELLS WITH HERPES SIMPLEX VIRUS
+HSV RNaseH I
−HSV RNaseH I
+HSV RNaseH II
−HSV RNaseH II
HOURS AFTER INFECTION
CELL CYCLE
MOUSE LYMPHOMA
RNaseH I
RNaseH II
HOURS
Phase S G2 M G1
S1 G2 M G1
REPAIR RAT LIVER
AFTER PARTIAL HEPATECTOMY
Control
Regener.
AFTER AMINO ACID DEPRIVATION
RNaseH A*
RNaseH B*
AA →
DAYS OF A. A. DEPRIVATION
CELL CYCLE
BHK CELLS
RNase HA1
RNase HA2
RNase HB1
RNase HB2
Go S

Development

The level and type of RNaseH activities observed during differentiation depend on the stage at which they are measured. Orloski et al. (1980) monitored RNaseH activity in mammary gland and mammary carcinoma tissue during the entire lactation cycle from the onset of pregnancy to involution. Changes in RNaseH activity that parallel differentiation of the mammary gland were found. The striking increase in the activity of the enzyme observed during pregnancy and lactogenesis ceases 2 days after birth when the cells have attained a wholly differentiated state. Enzyme activity continually declines to virgin levels during lactation and involution. In virgin animals, RNaseH activity is very low in normal mammary tissue but strikingly high in that of implanted mammary carcinoma, which mimics the stage of differentiation of the mammary gland in late pregnancy. During pregnancy and lactation of the host animal, changes in RNaseH activity of the tumor are slight but follow the same pattern seen in the normal mammary gland. Two days after birth, the peak levels of activity of the mammary gland are higher than those of the tumor. Orloski et al. 1980 suggest this difference in mammary gland and tumor RNaseH activities may reflect the wholly differentiated state of the mammary cells and partial differentiation in the tumor cells.

Results in congruence with this view were reported by Sawai et al. (1977, 1980a), who followed RNaseH activities in the rat brain from birth to adulthood. During postnatal neurogenesis of the cerebellum, RNaseH activity in that tissue increases after birth to peak levels on the sixth day and then decreases to low adult levels. In contrast, in noncerebellar brain where nerve cell division is essentially a prenatal event, the enzyme activity continually declines from the elevated level at birth to the lower adult level. Similar to the noncerebellar brain is rat liver where the RNaseH activity is high at birth and then declines steadily to adult levels (Sawai and Tsukada 1977).

Figure 3 Cell changes and RNaseH activity. Data compiled from the original literature are shown in unified format. A value of one was assigned to all controls and relative activities were computed as multiples of the control. In the interpretation of these results, each investigation should be treated as an independent unit. Any generalized conclusion drawn from the data as presented in this figure has to be restricted to a qualitative level because of the diversity of the systems studied, reference points used to establish control levels, and enzymatic assays. The ordinate of all figures gives relative RNaseH activities as defined above. The other parameters for a given set of data are indicated on the abscissa. For multiple RNaseH activities, numerical or letter designations given in the original publications are used. Otherwise, RNaseH A* or RNaseH B* are used to indicate different activities within a particular system. The assignment of A* or B* is based on the order of elution during separation of the two species by chromatography. References: Doenecke et al. (1972); Miller and Gill (1975); Büsen et al. (1977); Duff and Keir (1977); Sawai et al. (1977, 1980a,b); Sawai and Tsukada (1977); Tsukada et al. (1978); Müller et al. (1980a,b); Orloski et al. (1980).

A surge of RNaseH activity at the onset of differentiation was found by Doenecke et al. (1972) during metamorphosis of blowfly larvae. When added to hormone-depleted animals, β-ecdysone, which is elaborated during metamorphosis just prior to the increase in RNaseH activity, elicits the same type of increase in RNaseH activity.

Cell Proliferation: Cell Cycle

RNaseH activity during the complete cell cycle of L5178Y mouse lymphoma cells has been determined by Müller et al. (1980b) and at mid-S phase of BHK-21/C13 cells by Duff and Keir (1977). During S phase of synchronously growing mouse lymphoma cells, the activities of RNaseH II (30K) and of RNaseH I (70K) increase to maximum levels, a level that is maintained throughout most of the cycle by RNaseH I. The activity of RNaseH II, on the other hand, decreases during G_2 and M phases to a minimum below that found in the resting stage. At the end of M phase, the cells have doubled. After synchronous growth of BHK cells to mid-S phase, the activities of RNaseH A_1 and RNaseH B_1 (Cooper et al. 1974) are, respectively, 17 and 2 times higher than the corresponding activities at the quiescent (G_0) state.

Regenerating Rat Liver

Sawai and Tsukada (1977) showed changes in RNaseH activity following partial hepatectomy of rat liver. They observed an approximately twofold increase in RNaseH activity between 12 and 24 hours after the operation.

Stimulation of Bovine Lymphocytes by Concanavalin A

In a study of the temporal sequence of changes in multiple RNaseH activities during concanavalin A (Con A)-induced growth of bovine lymphocytes, Büsen et al. (1977) found that RNaseH activity increases in the stimulated cells after a lag period and that the lag period is different for individual RNaseH activities. The most pronounced difference in lag period is seen between RNaseH II_b (30K) and RNaseH I_a and I_b (70K).

Administration of Thioacetamide to Rats (Liver RNaseH)

Stimulation of RNaseH activities in the liver of rats that had received a single dose of thioacetamide was tested by Tsukada et al. (1978). They found that the RNaseH activities of the nuclear fraction increased after a lag period whereas the activity in the cytosol remained constant and only one activity RNaseH B (35K) in the nuclear fraction showed substantial elevation at 24 hours after treatment. It returned to control levels by 48 hours.

Rat Liver RNaseH Activity after Amino Acid Deprivation

The activity of two nuclear RNaseH species from hepatocytes was measured as a function of time of amino acid deprivation in rats by Sawai et al. (1980b).

During the period of dietary deficiency, they observed a gradual increase in only one, RNaseH B, of the two activities assayed, but an increase in both 20 hours after one feeding of high protein content and thyroid hormone to the deprived rats. Sawai et al. (1980b) also includes a table of the distribution of two RNaseH activities in tissues from 10 organs of the rat. The highest activity levels are found in thymus and bone marrow and the lowest in kidney and heart. The ratio of the two activities varies from 1.41 to 1.68.

Administration of ACTH

It has been reported by Miller and Gill (1975) that administration of ACTH to guinea pigs has no effect on the adrenocortical RNaseH activity measured 14 hours later. Similarly, no change in RNaseH activity in nuclear or cytoplasmic fractions of Y1 mouse adrenal tumor cells was found 24 hours after addition of ACTH to the tissue culture medium.

Infection of BHK Cells with Herpes Simplex Virus

Müller et al. (1980a) followed the activities of RNaseH I (75K) and RNaseH II (30K) after infection of BHK cells with herpes simplex virus and found a substantial increase only for RNaseH I. This activity reached a maximum between 8 and 10 hours after infection and returned by 14 hours to a level close to that of the uninfected cells. The activity of RNaseH II changed very slightly between 14 and 16 hours after infection.

Independence of Multiple RNaseH Activities

Independent behavior of multiple RNaseH activities is clearly seen in cell cultures where synchrony of cell activity is predominant (e.g., cell cycle of mouse lymphoma cells, stimulation of bovine lymphocytes by Con A, infection of BHK cells with HSV). It can also be discerned in whole tissues where an external stimulus probably leads to conditions that simulate synchrony. During spontaneous differentiation of a whole organ (rat brain), the two RNaseH activities assayed seemed to change in concert. These are average values of a given RNaseH activity in the heterogeneous population of a pool of whole organs from several animals. In individual cells the two activities may vary independently.

Is There a Correlation between RNaseH Activity and Transcription or DNA Replication?

RNaseH levels correlate with transcription or DNA replication sometimes—but not always.

Correlations with DNA Synthesis

Concomitant increases in DNA synthesis, DNA polymerase α, and RNaseH I_a and I_b (70K, Mg^{2+} or Mn^{2+} activated) activities were established by Büsen et al. (1977) in a kinetic analysis of RNaseH activities, DNA, RNA, and protein synthesis, after stimulation of lymphocytes by Con A. Upon infection of BHK cells with HSV, Müller et al. (1980a) found that selective induction of one RNaseH activity (RNaseH I, 75K) parallels that of HSV-coded DNA polymerase and alkaline DNase as well as HSV-DNA synthesis. The coincidence of DNA synthesis and the rise and decline of RNaseH II activity during the lymphoma cell cycle is not very striking, although the peak of DNA synthesis is seen during the period of maximal RNaseH II activity (Müller et al. 1980a). During regeneration after partial hepatectomy, a rise in RNaseH and DNA ligase activity coincides with DNA synthesis in the rat liver (Sawai and Tsukada 1977). A feeding of protein to amino acid-deprived rats results in simultaneous stimulation of mitosis, DNA synthesis, and two RNaseH activities (Sawai et al. 1980b). In the development of the rat liver a parallel decline from birth to adulthood of DNA synthesis and the activities of RNaseH, DNA ligase, DNA polymerase, and thymidine kinase (TK) were found by Sawai et al. (1977). In the postnatal development of the rat cerebellum, a parallel rise and decline of DNA synthesis and the activities of RNaseH, DNA polymerase, DNA ligase, and TK were reported (Sawai et al. 1977, 1980a). A correlation between DNA synthesis and RNaseH activity in synchronously growing BHK 21/C13 cells was reported by Duff and Keir (1977). At the quiescent (G_0) state, neither DNA synthesis nor DNA polymerase α activity was detectable. One (RNaseH A_1) out of four RNaseH activities was very low. After synchronous growth to mid-S phase, DNA synthesis reached peak levels and the activities of RNaseH A_1 and B_1 were, respectively, 17 and 2 times higher than the corresponding activities at G_0. RNaseH activity and DNA synthesis in rat mammary glands were found to increase and decrease in concert during the lactation cycle (Orloski et al. 1980). At a time during lactation when both decline continuously, RNA and protein synthesis proceed at accelerated rates. In addition, the activity of RNaseH and the rate of DNA synthesis in mammary tumor implants in virgin rats were both found to be high.

Correlations with RNA Synthesis

The stage in the development of the blowfly, chosen by Doenecke et al. (1972) to examine the RNaseH activities, was based on reports of dramatic increases of RNA at that point. They found a selective increase in one RNaseH activity. After stimulation of lymphocytes by Con A, Büsen et al. (1977) observed the selective increase of one RNaseH activity (30K, Mg^{2+} activated) concomitant with RNA synthesis. Sawai et al. (1980b) found a selective increase in one RNaseH activity (30K, Mg^{2+} activated) in parallel with an elevation in RNA polymerase levels and accumulation of nucleolar RNA (see also Tsukada et al. 1978).

No Correlation with DNA or RNA Synthesis

Miller and Gill (1975) examined the RNaseH activity in guinea pig cortices 14 hours after adrenocorticotropin hormone (ACTH) adminstration, because a doubling in DNA synthesis had been reported to occur at that time point. Similarly, the above authors determined the relative RNaseH activity in adrenal tumor cells 24 hours after addition of ACTH to the tissue culture medium since it had been reported that 90% inhibition of DNA synthesis occurred at that time. In both cases, a lack of effect of ACTH on RNaseH activity was found.

MEASUREMENT OF HYDROLYSIS OF RNA

Changes in Solubility of Substrate

Determination of the extent of hydrolysis of RNA is generally based on radioactivity recovered in remaining substrate or the lower-molecular-weight products. Trichloroacetic acid, perchloric acid, or ethanol are used to precipitate the high-molecular-weight material in the reaction mixture. After centrifugation, the radioactivity of the supernatant is determined as a direct measure of the hydrolysis of the RNA. By far the most widely used approach is an indirect one, where the difference between the total radioactivity and that of the remaining substrate gives the amount of RNA hydrolyzed. The remaining substrate is recovered either as the acid- or alcohol-insoluble fraction or is selectively retained on Whatman DE81 filter paper or on nitrocellulose membranes (Doenecke et al. 1972; Wyers et al. 1973; Büsen and Hausen 1975; Miller and Gill 1975; Orloski et al. 1980).

Determination of Phosphodiester Bond Cleavage

Two assays for the scission of phosphodiester bonds by RNaseH have been described in the literature. The rationale for each procedure is outlined in Figure 4. In method I, the 5′-terminal nucleotide of the enzymatic hydrolysis product is distinguishable from the others after alkaline hydrolysis whereas in method II the 3′-terminus is altered by periodate oxidation.

Simultaneous Quantitative and Qualitative Assay of the Digestion Products

PEI electrophoresis of the products of the digestion of homopolymers by RNaseH (Fig. 4) yields an array of di- and oligonucleotides (n_3 to n_9). This qualitative analysis can serve as an assay of the rate of degradation of the substrate as well. Application of the sample onto PEI terminates the reaction.

Fluorimetric Assay

A fluorimetric assay of RNaseH activity that has been developed by Morgan et al. (1979) is based on the enhancement of ethidium fluorescence upon intercalation with duplex nucleic acids. Thus, the destruction of the hybrid by RNaseH is accompanied by a drop in fluorescence. Since the substrate is not radioactively labeled, higher levels than in other assays are required.

Selective Assay of Endonucleolytic Action

To determine the mode of action of a given RNaseH, closed-circular superhelical DNA of the plasmid ColE1 containing covalently inserted ribonucleotides has been used as substrate. In gradient centrifugation of the digest, a shift in the S value to a lower value characteristic of open circles indicates endonucleolytic action. RNaseH with exonuclease specificity is unable to attack this substrate (Keller and Crouch 1972). This technique has been updated by Büsen (1980a) by taking advantage of the difference in a mobility on agarose gel electrophoresis of the cleaved forms of ColE1 DNA.

Electrophoresis of Digestion Products

To monitor the size distribution of products as well as the extent of hydrolysis during enzymatic digestion, the reaction mixture is subjected to electrophoresis in polyacrylamide gels in 7 M urea. This method is similar to the electrophoresis in 7 M urea on PEI sheets but since its resolution includes considerably larger fragments (Grandgenett and Green 1974), it can be used to detect enzymatic activities that yield long oligonucleotides as end products.

In Situ Assay of RNaseH Activity after SDS Gel Electrophoresis

After electrophoresis in SDS polyacrylamide gels containing macromolecular substrate, some nucleases can regain enzymatic activity upon removal of the SDS. Gels containing DNA-RNA hybrids are prepared according to Rosenthal and Lacks (1977) and sample preparation is identical with that for analysis of proteins in SDS-PAGE. After electrophoresis, SDS is allowed to diffuse out of the gel by prolonged standing in several changes of assay buffer. Since enzyme and substrate are immobilized in the matrix of the gel, the reactivated enzyme degrades only the substrate within reach. As the low-molecular-weight products diffuse out, the area occupied by the enzyme becomes devoid of RNA. Thus, when the substrate is radioactive (Huet et al. 1978), autoradiography of the gel reveals the position of the enzyme as a clear area (Fig. 4C). Note that a negative of the actual autoradiogram is shown and RNaseH appears as a dark band. It is possible to stain the same gel with Coomassie Blue to ascertain the position and amount of protein applied to the gel.

RNaseH AS A TOOL FOR STUDYING NUCLEIC ACIDS

A review of the application of RNaseH to the study of nucleic acid structure has appeared (Crouch 1981) and only a brief summary will be presented here.

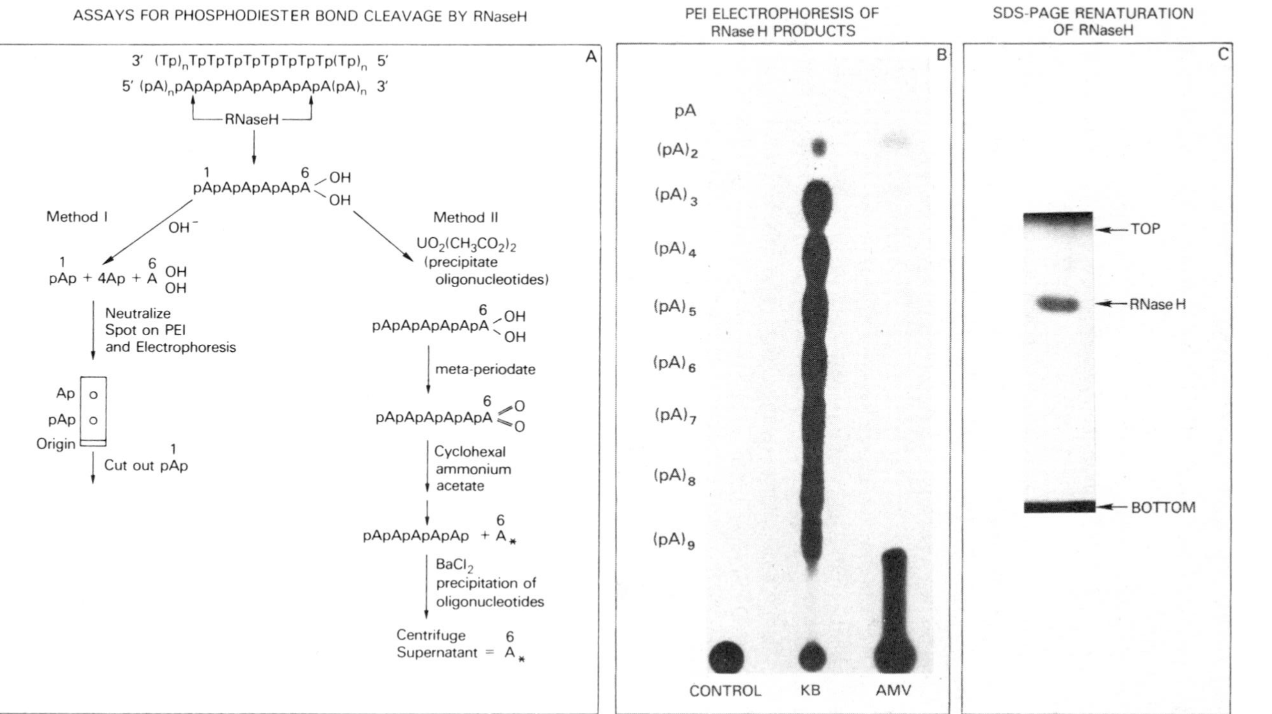

Figure 4 Assays for RNaseH activity used to detail important characteristics of the enzymes. (*A*) Two assays that measure cleavage of a single diester bond. In method I (Dirksen and Crouch 1981), the 5′ phosphate generated by RNaseH is detected, and, in method II (Stavrianopoulos and Chargaff 1976), the formation of 3′-OH groups is measured. (*B*) An autoradiogram of RNaseH products separated by PEI electrophoresis (e.g., KB RNaseH [high molecular weight] gives oligonucleotides 2 to 9 in length; AMV reverse transcriptase produces dinucleotides and larger oligonucleotides greater than 9 in length). The substrate was poly(rA)·poly(dT) labeled with ^{32}P in the poly(A). (*C*) Negative of an autoradiogram of a SDS-PAGE renaturation assay of *E. coli* RNaseH. The dark band indicating RNaseH activity appears as an unexposed area on the original autoradiogram.

RNaseH has been used to demonstrate DNA-RNA hybrids (e.g., the RNA in ColE1 [Keller and Crouch 1972]). A rather sensitive assay for such hybrids was developed by Jacquet et al. (1974) in which labeled DNA is hybridized to mRNA and unhybridized DNA is removed. RNaseH treatment exposes the labeled DNA to a single-strand-specific nuclease. Signal-to-noise ratios are significantly improved by this multistep procedure.

Removal of poly(A) sequences from mRNA by digestion with RNaseH in the presence of oligo(dT) has been employed to examine the function of poly(A) sequences (Sippel et al. 1974; Spector and Baltimore 1974) and to produce sharper bands of RNA during polyacrylamide gel electrophoresis (Vournakis et al. 1975).

Specific fragmentation of RNA has been achieved for detection of differences in termini of α-amylase mRNA (Hagenbüchle et al. 1980), for sequencing RNAs (Donis-Keller 1979; Harris 1979; Reddy et al. 1980; Darlix et al. 1981; Donis-Keller et al. 1981) and for genomic mapping (Darlix et al. 1977a).

A new procedure for obtaining high yields of full-length cDNA clones has been developed that takes advantage of the endonucleolytic nature of *E. coli* RNaseH. By an ingenious technique, a cDNA-mRNA hybrid is formed that is then covalently bound into a circular plasmid DNA. The DNA-RNA hybrid is converted to duplex DNA by simultaneous digestion of the RNA with RNaseH and resynthesis with *E. coli* DNA polymerase I (Okayama and Berg 1982). RNaseH acts to cleave the RNA and to produce primers for DNA synthesis.

POSSIBLE FUNCTIONS OF RNaseH

Removal of RNA Primers

Keller (1972) suggested RNaseH could be involved in the removal of RNA primers during DNA replication. To function effectively, RNaseH should remove all ribonucleotides in a DNA-RNA hybrid, particularly the ribonucleotide to which the first deoxyribonucleotide is joined. Darlix (1975) reported that *E. coli* RNaseH preparations were capable of breaking that bond, thus completely removing RNA primers. Unpublished experiments by W. Keller and R.J. Crouch indicate that KB RNaseH (high molecular weight) can also remove the final ribonucleotide but that high amounts of enzyme are required. Other RNasesH (Banks 1974) fail to cleave the ribo-deoxyribonucleotide junction.

Generation of Primers

Miller et al. (1973) suggested that RNaseH could be used to generate primers for DNA replication since the products of RNaseH digestion have 3′-hydroxyl groups. The same suggestion was made (Itoh and Tomizawa 1979, 1980;

Tomizawa and Selzer 1979) after finding that ColE1 plasmid DNA requires RNaseH for in vitro initiation of DNA replication (Itoh and Tomizawa 1979, 1980). Conclusive proof of this role awaits isolation of a mutant completely lacking in RNaseH activity, since a mutant with lowered levels of RNaseH can still support the growth of ColE1 plasmids (Carl et al. 1980).

In a highly purified DNA synthesizing system, ϕX174 DNA can be replicated with RNA polymerase as the enzyme responsible for primer production. Sumida-Yasumoto et al. (1978) and Vicuna et al. (1977a,b) have shown that ϕX174 DNA is not replicated in this in vitro system in the presence of RNaseH.

Transcription

RNaseH might participate in transcription as an editing function (Iborra et al. 1979) or as a terminator. Yeast RNaseH I, which is found closely associated with RNA polymerase A (see above), is inhibited if the polymerase is actively transcribing (Huet et al. 1978) and transcription of either DNA or chromatin in vitro is reportedly stimulated by RNaseH (Sekeris et al. 1972).

Chromatin Structure

Worcel and Burgi (1972) and Drlica and Worcel (1975) demonstrated the sensitivity of bacterial nucleoids to pancreatic RNase and, thus, a role for RNA in the maintenance of nucleoid structure. Wyers et al. (1976b) suggested that DNA-RNA hybrids may be responsible for some part of such structures and that folding and unfolding of nucleoids could be regulated by RNaseH.

Requirement for Retroviral DNA Synthesis

The presence of RNaseH on retroviral reverse transcriptase implies a role for this RNaseH in DNA replication. Most models of DNA synthesis of RNA tumor viruses require RNaseH activity to remove the 5′-portion of the genomic RNA after the initial synthesis of DNA from a tRNA primer. Exposure of the single-stranded DNA by RNaseH permits annealing of the DNA with a complementary region at the 3′-end of the genomic RNA. Further extension of the DNA produces a complete copy of the genomic RNA (see Moelling [1975], Verma [1977], and Bishop [1978] for details of these models). These models require RNaseH to be active under conditions in which DNA synthesis is occurring (Watson et al. 1979). Degradation of RNA following reverse transcription supports this proposal (Darlix et al. 1977a,b; Collett et al. 1978; Friedrich and Moelling 1979).

CONCLUDING REMARKS

RNasesH are present in a wide variety of cells, and work on these enzymes has been directed toward understanding their in vivo function(s). This has proved difficult because multiple RNasesH are often found in single cell types and they apparently occur in both the nucleus and cytoplasm. In the case of retroviral reverse transcriptase RNasesH, proteolytic cleavages of the enzyme can produce smaller proteins that still possess RNaseH activity. Recent evidence suggests that proteolysis of cellular RNasesH may occur that could lead to the multiple species observed in single cell types (Wada et al. 1980). Additionally, RNaseH may be a strictly nuclear enzyme, and cytoplasmic activities may result from leakage during preparation.

Most RNasesH yield small oligonucleotides as end products. However, some forms of retroviral RNasesH generate rather large oligomeric products. These would normally not be seen in assays based on differential solubility of the substrate and products. This leaves open the possibility that other RNasesH remain to be discovered.

Changes in RNaseH activities accompanying changes in cell activity have been considered by many authors as evidence supporting the participation of RNaseH in one or another metabolic event. Some correlation with DNA synthesis or transcription of RNA has been found in several instances, but no definitive role for an RNaseH has yet been found. In vitro replication of ColE1 DNA requires *E. coli* RNaseH for proper initiation, although its participation in vivo remains to be demonstrated. The recent isolation of a clone that apparently carries the *E. coli* RNaseH gene (Horiuchi et al. 1981) should aid in both the in vivo and the in vitro characterization of this interesting nuclease.

ACKNOWLEDGMENTS

We wish to thank Werner Büsen, Gary Gerard, Karin Moelling, Yasuko Sawai, Dennis Schwartz, and Andre Sentenac for communication of materials prior to publication. Thanks also go to C. and M. Chrambach for aid in organization. Terri Broderick deserves special thanks for her tireless efforts in typing this review.

REFERENCES

Arrendes, J., P.L. Carl, and A. Sugino. 1982. A mutation in the *rnh* locus of *Escherichia coli* affects the structural gene for RNaseH. *J. Biol. Chem.* **257**:4719.

Avramova, Z.V. and Z.I. Galcheva-Gargova. 1980. A ribonuclease specific for double-stranded RNA and two distinct ribonuclease H activities from the ribosomal salt wash fraction of Ehrlich ascites tumor cells. *Arch. Biochem. Biophys.* **204**:167.

Baltimore, D. and D.F. Smoler. 1972. Association of an endoribonuclease with the avian myeloblastosis virus deoxyribonucleic acid polymerase. *J. Biol. Chem.* **247**:7282.

Baltimore, D., I.M. Verma, D.F. Smoler, and N.L. Meuth. 1972. Avian myeloblastosis virus DNA polymerase. In *DNA synthesis* in vitro (ed. R.D. Wells and R.B. Inman), p. 333. University Park Press, Baltimore, London, Tokyo.

Banks, G.R. 1974. A ribonuclease H from *Ustilago maydis*. Properties and substrate specificity of the enzyme. *Eur. J. Biochem.* **47**:499.

Berkower, I., J. Leis, and J. Hurwitz. 1973. Isolation and characterization of an endonuclease from *Escherichia coli* specific for ribonucleic acid in ribonucleic acid deoxyribonucleic acid hybrid structures. *J. Biol. Chem.* **248:**5914.

Bishop, J.M. 1978. Retroviruses. *Annu. Rev. Biochem.* **47:**35.

Blair, D.G., D.J. Sherratt, D.B. Clewell, and D. Helinski. 1972. Isolation of supercoiled colicinogenic factor E_1 DNA sensitive to ribonuclease and alkali. *Proc. Natl. Acad. Sci.* **69:**2518.

Brewer, L.C. and R.D. Wells. 1974. Mechanistic independence of avian myeloblastosis virus DNA polymerase and ribonuclease H. *J. Virol.* **14:**1494.

Büsen, W. 1980a. Purification, subunit structure, and serological analysis of calf thymus ribonuclease H I. *J. Biol. Chem.* **255:**9434.

———. 1980b. Biochemical, serological and functional differences between the various ribonuclease H activities of bovine tissue. In *Biological implications of protein-nucleic acid interactions* (ed. J. Augustyniak), p. 571. Elsevier/North-Holland Biomedical Press, New York.

———. 1982. The subunit structure of calf thymus ribonuclease HI as revealed by immunological analysis. *J. Biol. Chem.* **257:**7106.

Büsen, W. and P. Hausen. 1975. Distinct ribonuclease H activities in calf thymus. *Eur. J. Biochem.* **52:**179.

Büsen, W., J.H. Peters, and P. Hausen. 1977. Ribonuclease H levels during the response of bovine lymphocytes to concanavalin A. *Eur. J. Biochem.* **74:**203.

Carl, P.L., L. Bloom, and R.J. Crouch. 1980. Isolation and mapping of a mutation in *Escherichia coli* with altered levels of ribonuclease H. *J. Bacteriol.* **144:**28.

Cathala, G., J. Rech, J. Ruet, and P. Jeanteaur. 1979. Isolation and characterization of two types of ribonucleases H in Krebs II ascites cells. *J. Biol. Chem.* **254:**7353.

Collett, M.S., P. Dierks, and J.T. Parsons. 1978. RNaseH hydrolysis of the 5′-terminus of the avian sarcoma virus genome during transcription. *Nature* **272:**181.

Cooper, R.J., P.M. Duff, A. Olivier, R.K. Craig, and T.I. Keir. 1974. Multiple ribonuclease H activities from BHK-21-C13. *FEBS Lett.* **45:**38.

Copeland, T.D., D.P. Grandgenett, and S. Oroszlan. 1980. Amino acid sequence analysis of reverse transcriptase subunits from avian myeloblastosis virus. *J. Virol.* **36:**115.

Crouch, R.J. 1974. Ribonuclease III does not degrade deoxyribonucleic:ribonucleic acid hybrids. *J. Biol. Chem.* **249:**1314.

———. 1981. Analysis of nucleic acid structure by RNase. In *Gene amplification and analysis* (ed. J.G. Chirikjian and T.S. Papas), vol. 2, p. 217. Elsevier/North Holland, New York.

Darlix, J.-L. 1975. Simultaneous purification of *Escherichia coli* termination factor rho, RNase III and RNaseH. *Eur. J. Biochem.* **51:**369.

Darlix, J.-L., P.A. Bromley, and P.F. Spahr. 1977a. New procedure for the direct analysis of *in vitro* reverse transcription of Rous sarcoma virus RNA. *J. Virol.* **22:**118.

———. 1977b. Extensive *in vitro* transcription of Rous sarcoma virus RNA by avian myeloblastosis virus DNA polymerase and concurrent activation of the associated RNaseH. *J. Virol.* **23:**659.

———. 1981. A model structure for Rous sarcoma virus genomic RNA and its implications for various functions of the viral RNA. *Mol. Biol. Rep.* **7:**127.

Dezelee, S., F. Wyers, J.-L. Darlix, A. Sentenac, and P. Fromgeot. 1977. DNA binding, condensing and unwinding properties of yeast RNaseH. *J. Biol. Chem.* **232:**8935.

Dion, A.S., C.J. Williams, and D.H. Moore. 1977. RNaseH and RNA-directed DNA polymerase: Associated enzymatic activities of murine mammary tumor virus. *J. Virol.* **22:**187.

Dirksen, M.-L. and R.J. Crouch. 1981. Selective inhibition of RNaseH by dextran. *J. Biol. Chem.* **256:**11569.

Doenecke, D., V.J. Marmaras, and C.E. Sekeris. 1972. Increased RNase H (hybridase) activity in the integument of blowfly larvae during development and under the influence of β-ecdysone. *FEBS Lett.* **22:**261.

Donis-Keller, H. 1979. Site specific enzymatic cleavage of RNA. *Nucleic Acids Res.* **7:**179.

Donis-Keller, H., K.S. Browning, and J.M. Clark, Jr. 1981. Sequence heterogeneity in satellite tobacco necrosis virus RNA. *Virology* **110:**43.

Drlica, K. and A. Worcel. 1975. Conformational transitions in the *Escherichia coli* chromosome: Analysis by viscometry and sedimentation. *J. Mol. Biol.* **98:**393.

Duff, P.M. and H.M. Keir. 1977. The relationship between the activity of ribonuclease H and synthesis of deoxyribonucleic acid in synchronously growing BHK-21/C13 cells. *(Proceedings) Biochem. Soc. Trans.* **5:**676.

Ferrari, S., C.O. Yehle, H.D. Robertson, and E. Dickson. 1977. Specific RNA-cleaving activities from HeLa cells. *Proc. Natl. Acad. Sci.* **77:**2395.

Flügel, R.M., J.E. Larson, P.F. Schendel, R.W. Sweet, T.R. Tamblyn, and R.D. Wells. 1972. RNA-DNA bonds formed by DNA polymerases from bacteria and RNA tumor viruses. In *DNA synthesis* in vitro (ed. R.D. Wells and R.B. Inman), p. 309. University Park Press, Baltimore, London, Tokyo.

Friedrich, R. and K. Moelling. 1979. Effect of viral RNaseH on the avian sarcoma viral genome during early transcription *in vitro*. *J. Virol.* **31:**630.

Galcheva-Gargova, Z.I., Z.V. Avramova, and G.I. Milchew. 1980. Three distinct RNaseH activities isolated from the cytosol of Ehrlich ascites tumor cells. *C. R. Acad. Bulg. Sci.* **33:**551.

Gerard, G.F. 1978. Multiple RNaseH activities in mammalian type C retrovirus lysates. *J. Virol.* **26:**16.

———. 1981a. Mechanism of action of Moloney murine leukemia virus RNase III. *J. Virol.* **37:**748.

———. 1981b. Mechanism of action of Moloney murine leukemia virus RNA-directed DNA polymerase associated RNase (RNase HI). *Biochemistry* **20:**256.

Gerard, G.F. and D.P. Grandgenett. 1975. Purification and characterization of the DNA polymerase and RNaseH activities in Moloney murine sarcoma-leukemia virus. *J. Virol.* **15:**785.

Gibson, W. and I.M. Verma. 1974. Studies on the reverse transcriptase of RNA tumor viruses. Structural relatedness of two subunits of avian RNA tumor viruses. *Proc. Natl. Acad. Sci.* **71:**4991.

Gorecki, M. and A. Panet. 1978. Discrimination of DNA polymerase and RNaseH activities in reverse transcriptase of avian myeloblastosis virus. *Biochemistry* **17:**2438.

Grandgenett, D. 1975. Correlation of the binding properties of alpha and alpha beta DNA polymerase of avian myeloblastosis virus with their different mode of ribonuclease H activity. In *DNA synthesis and its regulation* (ed. M. Goulian and P. Hanawalt), p. 760. Benjamin, Menlo Park, California.

Grandgenett, D.P. and M. Green. 1974. Different mode of action of ribonuclease H in purified alpha beta ribonucleic acid-directed deoxyribonucleic polymerase from avian myeloblastosis virus. *J. Biol. Chem.* **249:**5148.

Grandgenett, D.P. and H.M. Rho. 1975. Binding properties of avian myeloblastosis virus DNA polymerases to nucleic acid affinity columns. *J. Virol.* **15:**526.

Grandgenett, D.P., G.F. Gerard, and M. Green. 1972. Ribonuclease H: A ubiquitous activity in virions of ribonucleic acid tumor viruses. *J. Virol.* **10:**1136.

———. 1973. A single subunit from avian myeloblastosis virus with both RNA-directed DNA polymerase and ribonuclease H activity. *Proc. Natl. Acad. Sci.* **70:**230.

Green, M., D.P. Grandgenett, G.F. Gerard, H.M. Rho, M.C. Loni, M. Robins, S. Salzberg, G. Shanmugam, S. Bhaduri, and G. Vecchio. 1974. Properties of oncornavirus RNA-directed DNA polymerase, the RNA template, and the intracellular products formed early during infection and cell transformation. *Cold Spring Harbor Symp. Quant. Biol.* **39:**975.

Grossman, L.I., R. Watson, and J. Vinograd. 1973. The presence of ribonucleotides in mature closed-circular mitochondrial DNA. *Proc. Natl. Acad. Sci.* **70:**3339.

Haberkern, R.C. and G.L. Cantoni. 1973. Studies on a calf thymus ribonuclease specific for ribonucleic acid-deoxyribonucleic acid hybrids. *Biochemistry* **12:**2389.

Hagenbüchle, D., R. Bovey, and R.A. Young. 1980. Tissue specific expression of mouse α-amylase genes: Nucleotide sequence of isoenzyme mRNAs from pancreas and salivary gland. *Cell* **21:**179.

Hall, S.H. and R.J. Crouch. 1977. Isolation and characterization of two enzymatic activities from chick embryos which degrade double-stranded RNA. *J. Biol. Chem.* **252:**4092.

Harris, T.J. 1979. The nucleotide sequence at the 5′ end of foot and mouth disease virus RNA. *Nucleic Acids Res.* **11:**1765.

Hausen, P. and H. Stein. 1970. Ribonucleic acid: An enzyme degrading the RNA moiety of DNA-RNA hybrids. *Eur. J. Biochem.* **14:**278.

Henry, C.M., F.-J. Ferdinand, and R. Knippers. 1973. A hybridase from *Escherichia coli. Biochem. Biophys. Res. Commun.* **50:**603.

Hizi, A. and W.K. Joklik. 1977. RNA-dependent DNA polymerase of avian sarcoma virus B77. I. Isolation and partial characterization of the α, β_2 and $\alpha\beta$ forms of the enzyme. *J. Biol. Chem.* **252:**2281.

Hizi, A. and A. Yaniv. 1980. RNA-dependent DNA polymerase of an endogenous type C virus of mice: Purification and partial characterization. *J. Virol.* **34:**795.

Hizi, A., J.P. Leis, and W.K. Joklik. 1977. RNA-dependent DNA polymerase of avian sarcoma virus B77. II. Comparison of the catalytic properties of the α, β_2 and $\alpha\beta$ enzyme forms. *J. Biol. Chem.* **252:**2290.

Horiuchi, T., H. Maki, M. Maruyana, and M. Sekiguchi. 1981. Identification of the *dnaQ* gene product and location of the structural gene for RNaseH of *Escherichia coli* by cloning of the genes. *Proc. Natl. Acad. Sci.* **78:**3770.

Huet, J., A. Sentenac, and P. Fromageot. 1978. Detection of nucleases degrading double helical RNA and of nucleic acid binding proteins following SDS-gel electrophoresis. *FEBS Lett.* **94:**28.

Huet, J., J.-M. Buhler, A. Sentenac, and P. Fromageot. 1977. Characterization of ribonuclease H activity associated with yeast RNA polymerase A. *J. Biol. Chem.* **252:**8848.

Huet, J., T. Wyers, J.-M. Buhler, A. Sentenac, and P. Fromageot. 1976. Association of RNaseH activity with yeast RNA polymerase A. *Nature* **261:**431.

Iborra, F., J. Huet, B. Breant, A. Sentenac, and P. Fromageot. 1979. Identification of two different RNaseH activities associated with yeast RNA polymerase A. *J. Biol. Chem.* **254:**10920.

Itoh, T. and J. Tomizawa. 1979. Initiation of replication of plasmid ColE1 DNA by RNA polymerase ribonuclease H and DNA polymerase I. *Cold Spring Harbor Symp. Quant. Biol.* **43:**409.

———. 1980. Formation of an RNA primer for initiation of replication of ColE1 DNA by ribonuclease H. *Proc. Natl. Acad. Sci.* **77:**2450.

Jaquet, M., Y. Groner, G. Monroy, and J. Hurwitz. 1974. The *in vitro* synthesis of avian myeloblastosis viral RNA sequences. *Proc. Natl. Acad. Sci.* **71:**3045.

Kacian, D.L., K.F. Watson, A. Burney, and S. Spiegelman. 1971. Purification of the DNA polymerase of avian myeloblastosis virus. *Biochim. Biophys. Acta* **246:**365.

Keller, W. 1972. RNA-primed DNA synthesis *in vitro*. *Proc. Natl. Acad. Sci.* **69:**1560.

Keller, W. and R. Crouch. 1972. Degradation of DNA RNA hybrids by ribonuclease H and DNA polymerases of cellular and viral origin. *Proc. Natl. Acad. Sci.* **69:**3360.

Kleinicke, C., H. Fischer, and R.M. Flügel. 1974. Ribonuclease H activities associated with herpes simplex virus. *Biochem. Biophys. Res. Commun.* **60:**1491.

Lai, M.-H. and I.M. Verma. 1978. Reverse transcriptase of RNA tumor viruses. V. *In vitro* proteolysis of reverse transcriptase from avian myeloblastosis virus and isolation of a polypeptide manifesting only RNaseH activity. *J. Virol.* **25:**652.

Lai, M.-H., I.M. Verma, S.R. Tronic, and S.A. Aaronson. 1978. Mammalian retrovirus-associated RNaseH is virus coded. *J. Virol.* **27:**823.

Leis, J., I. Berkower, and J. Hurwitz. 1972. RNA-dependent DNA polymerase activity of RNA tumor viruses. IV. Characterization of AMV stimulatory protein and RNaseH-associated activity. In *DNA synthesis* in vitro (ed. R.D. Wells and R.P. Inman), p. 287. University Park Press, Baltimore, London, and Tokyo.

Leis, J.P., I. Berkower, and J. Hurwitz. 1973. Mechanism of action of ribonuclease H isolated from avian myeloblastosis virus and *Escherichia coli*. *Proc. Natl. Acad. Sci.* **70:**466.

Leis, J., J. Hurwitz, A.L. Schincariol, M. Stone, and W.K. Joklik. 1976. The avian myeloblastosis virus RNA-dependent DNA polymerase. In *The biology of tumor viruses* (ed. G. Beaudreau), p. 77. Oregon State University Press, Corvallis.

Marcus, S.L., G.W. Smith, and M.J. Modak. 1978. Reverse transcriptase associated RNaseH activity. II. Inhibition by natural and synthetic RNA. *J. Virol.* **27:**576.

Miller, H.I. and G.N. Gill. 1975. Adrenal cortical ribonuclease H (hybrid) (38494). *Proc. Soc. Exp. Biol. Med.* **148:**151.

Miller, H.I., A.D. Riggs, and G.N. Gill. 1973. Ribonuclease H (hybrid) in *Escherichia coli*. Identification and characterization. *J. Biol. Chem.* **248:**2621.

Modak, M.J. 1976. Pyridoxal 5′-phosphate: A selective inhibitor of oncornaviral DNA polymerases. *Biochem. Biophys. Res. Commun.* **71:**180.

Modak, M.J. and S.L. Marcus. 1977. Specific inhibition of DNA polymerase-associated RNaseH by DNA. *J. Virol.* **22:**243.

Modak, M.J. and A. Srivastava. 1979. Reverse transcriptase-associated ribonuclease H does not require zinc for catalysis. *J. Biol. Chem.* **254:**4756.

Moelling, K. 1974a. Reverse transcriptase and RNaseH present in a murine virus and in both subunits of an avian virus. *Cold Spring Harbor Symp. Quant. Biol.* **39:**969.

__________. 1974b. Characterization of reverse transcriptase and RNaseH from Friend-murine leukemia virus. *Virology* **62:**46.

__________. 1975. Ribonuclease H of viral and cellular origin. In *Recent advances in cancer research: Cell biology, molecular biology, and tumor biology* (ed. R.C. Gallo), vol. II, p. 121. CRC Press, Inc., Cleveland, Ohio.

__________. 1976. Further characterization of the Friend leukemia virus reverse transcriptase-RNaseH complex. *J. Virol.* **18:**418.

Moelling, K. and R.R. Friis. 1979. Two avian sarcoma virus mutants with defects in the DNA polymerase-RNaseH complex. *J. Virol.* **32:**370.

Moelling, K., H. Gelderblom, G. Pauli, R. Friis, and H. Bauer. 1975. A comparative study of the avian reticuloendotheliosis virus: Relationship to murine leukemia virus and viruses of the avian sarcoma-leukosis complex. *Virology* **65:**546.

Moelling, K., D.P. Bolognesi, H. Bauer, W. Büsen, H.W. Plassmann, and P. Hausen. 1971. Association of viral reverse transcriptase with an enzyme degrading the RNA moiety of RNA:DNA hybrids. *Nature New Biol.* **234:**240.

Morgan, A.R., D.H. Evans, J.S. Lee, and D.E. Pulleyblank. 1979. Review: Ethidium fluorescence assay. Part II. Enzymatic studies and DNA-protein interactions. *Nucleic Acids Res.* **7:**571.

Müller, W.E.G., D. Falke, R.K. Zahn, and J. Arendes. 1980a. Ribonuclease H levels in herpes simplex virus-infected cells. *Arch. Virol.* **64:**269.

Müller, W.E.G., W. Geurtsen, R.J. Zahn, and J. Arendes. 1980b. Cell cycle-dependent alterations of the two types of ribonucleases H in L5178y cells. *FEBS Lett.* **110:**119.

Natori, S., K. Takeuchi, and D. Mizuno. 1973. DNA-dependent RNA polymerase from Ehrlich ascites tumor cells. 3. Ribonuclease H and elongating activity of stimulatory factor S-II. *J. Biochem.* (Tokyo) **74:**1177.

Nowinski, R.C., K.F. Watson, A. Yaniv, and S. Spiegelman. 1972. Serological analysis of deoxyribonucleic acid and polymerase of avian oncornaviruses. II. Comparison of avian deoxyribonucleic acid polymerases. *J. Virol.* **10:**959.

O'Cuinn, G., F.J. Persico, and A.A. Gottlieb. 1973. Two ribonuclease H activities from the murine myeloma, MOPC-21. *Biochim. Biophys. Acta* **324:**78.

Okayama, H. and P. Berg. 1982. High-efficiency cloning of full-length cDNA. *Mol. Cell Biol.* **2:**161.

Orloski, J.M., P.J. Fritz, and D.K. Liu. 1980. Ribonuclease H activity in rat mammary gland during the lactation cycle and in R3230AC mammary adenocarcinoma. *Biochim. Biophys. Acta* **632:**1.

Panet, A., I.M. Verma, and D. Baltimore. 1974. Role of the subunits of avian RNA tumor virus reverse transcriptase. *Cold Spring Harbor Symp. Quant. Biol.* **39:**919.

Papas, T.S., G.R. Renzi, and W.J. Martin. 1977. Immunological distinction between ribonuclease H activity of alpha and alpha beta forms of avian myeloblastosis virus (AMV) DNA polymerase. *Virology* **76:**882.

Reddy, R., D. Henning, and H. Busch. 1980. Substitutions, insertions, and deletions in two highly conserved U3 RNA species. *J. Biol. Chem.* **255:**7029.

Rho, H.M. and R.C. Gallo. 1980. Biochemical and immunological properties of the DNA polymerase and RNaseH activities of purified feline leukemia virus reverse transcriptase. *Cancer Lett.* **10:**207.

Rho, H.M., D.P. Grandgenett, and M. Green. 1975. Sequence relatedness between the subunits of avian myeloblastosis virus reverse transcriptase. *J. Biol. Chem.* **250:**5278.

Robertson, H.D. 1971. Enzymatic synthesis of bacteriophage f1 DNA:RNA hybrid and double-stranded RNA. *Nature New Biol.* **229:**169.

Robertson, H.D. and J.J. Dunn. 1975. Ribonucleic acid processing activity of *Escherichia coli* by ribonuclease III. *J. Biol. Chem.* **250:**3050.

Robertson, H.D. and N. Zinder. 1968. Selective digestion of a DNA:RNA hybrid by *E. coli* ribonuclease III. *Fed. Proc.* **27:**296.

Robertson, H.D., R.E. Webster, and N. Zinder. 1968. Purification and properties of ribonuclease III from *Escherichia coli*. *J. Biol. Chem.* **243:**82.

Roewekamp, W. and C.E. Sekeris. 1974. Purification and characteristics of hybridase (ribonuclease H) from rat-liver cytosol. *Eur. J. Biochem.* **43:**405.

Rosenthal, A.L. and S. Lacks. 1977. Nuclease detection in SDS-polyacrylamide gel electrophoresis. *Anal. Biochem.* **80:**76.

Rucheton, M., M.N. Lelay, and P. Jeanteur. 1979. Evidence from direct visualization after denaturing and electrophoresis that RNaseH is associated with MSV-MULV reverse transcriptase. *Virology* **97:**221.

Sarngadharan, M.G., J.P. Leis, and R.C. Gallo. 1975. Isolation and characterization of a ribonuclease from human leukemic blood cells specific for ribonucleic acid of ribonucleic acid-deoxyribonucleic acid hybrid molecules. *J. Biol. Chem.* **250**:365.

Sarngadharan, M.G., V.S. Kalyanaraman, and R.C. Gallo. 1978. Inhibition by RNA of RNaseH activity associated with reverse transcriptase in Rauscher murine leukemia virus cores. *J. Virol.* **27:**568.

Sawai, Y. and K. Tsukada. 1977. Change of ribonuclease H activity in developing and regenerating rat liver. *Biochim. Biophys. Acta* **479:**126.

Sawai, Y., J. Saito, and K. Tsukada. 1980a. Developmental changes of two distinct ribonuclease H activities from rat brain and characterization of these enzymes. *Biochim. Biophys. Acta* **630:**386.

Sawai, Y., Y. Sawasaki, and K. Tsukada. 1977. Ribonuclease H activity in developing rat brain. *Life Sci.* **21:**1351.

Sawai, Y., N. Sugano, and K. Tsukada. 1978. Ribonuclease H activity in cultured plant cells. *Biochim. Biophys. Acta* **518:**181.

Sawai, Y., M. Unno, and K. Tsukada. 1978. Two ribonuclease H activities from rat liver nuclei. *Biochem. Biophys. Res. Commun.* **84:**313.

Sawai, Y., K. Wada, and K. Tsukada. 1980b. Change of nuclear ribonuclease H activity following deoxyribonucleic acid synthesis in liver of protein-deficient rat. *Life Sci.* **26:**1497.

Sawai, Y., M. Yanokura, and K. Tsukada. 1979a. Multiple forms of ribonuclease H from rat liver cytosol. *J. Biochem.* (Tokyo) **86:**757.

Sawai, Y., N. Kitahara, W.L. Thung, M. Yanokura, and K. Tsukada. 1981. Nuclear location of ribonuclease H and increased levels of magnesium-dependent ribonuclease H from rat liver by thioacetamide. *J. Biochem.* (Tokyo) **90:**11.

Sawai, Y., S. Uchida, J. Saito, N. Sugano, and K. Tsukada. 1979b. Two ribonucleases H from cultured plant cells. *J. Biochem.* (Tokyo) **85:**1301.

Sekeris, C.E. and W. Roewekamp. 1972. Inhibitory effects of rifampicin and some derivatives on ribonuclease H (hybridase) from rat liver. *FEBS Lett.* **23:**34.

Sekeris, C.E., W. Schmid, and W. Roewekamp. 1972. Stimulation of *in vitro* transcription by ribonuclease H (hybridase). *FEBS Lett.* **24:**27.

Sippel, A.E., J.G. Stavrianopoulos, G. Schutz, and P. Feigelson. 1974. Translational properties of rabbit globin mRNA after specific removal of poly(A) with ribonuclease H. *Proc. Natl. Acad. Sci.* **71:**4635.

Soriano, L., J. Smith, Y. Croisille, and B. Dastugue. 1974. Mitochondrial DNA polymerase, deoxyribonuclease and ribonuclease H activities from brain of chick embryo. *Nucleic Acids Res.* **6:**1085.

Spector, D.H. and D. Baltimore. 1974. Requirement of 3′-terminal poly(adenylic acid) for the infectivity of poliovirus RNA. *Proc. Natl. Acad. Sci.* **71:**2983.

Srivastava, A. and M.J. Modak. 1979. Reverse transcriptase-associated RNaseH. Part IV. Pyrophosphate does not inhibit RNaseH activity of AMV DNA polymerase. *Biochem. Biophys. Res. Commun.* **91:**892.

__________. 1980. Enzymatic activities associated with avian and murine retroviral DNA polymerases. Catalysis of and active site involvement in pyrophosphate exchange and pyrophosphorolysis reactions. *J. Biol. Chem.* **255:**2000.

Stavrianopoulos, J.G. and E. Chargaff. 1973. Purification and properties of ribonuclease H of calf thymus. *Proc. Natl. Acad. Sci.* **70:**1959.

__________. 1976. An assay of ribonuclease H, endoribonucleases, and phosphatases. *Proc. Natl. Acad. Sci.* **73:**1556.

__________. 1978. Simplified method for purification of ribonuclease H from calf thymus. *Proc. Natl. Acad. Sci.* **75:**4140.

Stavrianopoulos, J.G., A. Gambino-Giuffrida, and E. Chargaff. 1976. Ribonuclease H of calf thymus: Substrate specificity, activation, inhibition. *Proc. Natl. Acad. Sci.* **73:**1087.

Stein, H. and P. Hausen. 1969. Enzyme from calf thymus degrading the RNA moiety of DNA:RNA hybrids: Effect on DNA dependent RNA polymerase. *Science* **166:**393.

__________. 1970. Factors influencing the activity of mammalian RNA polymerase. *Cold Spring Harbor Symp. Quant. Biol.* **35:**709.

Sumida-Yasumoto, C., J.-E. Ikeda, E. Benz, K.J. Marians, R. Vicuna, S. Sugrue, S.L. Zipursky, and J. Hurwitz. 1978. Replication of ϕX174 DNA: In vitro synthesis of ϕX RFI DNA and circular, single-stranded DNA. *Cold Spring Harbor Symp. Quant. Biol.* **43:**311.

Tashiro, F. and Y. Ueno, 1978a. Ribonuclease H from rat liver. I. Partial purification and characterization of nuclear ribonuclease H1. *J. Biochem.* (Tokyo) **84:**385.

__________. 1978b. Ribonuclease H from rat liver. II. Partial purification and characterization of nuclear ribonuclease H1. *J. Biochem.* (Tokyo) **84:**395.

Tashiro, F., K. Hirai, and Y. Ueno. 1979. Inhibitory effects of carcinogenic mycotoxins on deoxyribonucleic acid-dependent ribonucleic acid polymerase and ribonuclease H. *Appl. Environ. Microbiol.* **38:**191.

Tashiro, F., T. Mita, and T. Higashinakagawa. 1976. Multiple forms of nuclear ribonuclease H from *Tetrahymena pyriformis*. *Eur. J. Biochem.* **65:**123.

Tomizawa, J. and G. Selzer. 1979. Initiation of DNA synthesis in *Escherichia coli*. *Annu. Rev. Biochem.* **48:**999.

Tsukada, K., Y. Sawai, J. Saito, and F. Sato. 1978. Change of ribonuclease H activity from rat liver nuclei by thioacetamide: Preferential increase in the activity of magnesium-dependent ribonuclease H. *Biochem. Biophys. Res. Commun.* **85:**280.

Verma, I.M. 1975a. Studies on reverse transcriptase of RNA tumor viruses. I. Localization of thermolabile DNA polymerase and RNaseH activities on one polypeptide. *J. Virol.* **15:**121.

__________. 1975b. Studies on reverse transcriptase of RNA tumor viruses. III. Properties of purified Moloney murine leukemia virus DNA polymerase and associated RNaseH. *J. Virol.* **15:**843.

———— . 1977. The reverse transcriptase. *Biochim. Biophys. Acta* **473:**1.

Verma, I.M., M.L. Meuth, H. Fan, and D. Baltimore. 1974. Hamster leukemia virus: Lack of endogenous DNA synthesis and unique structure of its DNA polymerase. *J. Virol.* **13:**1075.

Vicuna, R., J.E. Ikeda, and J. Hurwitz. 1977a. Selective inhibition of ϕXRFII compared with fdRFII DNA synthesis *in vitro. J. Biol. Chem.* **252:**2534.

Vicuna, R., H Hurwitz, S. Wallace, and M. Girard. 1977b. Selective inhibition of *in vitro* DNA synthesis dependent on ϕX174 compared with fd DNA. I. Protein requirements for selective inhibition. *J. Biol. Chem.* **252:**2524.

Vournakis, J.N., A. Efstratiadis, and F.C. Kafatos. 1975. Electrophoretic patterns of deadenylated chorion and globin mRNAs. *Proc. Natl. Acad. Sci.* **72:**2959.

Wada, K., Y. Sawai, and K. Tsukada. 1980. A protease from rat intestine. *Biochim. Biophys. Acta* **612:**253.

Wang, L.-H. and P.H. Duesberg. 1973. DNA polymerase of murine sarcoma-leukemia virus: Lack of detectable RNaseH and low activity with viral RNA and natural DNA templates. *J. Virol.* **12:**1512.

Wang, L.H. and W. Levinson. 1978. *N*-Methylation of beta-thiosemicarbazone-copper complex inhibits RNA-dependent DNA polymerase but not ribonuclease H of Rous sarcoma virus. *Bioinorg. Chem.* **8:**535.

Wassarman, P.M., T.G. Hollinger, and L.D. Smith. 1974. Ribonuclease H activity in germinal vesicles of oocytes from *Rana pipiens. Exp. Cell Res.* **89:**410.

Watson, K.F., K. Moelling, and H. Bauer. 1973. Ribonuclease H activity present in purified DNA polymerase from avian myeloblastosis virus. *Biochem. Biophys. Res. Commun.* **51:**232.

Watson, K.F., P.L. Schendel, M.J. Rosok, and L.R. Ramsey. 1979. Model RNA-directed DNA synthesis by avian myeloblastosis virus DNA polymerase and its associated RNaseH. *Biochemistry* **18:**3210.

Weatherford, S.C., L.S. Weisberg, D.T. Achord, and D. Apirion. 1972. Separation of *Escherichia coli* ribonucleases on a DNA agarose column and the identification of an RNaseH activity. *Biochem. Biophys. Res. Commun.* **498:**1307.

Weimann, B.J., J. Schmidt, and D.I. Wolfrum. 1974. RNA-dependent DNA polymerase and ribonuclease H from virions. *FEBS Lett.* **43:**37.

Weiss, B., S.G. Rogers, and A.F. Taylor. 1978. The endonuclease activity of exonuclease III and the repair of uracil-containing DNA in *Escherichia coli.* In *DNA repair mechanisms* (ed. P.C. Hanawalt, E.C. Freidberg, and C.F. Fox), p. 191. Academic Press, New York.

Worcel, A. and E. Burgi. 1972. On the structure of the folded chromosome of *Escherichia coli. J. Mol. Biol.* **71:**127.

Wu, A.M., M.G. Sarngadharan, and R.C. Gallo. 1974. Separation of ribonuclease H and RNA directed DNA polymerase (reverse transcriptase) of murine type-C RNA tumor viruses. *Proc. Natl. Acad. Sci.* **71:**1871.

Wyers, F., A. Sentenac, and P. Fromageot. 1973. Role of DNA: RNA hybrids in eukaryotes. Ribonuclease H in yeast. *Eur. J. Biochem.* **35:**270.

———— . 1976a. role of DNA: RNA hybrids in eukaryotes. *J. Biochem.* **69:**377.

Wyers, F., J. Huet, A. Sentenac, and P. Fromageot. 1976b. Role of DNA:RNA hybrids in eukaryotes. Characterization of yeast ribonucleases H_1 and H_2. *Eur. J. Biochem.* **69:**385.

RNA-processing Nucleases

Sidney Altman,* Cecilia Guerrier-Takada,*
Howard M. Frankfort,† and Hugh D. Robertson†

*Department of Biology
Yale University
New Haven, Connecticut 06520

†Rockefeller University
New York, New York 10021

In all living organisms there are biosynthetic pathways by which gene transcripts are converted step-by-step into their final functional form in cells. The several classes of functional RNAs, like mRNA, tRNA, or rRNA, each undergo processing events during their intracellular metabolism (Perry 1982). The evidence is overwhelming that for each type of RNA there are ribonucleases that systematically alter the size of the original gene transcript. In addition to size changes, other modifications also occur in RNA biosynthetic pathways (e.g., methylation, polyadenylation, CCA addition, and the acquisition of cap structures at 5′ termini). This article is concerned solely with the nature and mode of action of enzymes that alter the size of primary RNA transcripts and their processing intermediates.

Since every gene transcript contains RNA, the main distinguishing feature among transcripts is nucleotide sequence. It is probably sequence, therefore, that determines which biosynthetic pathway an RNA molecule enters—that for rRNA, mRNA, tRNA, or some other RNA species. It must also be sequence that ultimately determines which set of processing RNases acts on a particular gene transcript, although it may not be sequence alone that directly defines a cleavage site.

RNA-PROCESSING EVENTS

The involvement of RNases in RNA-processing events can be categorized as follows: (1) specific endonucleolytic cleavage of gene transcripts to alter their size; (2) specific exonucleolytic cleavage to perform small size reductions at either end of a molecule; or (3) specific rejoining of RNA molecules (ligation) associated with splicing of transcripts. The first two nucleolytic mechanisms are known to occur widely in both prokaryotes and eukaryotes, and the nature of the events in the various classes of RNA are similar in both kinds of organisms. Cleavages associated with splicing reactions are known to occur only in eukaryotes (Abelson 1979). RNA ligation has been reported in both prokaryotes and eukaryotes but its role in RNA processing has yet to be worked out (Gumport and Uhlenbeck 1981).

Our knowledge of the details of these various kinds of RNA-processing events is not very profound. However, it does seem clear that most specific endonucleolytic cleavage events are governed, in terms of site specificity, by combinations of sequence and higher-order structure or by higher-order structure alone in the substrates (Robertson 1977). There is no clear case of an endonucleolytic cleavage site being determined solely by primary sequence in the substrate.

The action of specific exonucleases is not known to be controlled by such sophisticated mechanisms. Rather, in those cases where something is known about an exonucleolytic mechanism, trimming at the single-stranded end of an RNA molecule is carried out by a specific nuclease which stops its action when it reaches an obstacle such as secondary or tertiary structure (Cudny and Deutscher 1980). It is probably true, however, that the initial recognition of substrates by processing exonucleases is carried out, in part, by recognition of a solution structure of that substrate and a particular free end. Aside from specific recognition of 5′- or 3′-terminal end groups, however, little else is known about the parameters involved in exonuclease substrate recognition.

The factors controlling substrate recognition by endonucleases and RNA ligases involved in splicing reactions seem to vary in complexity, depending upon the kind of RNA transcript being spliced (Johnson and Abelson 1980; Akusjarvi and Persson 1981). That is, splicing of tRNA transcripts in eukaryotes appears to be controlled primarily by tertiary and secondary structure as is the action of other tRNA-processing endonucleases (Altman 1981; Colby et al. 1981). The recognition of hnRNA transcripts to be spliced by the enzymes involved in this process is very complex and is as yet poorly understood (Sharp

1981). The ensuing discussion will deal in some detail with the various pathways of RNA biosynthesis and the processing of specific classes of RNA.

tRNA BIOSYNTHESIS

As knowledge of gene structure increases, it becomes easier to infer what kinds of processing events must occur to transform a gene transcript into a functional RNA molecule. This point is illustrated by the pathway of tRNA biosynthesis. Sequences coding for tRNA genes can be found in at least two different genetic contexts in prokaryotes. Such sequences can be cotranscribed with the sequences coding for rRNAs (there are seven such cistrons in *Escherichia coli* [Nomura and Post 1980]) and they can also be found elsewhere in the genome of *E. coli* (Ozeki 1980). In two cases where tRNA gene transcripts have been fully characterized, it appears that there are also protein-coding sequences near the 3′-termini of the long transcripts containing one or several tRNA sequences (An and Friesen 1980; Altman et al. 1981; Hudson et al. 1981). Other gene transcripts of tRNA sequences from *E. coli* are known that contain up to seven tRNA sequences (Nakajima et al. 1981), but it is not known whether these other sequences also contain adjacent protein-coding sequences in the nascent transcript from the *E. coli* chromosome. In any case, it is clear that more often than not, tRNA sequences must somehow be rescued from the long multimeric transcripts. A schematic illustration of this process is shown in Figure 1 (see also Fig. 4 below). The nucleotide sequences bounding the mature tRNA sequence must be removed during the process of maturation of any particular tRNA. Whether or not adjacent sequences contain rRNA, tRNA, or protein-coding sequences is irrelevant for present considerations. There must be an enzyme or enzymes that remove sequences from beyond the 3′ end, and an enzyme or enzymes that remove sequences upstream from the 5′ end of the mature tRNA se-

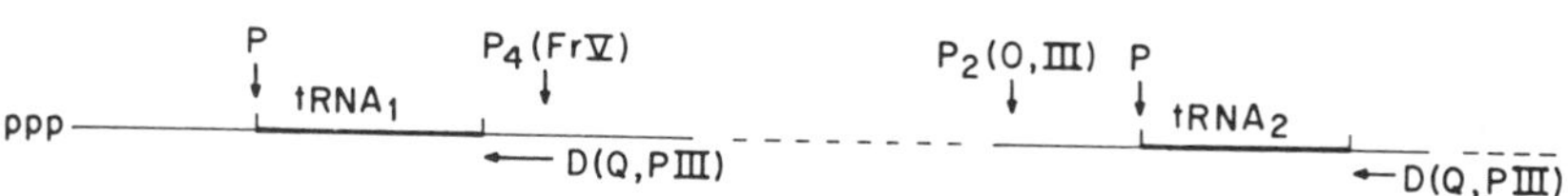

Figure 1 Schematic illustration of the interaction of tRNA-processing enzymes with tRNA precursor molecules in *E. coli*. The sketch shows potential sites of action of various enzymatic activities (see text) on either mono- or polycistronic tRNA precursor molecules. Endoribonuclease action is indicated by an arrow perpendicular to the line showing the gene transcript and exoribonuclease action is indicated by a parallel line. There may exist more events than those shown; also some of the enzymatic activities indicated may be identical to each other and these are indicated as such by the parentheses. The labels in parentheses designate less well-characterized activities and are not discussed extensively in the text.

quence. From the extensive knowledge of tRNA coding regions in *E. coli* and tRNA gene transcripts, it is also apparent that the enzymes involved in tRNA biosynthesis cannot have any great sequence specificity for their substrates because there is no nucleotide sequence homology around the expected cleavage sites (Altman 1978; Apirion et al. 1980). Since no one has successfully characterized an exonuclease that works 5′→3′ on gene transcripts in *E. coli*, cleavage near the 5′ end of a mature tRNA sequence is probably carried out endonucleolytically, whereas cleavage near the 3′ end could be carried out by either endo- and/or exonucleases.

RIBONUCLEASE P

RNase P is the best characterized endoribonuclease involved in tRNA biosynthesis (Kole and Altman 1979, 1981; Kole et al. 1980). This enzyme is essential for tRNA biosynthesis in *E. coli*, being responsible for the endonucleolytic cleavages that generate mature 5′ termini, as shown by the isolation of conditional lethal mutants that have temperature-sensitive enzyme activity (Schedl and Primakoff 1973; Sakano et al. 1974). Only a single RNase-P species is needed to process the approximately 60 tRNA precursor molecules that must exist in *E. coli*. Similar activities have been identified in extracts of other bacteria, yeast, frogs, silkworms, and various mammalian sources (Altman et al. 1980; Kline et al. 1981). A remarkable feature of RNase P is the precision with which it recognizes the correct site for cleavage in all its substrates—sites around which there is no nucleotide sequence homology. Furthermore, RNase-P enzymes from various sources appear to have interchangeable substrate specificities. RNase P from *E. coli* can recognize tRNA precursors from eukaryotic sources correctly and, in a reciprocal fashion, eukaryotic RNase-P preparations can process prokaryotic substrates accurately. There are many mutant tRNAs from *E. coli* (Altman 1978), T4 (McClain 1977), and yeast (Koski et al. 1980) that cannot be processed efficiently by RNase P. The mutations invariably disturb secondary and/or tertiary structure of the tRNA precursor molecules. They do not induce aberrant processing but merely reduce the levels of correct processing. Such observations have led to the postulate that RNase P recognizes the tertiary structure in the tRNA region of the precursors. Assuming that the tertiary structure of all tRNAs are similar, this provides an explanation for the reciprocity of substrate recognition across the evolutionary scale.

A striking aspect of the structure of RNase P may lend a clue to this recognition process. RNase P from *E. coli* is a ribonucleoprotein. The purified enzyme has a small protein component of 18.5K and an RNA component about 365 bases in length (120K). In yeast RNase P, the RNA moiety is about 250 nucleotides long (S. Nishikawa et al., pers. comm.) and in mammalian cells it may be even shorter (E. Akaboshi and S. Altman, unpubl.). The essential role in vitro of the RNA has been demonstrated by observing the loss of activity of RNase P after pretreatment with another ribonuclease. The purified components cannot, when separated, carry out the RNase-P reaction. Upon reconstitution under the appropriate conditions, activity is restored. By examining the reconstitution properties of various components from temperature-sensitive mutants in RNase P function, it can be shown that the RNA is absolutely required for action in vivo. Since the protein component has no apparent ribonuclease activity by itself, the RNA is not just acting as a cofactor that alters the specificity of a nonspecific ribonuclease.

What is the function of this RNA? One function could be to facilitate substrate recognition by nucleotide-nucleotide interaction with the invariant nucleotides of the tRNA moiety of precursor molecules. Moreover, the RNA might activate the catalytic activity of the protein component. This situation would be analogous to the complementary role of rRNA and protein in the ribosome.

The protein and RNA components of RNase P may also show regulatory interaction during their biosynthesis. The structural gene for the protein moiety has been identified as the *A*49 locus near 81 minutes on the *E. coli* map (Stark 1977). The exact position of the gene coding for the RNA component remains to be determined, although the *ts*709 locus (near 69 minutes on the map) produces an RNA component with a molecular weight that differs from the one measured in wild-type strains. Recent data from F. Schmidt's group (H. Motamedi et al., pers. comm.) indicate that the control of the biosynthesis of the protein and RNA components may be intimately related.

Turning to mechanistic considerations, three important reactions are shown below.

$$\text{RNA} + \text{protein} \rightleftharpoons \text{E (RNase P)} \tag{1}$$

$$\text{E} + \text{S (tRNA precursor) ES} \leftrightarrows \text{P}_1 \text{ (tRNA)} + \text{P}_2 \text{ (extra fragment)} \tag{2}$$

$$\text{E} + \text{I (tRNA)} \rightleftharpoons \text{EI} \tag{3}$$

The K_d of the complex shown in Equation (1) is approximately 10^{-11} M as measured by the recovery of this complex from Cs_2SO_4 gradients after reconstitution from the individual components. The K_m for the formation of ES in Equation (2) is 10^{-8} M and the K_i for Equation (3) is 10^{-6} M (Stark 1977). The tightest association then is between the components of the enzyme complex itself, which must stay together under conditions of interaction with the substrate. tRNA, the product of the RNase-P reaction, associates 100-fold less strongly with the enzyme than does the precursor substrate. These numbers, which are accurate with respect to order of magnitude, are satisfying from the point of view of the biological function of the enzyme and its components.

To examine these processes more deeply, an RNaseP-$tRNA^{Tyr}$ precursor complex has recently been isolated. The precursor tRNA can be detected in Cs_2SO_4 gradients at a density of about 1.4 g/ml whereas the E-S complex, containing intact precursor, bands at a density of about 1.5 g/ml (Fig. 2). When diluted out of Cs_2SO_4, the complex still manifests enzymatic activity (Fig. 2). Photochemically activated cross-linking agents and partial digests of the complex with certain RNases might now be used to identify the regions of both the substrate and the RNA component of the enzyme that are involved both in binding to the protein and to each other. Obviously, sequence data concerning the RNA will be invaluable in interpreting these experiments.

tRNA PROCESSING AT THE 3′ END

Studies in vitro of the processing of transcripts of both a natural and a synthetic *E. coli* $tRNA^{Tyr}$ gene have revealed the existence of an endoribonuclease, RNase P4, which appears to be responsible for cutting the gene transcript eight nucleotides downstream from the 3′ terminus of the mature tRNA sequence (Bikoff et al. 1975; Sekiya et al. 1976). Mutations in this tRNA gene that cause accumulation of unusually long transcripts in vivo, may affect this 3′-distal processing site (Altman 1971; McCorkle and Altman

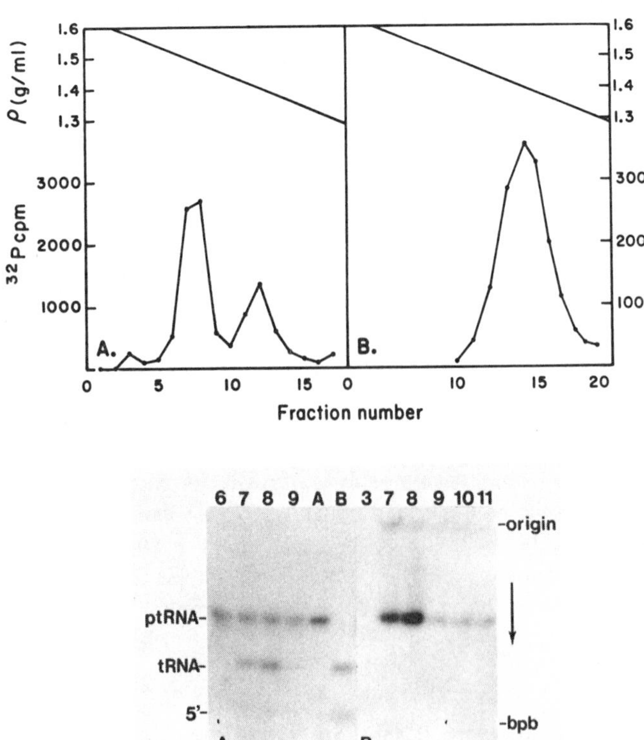

Figure 2. (See facing page for legend.)

1982). The existence of a 3′-processing endonuclease in eukaryotes has been postulated from the nature of gene transcripts from silkworm tRNA genes (Garber and Gage 1979). Indeed, an endoribonuclease that processes transcripts produced by RNA polymerase III near their 3′ termini has been identified recently in yeast cell extracts (Tekamp et al. 1980), but a necessary role for it in tRNA biosynthesis remains to be demonstrated. In addition to these endonucleases, an exoribonuclease must exist to remove any extra nucleotides left at the 3′ termini of precursors after cleavage by RNase P4 or similar enzymes. RNase D is a good candidate for this enzyme in *E. coli* (Cudny and Deutscher 1980).

RNase D has been purified to homogeneity by Deutscher and his colleagues (Cudny et al. 1981b). The enzyme is physically and genetically distinct from RNase II, another 3′–5′ exonuclease that does not mimic the action of RNase D on synthetic or natural precursor molecules. Current attempts to isolate mutants of *E. coli* completely lacking in RNase D activity (M. Deutscher et al., pers. comm.) will reveal if this enzyme is essential for tRNA biosynthesis. Other enzymes that act exonucleolytically on *E. coli* tRNA precursor have been called RNase P3 (Bikoff et al. 1975) and RNase Q (Sakano and Shimura 1975).

Figure 2 Isolation of RNase-P enzyme-substrate complex using Cs_2SO_4 buoyant density gradients. $tRNA^{Tyr}$ precursor was mixed on ice with *E. coli* RNase P, purified through the Sepharose-4B step (Kole and Altman 1981), in 15 μl of standard RNase-P reaction mix buffer. The mixture was then immediately diluted with 40 μl of 50 mM Tris-HCl (pH 7), 500 mM NH_4Cl, 10 mM EDTA-Na_2, 6 mM mercaptoethanol, and loaded onto a preformed Cs_2SO_4 gradient. The gradient was made by layering 190 μl each of 45%, 37%, and 31% (w/w) of Cs_2SO_4 in the dilution buffer noted above. The centrifuge tubes were of cellulose nitrate and 0.65-ml capacity. Centrifugation was carried out in SW 65 rotor fitted with tube adapters for 6 hours at 3°C and 59,000 rpm. One-drop fractions were collected from the bottom of the tubes after puncturing with a 27-gauge hypodermic needle. 3 μl of each fraction were assayed for RNase-P activity in standard fashion. Selected fractions were also extracted with phenol, the RNA in the aqueous phase precipitated with ethanol, and the ensuing precipitate loaded on a standard 10% polyacrylamide gel used for RNase-P assays. (*A; upper panel*) Distribution of radioactivity (•) in the gradient containing the enzyme substrate complex. (———) buoyant density as measured by refractive index. (*B; upper panel*) Distribution of radioactivity (•) in the gradient containing precursor tRNA substrate only. (*A; lower panel*) Selected fractions of A (*upper panel*), assayed for RNase-P activity. Fractions numbered as in upper panel except lane A, which contains untreated precursor tRNA, and lane B, which is a control reaction with a known quantity of active RNase P. (*B; lower panel*) Radioactive RNA in selected fractions from *A (upper panel)*, phenol extracted and rerun to examine molecular weight. Numbering as in *A (upper panel)*. (*Lower panel*) The position of intact precursor tRNA (ptRNA), the RNase-P cleavage fragment containing the mature tRNA sequence (tRNA), and the 5′ fragment (5′) are marked. (bpb) The position of the bromophenol blue dye marker.

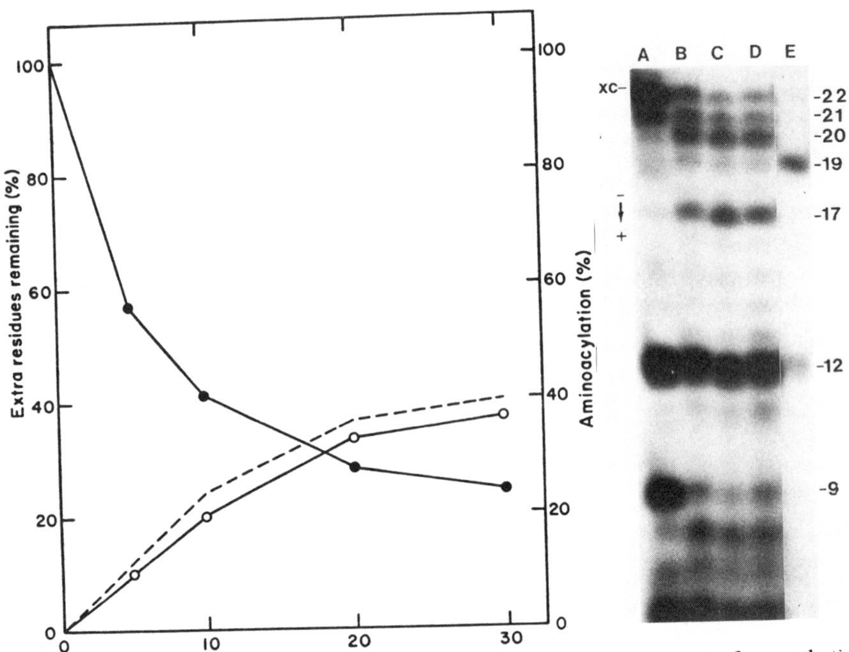

Figure 3 (*A*) Generation of tyrosine-acceptor activity upon treatment of a synthetic tRNA precursor with RNase D. Assays were carried out in reaction mixtures (0.1 ml) containing: 10 mM Tris-Cl (pH 7.5); 5 mM $MgCl_2$; 0.1 M KCl; 1 mM ATP; 10 μM [^{3}H]tyrosine (400 cpm per pmole); 50 μg tRNA-C-C-A-[^{14}C]C_n; 0.1 unit RNase D and excess tyrosyl-tRNA synthetase. Samples were incubated for the indicated times at 37°C and acid-precipitable material collected on Whatman GF/C filters (Cudny and Deutscher 1980). Nucleotide release was determined from the decrease in ^{14}C cpm (•), and aminoacylation from the level of ^{3}H cpm (○). Hydrolysis is measured relative to a zero time sample, and aminoacylation relative to the original tRNA used to synthesize the precursor. The dashed line is the theoretical level of aminoacylation expected, assuming random removal of nucleotides from the 3′ terminus of the synthetic precursor. (*B*) Oligonucleotides generated by RNase-T_1 digestion of ptRNA treated with RNase D (0.045 units) and RNase P (Sepharose 4B step; 2 μl). Reactions were carried out according to Cudny and Deutscher (1980) and Kole and Altman (1981). After RNase-T digestion of the reaction products, the oligonucleotides were electrophoresed in a 20% ultrathin polyacrylamide gel (40 cm long) as described by Sanger and Coulson (1978). The top of the gel is not shown in the figure. The positions of various sizes of oligonucleotides are indicated. Oligonucleotides of 17 nucleotides or greater contain the 3′ terminal sequence of $tRNA^{Tyr}$; the dodecanucleotide contains the anticodon of the tRNA and the nonanucleotide is derived from the 3′ end of the 5′ proximal fragment released by RNase-P cleavage of ptRNA. Lane A: no enzyme added; incubation for 190 minutes. Lane B: RNase P and RNase D added; 80 minutes. Lane C: as in B, 145 minutes. Lane D: as in B, 190 minutes. Lane E: mature $tRNA^{Tyr}$ incubated for 190 minutes with no added enzyme. The mature $tRNA^{Tyr}$ yields a 3′ terminal oligonucleotide 19 nucleotides long that comigrates in the figure with material found in lanes B–D. We thank S. Altman, H. Cudny, and M. Deutscher for providing this figure. The position of the marker dye, xylene cyanol (xc) is indicated.

The ability of RNase D to trim extra nucleotides from the 3′ termini of tRNA precursor molecules was demonstrated with the synthetic substrate tRNA-CCACC. The enzyme is capable of removing the extra nucleotides from this substrate and stops when it reaches the terminal A of the common CCA sequence found in all tRNA gene transcripts in *E. coli*. It acts similarly on natural precursor molecules, but the rate at which the exonucleolytic cleavage decreases when the CCA sequence is reached may vary with individual precursor molecules. Furthermore, RNase D removes extra nucleotides more slowly from synthetic precursor molecules of the type tRNA-CU (which are analogs of certain bacteriophage T4-encoded tRNA precursors) than it does from precursors containing the CCA sequence.

When *E. coli* is infected by T4, two enzymes involved in making gene transcripts, RNA polymerase and RNase D, are modified. RNase D found after T4 infection has an increased molecular weight (Cudny et al. 1981a), but as yet no change in function is associated with this modified form of the enzyme. Curiously, in *E. coli* strain BN, there appears to be no intracellular capability to process T4 tRNA precursors (of the type tRNA-CU) that require the removal of 3′ nucleotides prior to the addition of the CCA sequence (McClain 1977). Since the RNase D extracted from this strain is identical in vitro with that extracted from wild-type cells, another nuclease is probably present in *E. coli* that processes the 3′ termini of tRNA precursors. An enzymatic activity, possibly an endoribonuclease, which is capable of cleaving tRNA-CUC to produce a shortened tRNA and an oligonucleotide, has been identified in extracts of *Xenopus laevis* oocytes (M. Deutscher, pers. comm.).

The approach taken by Deutscher and his colleagues in identifying RNase D and its function has been fruitful, but discrepancies remain about the presumed processing capability of RNase D when comparing synthetic and natural substrates. For example, RNase D processes synthetic precursor to $tRNA^{Tyr}$ very efficiently (see Fig. 3) so that about 50% of the precursor becomes acylatable by tyrosine. However, using a natural $tRNA^{Tyr}$ precursor as a substrate, S. Altman, H. Cudny, and M. Deutscher (pers. comm.) have found that most of the product has two more than the requisite number of nucleotides removed (Fig. 3). This discrepancy may be due to the fact that with different types of substrate, enzyme-substrate ratios varied considerably. Alternatively, the supposed natural substrate may not be the true substrate for RNase D in vivo. Finally, it should be pointed out that while RNase Q (Sakano and Shimura 1975) is not as well characterized as RNase D, its activity resembles that of RNase D and it was identified using natural substrates.

INTERCISTRONIC CLEAVAGE EVENTS

If a tRNA gene is found in a genetic context within which it is cotranscribed with other genes, such as rRNA genes or other tRNA genes, then several nucleases may be involved in cutting the monocistronic tRNA gene transcript out of the larger polycistronic transcript (Lund and Dahlberg 1977; Apirion et al. 1980; D. Apirion, pers. comm.). In principle, an enzyme like RNase P4, acting in concert with RNase P, may be sufficient to perform this excision if it is assumed that RNase P4 has a variety of cleavage sites in intercistronic regions. Such a role has also been postulated for RNase P2 (Schedl et al. 1974) and for a newly identified activity that acts on the large transcript containing T4 tRNA precursors. This new activity, RNase PC (Goldfarb and Daniel 1980 and pers. comm.), yields fragments containing both monomeric and dimeric precursor sequences. RNases P2 and PC may or may not be identical to RNase P4. No mutations have been found that directly affect any of these activities, and it is still uncertain whether any endoribonuclease specific for intercistronic regions is essential for tRNA biosynthesis.

While the involvement of an activity like RNase P4 in cleavage near the 3′ terminus of a specific tRNA gene transcript has been observed, it is less clear that such a ribonuclease is essential for all tRNA biosynthetic pathways. For example, in the biosynthesis of the tRNA sequences found in the spacer regions of rRNA cistrons, intermediates containing the tRNA sequences are found that appear to be generated by enzymes other than RNase P4 (possibly RNase E; see Ray and Apirion 1981). Since the unique identity of all these enzymes is not clearly established and detailed studies of these intermediates and their biosynthesis are lacking, it is not possible to rule out rigorously the role for RNase P4 in the biosynthesis of these tRNAs. Although, as mentioned above, tRNA biosynthesis events of the kinds discussed seem to be similar in eukaryotes and prokaryotes, cleavage events near the 3′ termini of eukaryotic tRNA gene transcripts are not well characterized. However, they do appear to involve specific endonucleolytic cleavages very close to the region where CCA must be added to the original gene transcripts.

SPLICING OF tRNA GENE TRANSCRIPTS

Removal of intervening sequences in certain tRNA transcripts is unique to eukaryotes. Not all tRNAs in eukaryotes contain intervening sequences and to date there is no rational understanding of why intervening sequences are found in certain tRNA species and not others. Studies from Abelson's group show that splicing is carried out by the successive or possibly concerted actions of an endonuclease and a ligase. In vitro, at least, it is possible to separate the endonuclease action into separate events so that products which result from only one of the two endonucleolytic cuts are retrievable (J. Abelson, pers. comm.). The splicing reaction, when necessary, appears to be a late step or the last step in tRNA processing in eukaryotes (Melton et al. 1980). That is, trimming at both 5′ and 3′ termini of gene transcripts has already occurred prior to the splicing event. This may lend credence to the idea that splicing is involved in transport of particular tRNA species from the nucleus to the cytoplasm. Further evidence comes from the observation that the splicing endonuclease is associated with a nuclear membrane fraction but can be separated from this fraction during biochemical purification (C. Peebles and J. Abelson, pers. comm.).

The splicing endonuclease recognizes secondary and/or tertiary conformation in its substrates. As with RNase P, mutations near the splice sites do not alter the accuracy of the cleavage events but do change the rate at which the cleavage occurs (Colby et al. 1981). One very interesting aspect of the splicing endonuclease is that it, unlike all other known processing endonucleases, produces 5′-hydroxyl and 3′-phosphate cleavage products whereas all known DNA or RNA ligases require 5′-phosphate and 3′-hydroxyl groups.

Recently, Abelson's group (C. Peebles and J. Abelson, pers. comm.) have purified a similar ligation activity from wheat germ extracts. Konarska et al. (1981) have also shown that a wheat germ ligase activity can join together RNAs with a 3′-phosphate and 5′-hydroxyl group through the intermediate formation of a cyclic 2′–3′ compound and the subsequent formation of a 2′-phosphate end group in the spliced product. Should these two wheat germ ligases prove to be identical and indeed needed for RNA-processing events, this would raise exciting possibilities for the role of the free 2′-phosphate group in the splicing reaction. One could imagine, for example, proteins essential for recognition of splice junctions, or for transport across nuclear membranes of spliced products, becoming transiently esterified to phosphate groups in the substrate before and after splicing.

ORDER OF tRNA-PROCESSING EVENTS

Although the maturation of some T4 tRNA precursors requires processing at the 3′ termini before processing can begin at the 5′ termini, such is not universally the case in prokaryotes. In eukaryotes, it is clear from experiments both in vivo and in vitro, that processing at 3′ and 5′ termini occurs before the removal of intervening sequences in those cases where splicing is necessary. Similar enzymes to those required for processing in prokaryotic cells are required for processing eukaryotic tRNA precursors. There is some evidence, from experiments in vitro, that some tRNA precursors are dimeric (Mao et al. 1980; Schmidt et al. 1980) but the vast majority are monomeric (i.e., contain only one tRNA sequence [Feldman 1976]). There is as yet, however, no evidence that dimeric precursors exist in vivo.

In the case of tRNA biosynthesis, knowledge of the nature of tRNA gene transcripts, or the DNA sequences from which these transcripts derive, allows strong predictions about the kinds of nucleases that must be involved in the processing events. These have turned out to be largely correct. These nucleases do not have strong sequence specificity, if any at all, but rather, tertiary structure plays the predominant role in determining substrate recognition.

PROKARYOTIC rRNA BIOSYNTHESIS

E. coli rRNA transcription units contain genes for one or more rRNAs interspersed with those for 16S, 23S, and 5S RNAs. Studies of cellular RNA from chloramphenicol-treated *E. coli* (Gegenheimer et al. 1977) showed that there were intermediates in the processing of the primary transcripts to give mature RNAs. In particular, immediate precursors to the three rRNA species have extra sequences attached to the 3′ and 5′ termini of the mature RNAs. An important advance was the discovery that an *E. coli* strain lacking RNase III (Kindler et al. 1973) was incapable of efficient rRNA processing and accumulated large amounts of a 30S "precursor" species (Dunn and Studier 1973b; Nikolaev et al. 1973). The RNAs accumulating in these mutant cells show that some processing still occurs, converting the 30S rRNA precursor to smaller species resembling immediate precursors of 16S, 23S, and 5S rRNA. Genetic studies implicate two enzymatic activities in these events (Gegenheimer and Apirion 1978), one of which involves the RNase-P gene. It would seem that several (at least three) enzymes are involved in cleaving the primary transcript, and that several (at least two) more must be involved in converting the immediate precursors to the mature rRNAs.

ENDORIBONUCLEASES

Several of the enzymes actually involved in *E. coli* RNA processing have been identified. The most prominent is RNase III, which was shown to be specific for cleavage within RNA-RNA duplex regions (Robertson et al. 1968). Dunn and Studier (1973a) showed that this enzyme could process bacteriophage T7 early mRNA precursor correctly, as well as the 30S rRNA. The cleavages occur at four locations along the 30S precursor (Fig. 4): (1) to the left of the 16S region; (2) to the right of this region; and (3,4) on both sides of the 23S region. That these cleavages actually occur within two limited regions of a nearly perfect double helix was concluded from the follwing observations: (1) electron microscopy by Wu and Davidson (1975) using *E. coli* DNA-encoding rRNA genes, suggested that regions on either side of the 16S and 23S coding regions could fold together; (2) DNA sequencing by Young and Steitz (1978) showed that sequences flanking the termini of the mature RNAs contained complementary regions that might interact in the precursor to produce RNA-RNA duplexes within which the RNase-III cleavages could occur; (3) Robertson and Barany (1978; Robertson et al. 1980) showed that stable, nuclease-resistant RNA segments could be isolated from the 30S rRNA precursor; these were shown by sequence analysis to contain the RNase-III cleavage sites and could be cleaved in vitro by RNase III at the correct locations, despite the

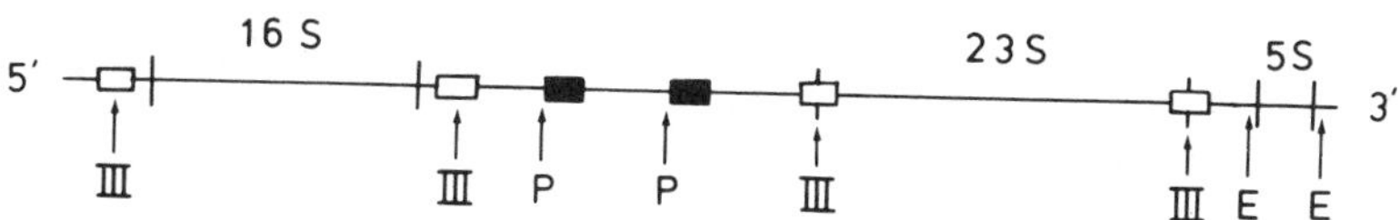

(b)

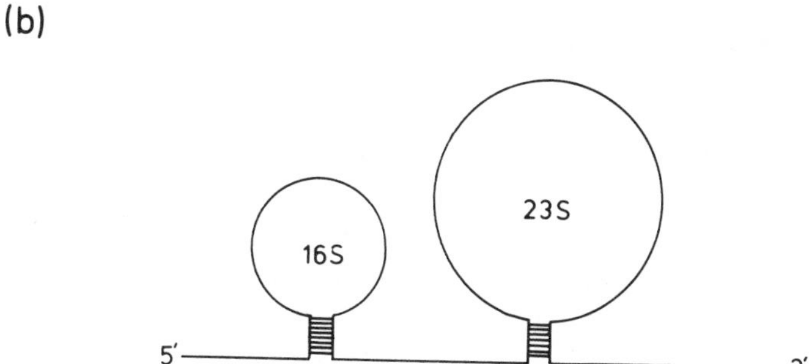

Figure 4 Schematic representations of the *E. coli* 30S rRNA precursor. (*A*) Linear depiction of the 30S rRNA precursor molecule. 16S, 23S, and 5S refer to the RNA regions containing the mature rRNA sequences, in order from the 5′ to 3′ end of the molecule. Open rectangles indicate those regions containing the four RNase-III cleavage sites (denoted by III and an arrow in each case); closed rectangles indicate the tRNA moieties (of which there can be either one or two in the seven different 30S rRNA transcripts produced from the seven distinct *E. coli* rRNA genes), while sites for RNase-P cleavage are indicated by P and an arrow in each case; sites for RNase-E cleavage are indicated by E and an arrow in each case. (*B*) Folded depiction of the 30S rRNA molecule. As described in the text, regions of stable RNA-RNA duplex are capable of forming between regions in the vicinity of the 16S and 23S rRNA termini while they are embedded within the 30S rRNA precursor molecule. The duplex region in the vicinity of the 16S rRNA region is formed by base-pairing of the leftward pair of RNA regions indicated by open rectangles in Fig. 4A, while the duplex region in the vicinity of the 23S rRNA region is formed by base-pairing of the rightward pair of RNA segments indicated by open rectangles in Fig. 4A.

absence of the extensive sequences connecting the sites (see Figs. 4, A and B); (4) studies of in vitro cleavage of *E. coli* 30S rRNA precursor by several enzymes have further defined the requirements for its processing. First, Lund and Dahlberg (1977; Dahlberg et al. 1978) have studied in vitro cleavage by RNase III and have shown accumulation of the expected RNA species. Second, these workers (Lund et al. 1980) have shown that, in the absence of RNase III, RNase P can cleave the 30S rRNA precursor, separating the 16S and 23S coding regions. In addition, following RNase-III cleavage, RNase P can cleave to form the mature 5′ termini of the tRNAs found in the 16S–23S spacer region.

The processing of 30S rRNA has been studied in mutants lacking RNase III, RNase P, or both (Robertson et al. 1980; Gegenheimer et al. 1977). The latter group has also isolated several other classes of mutants that change the in vivo pattern of rRNA metabolism. In order to explain the novel RNA species observed in some triple mutants of *E. coli*, Apirion et al. have postulated genes for two enzymes, RNase E and RNase F, although these have not yet been fully characterized enzymatically (Misra and Apirion 1979; Anderson et al. 1981). RNase E seems to be involved in processing the region of the 30S rRNA precursor that contains 5S RNA. The approximate sites of these cleavages are indicated in Figure 4, and the precise locations proposed by Apirion and his colleagues in Figure 5 (B.K. Ray et al., pers. comm.).

Most of the cleavages that occur during *E. coli* rRNA processing can be explained by invoking the action of RNases P, III, and E. Subsequent to the endonucleolytic cleavages by these enzymes within 30S rRNA, further events must occur that convert intermediates to mature RNA species. One or more of the activities involved in the maturation events of 16S and 23S rRNA may require the presence of ribosomal proteins (Nikolaev et al. 1973; Dahlberg et al. 1978). In a certain sense, therefore, one might consider specific ribosomal proteins as cofactors in rRNA-processing reactions.

The maturation of *Bacillus subtilis* 5S rRNA from its immediate precursor has been particularly well-studied. Pace and his colleagues (Sogin et al. 1977) have purified and characterized an enzyme, RNase m5, which is responsible for the two cleavages that form the 5′ and 3′ termini of the mature 5S RNA. The signal for this cleavage is neither primary sequence nor Watson-Crick secondary structure (Pace et al. 1980).

The RNase m5 enzymatic activity contains two components, alpha and beta, both of which are required for cleavage. The beta subunit has been recently purified and characterized (D.A. Stahl and N.R. Pace, pers. comm.). It is a basic protein of 15–17K that will retain both precursor and mature 5S RNAs on nitrocellulose filters. An RNase m5-like activity is also present in *E. coli*. The *E. coli* equivalent of the beta subunit, when added to the *B. subtilis* alpha subunit, will correctly process the *B. subtilis* 5S RNA precursor, demonstrating the similarity of these activities in the two species of bacteria (D.A. Stahl and N.R. Pace, pers. comm.).

EUKARYOTIC rRNA BIOSYNTHESIS

Compared with the fairly detailed studies available on the RNases involved in prokaryotic rRNA metabolism, there is only limited information on the enzymes involved in the analogous pathways for eukaryotes. One of the major differences between prokaryotic and eukaryotic rRNA transcription is that whereas the 5S RNA genes of eukaryotes are transcribed by RNA polymerase III, the 18S, 28S, and 5.8S RNA genes are transcribed by a second enzyme, RNA polymerase I. In the case of 5S RNA, there have been some suggestions that an RNA precursor with extra sequences at the 3′ end arises in some species (Rubin and Hogness 1975; Tekamp et al. 1980). The 5.8S rRNA species represents a component of the ribosome found in intimate association with 28S RNA and represents a second difference from prokaryotic rRNAs. An additional difference between prokaryotic and eukaryotic rRNA gene organization is the apparent absence of tRNA sequences embedded within the eukaryotic rRNA transcripts.

The biosynthetic pathway of the rRNA species (excluding 5S RNA) of mammalian cells was the first RNA metabolic pathway to be studied. It has

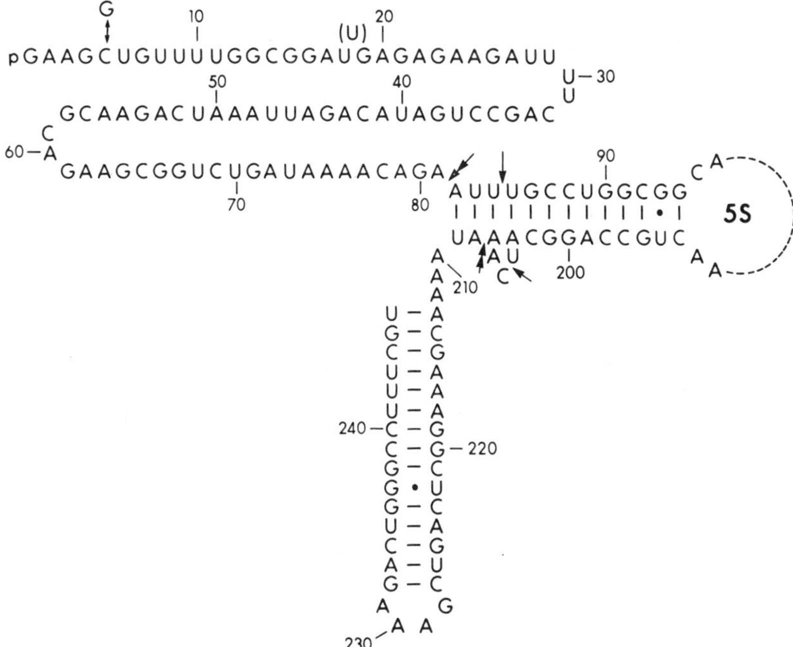

Figure 5 Sequence and partial secondary structure of an RNase-E substrate, a 9S rRNA of *E. coli* which is an intermediate in 5S rRNA maturation derived from the 30S rRNA in certain mutant cells. The sequence shown is from transcripts of the *rrnB* gene cluster and is arranged in two duplex structures. Double-headed arrows show the proposed RNase-E cleavage sites; single-headed arrows indicate the mature 5′ and 3′ termini of 5S rRNA. The structure shown is transcribed from the four rRNA gene clusters which do not contain trailer tRNAs. An almost identical RNA, up to nucleotide 209, is also transcribed from the other rRNA gene clusters which do contain trailer tRNAs. We thank D. Apirion for providing this figure.

been clear for some time (Darnell 1968) that the 45S rRNA precursor detected in nucleoli of mammalian cells is cleaved to (at least) 18S and 28S forms via a series of intermediates. At first it seemed likely that the 45S rRNA precursor, and the other abundant and easily obtainable species, would aid studies of their enzymatic processing. However, such studies have so far concentrated on the final steps of rRNA processing or upon the more recently identified splicing events necessary to remove intervening sequences. In part, this is due to: (1) the difficulty of preparing clean substrates and the lack of a gel system to assay their processing; (2) the absence of mutants in rRNA processing; and (3) the ubiquity of nonspecific nucleases in mammalian cell-free extracts. It is not clear whether the rRNA precursor is cleaved by a series of specific and reproducible endonucleolytic cuts, as in *E. coli*, or whether the specificity depends upon "packaging" of the RNA precursor with proteins, and/or altering it by nucleotide methylation so that the sequences to be preserved are protected from cleavage by nonspecific ribonuclease action. However, from what is known of *E. coli* rRNA processing, and of the intermediates of eukaryotic cytoplasmic rRNA synthesis, it is simplest to postulate that specific nucleolar endonucleases acting in sequence will be involved in eukaryotic rRNA processing. In addition, in those cases (principally in lower eukaryotes; see Carin et al. 1980) where intervening sequences are found in the larger rRNA precursors it is likely that splicing events will be carried out by nucleolar enzymes.

IN VITRO PROCESSING STUDIES

Since RNase III is active in processing the 30S rRNA precursor of *E. coli*, it seemed reasonable to test the effect of this enzyme on the mammalian 45S rRNA precursor. This *E. coli* enzyme, while yielding polyacrylamide gel bands from 45S rRNA whose mobilities sometimes correspond to 18S, 28S, or both RNAs, seems unable to duplicate eukaryotic rRNA metabolism in vitro as assayed by sequence analysis of the reaction products. An activity found in the postribosomal supernatant of HeLa cell extracts can convert ^{32}P-labeled 45S rRNA precursor reproducibly to bands with mobilities of 28S and 18S (Ferrari et al. 1980), but so far no fingerprinting or hybridization studies on these bands have been reported. Similarly, Grummt et al. (1979) have reported that a nuclease, RNase D2, capable of cleaving double-stranded RNA, can also cleave rRNA transcripts from nucleoli in a reproducible manner.

A possible new insight into the mechanism of eukaryotic rRNA processing has been provided by R.J. Crouch et al. (pers. comm.). They have found that the conversion of 32S rRNA to 28S and 5.8S rRNA first requires the formation of the 5.8S rRNA 3′ terminus and the 28S rRNA 5′ terminus. The 32S rRNA-processing intermediate, containing 5.8S and 28S rRNA, was cloned and the nucleotide sequence of the 650-base extra region determined. This region is composed of 85% G and C residues and can be drawn into a hairpin structure. Crouch and his colleagues suggest that RNase D2, as well as the nucleolar U3 RNA species (Ramachandra and Busch 1981) may be involved in the processing of the rRNA precursor mediated by changes in hypothetical base-paired regions between the U3 RNA and the RNA substrate.

SPLICING IN rRNA BIOSYNTHESIS

The macronuclear rRNA genes of *Tetrahymena* have a 26S rRNA coding region that is interrupted by a 400-bp intervening sequence. Some insight into the mechanism of the splicing of the rRNA precursors transcribed from this region has been provided by Cech and his colleagues (Grabowski et al. 1981). When transcription and splicing of rRNA are allowed to occur at 30°C in isolated nuclei, the excised intervening sequence can be identified as a linear species accumulating with time. At 39°C, however, a conversion of the linear intervening sequence to a circular form occurs. It is not known whether this conversion occurs in vivo. Further studies by P. Grabowski et al. (pers. comm.) focusing on the mechanism of this reaction, suggest that a single activity is involved in the excision of the intervening sequence, ligation of the spliced pre-rRNA and cyclization of the intervening sequence. The products of this reaction have been analyzed by polyacrylamide gel electrophoresis and RNase T1 fingerprinting. An extra guanosine residue, not encoded by the DNA of the intervening sequence, is present in the product of this reaction. This extra guanosine moiety is added to the linear intervening sequence during its excision. Grabowski et al. propose that the *Tetrahymena* splicing activity is a phosphoester transferase. Although the data of Carin et al. (1980) do not cover the circularization of the intervening sequence of the *Tetrahymena* rRNA transcript, they found by R-loop analysis, that the two halves of the 26S rRNA are joined together in vitro to form an RNA with the hybridization properties expected of the mature sequence. Further analysis of this RNA by sequencing methods will be required to determine whether this in vitro splicing event has occurred with the required fidelity.

Splicing of *Neurospora* mitochondrial rRNA has been studied by Grimm et al. (1981), who have isolated a series of mutants that fail to remove an intervening sequence from the 25S rRNA precursor. Their studies suggest that a structured region near one of the processing sites, together with prebinding of ribosomal proteins, could be important in the processing. The enzymatic components of this promising system have yet to be identified.

PROKARYOTIC mRNA PROCESSING

Of all of the major species of RNA to be studied, prokaryotic mRNA is the one about which there is the least evidence for systematic RNA processing. Such processing events would be difficult to detect and characterize because:

(1) mRNA transcription and translation are tightly coupled such that polycistronic mRNAs are attached to large polysomes in their 5′ regions while the 3′ regions are still undergoing synthesis. The overall half-life of many mRNA molecules is so short that the entire transcript never exists as such because mRNA breakdown begins within regions already translated before the RNA polymerase has reached the termination signal in the DNA. Thus, considering the speed of transcription and turnover in *E. coli*, and the presence of a large proportion of the mRNA as nascent fragments, the technical obstacles to the characterization of intermediates of RNA processing in vivo are formidable. (2) For similar reasons, the isolation of intact primary transcripts with which to study in vitro mRNA processing cannot be accomplished except with certain viral systems.

While precursors of rRNA species accumulate in RNase-III mutants of *E. coli* (Gegenheimer et al. 1977), repeated efforts to seek high-molecular-weight mRNA precursors in such cells have failed. This may reflect a lack of systematic processing steps utilizing RNase III, or it may indicate that the enzymes responsible for the rapid breakdown and destruction of prokaryotic mRNAs are so efficient that they do not allow such precursors to survive. Nevertheless, it has been suggested that RNase III might play a role in regulating the stoichiometry of gene products from certain polycistronic transcripts in *E. coli* (Barry et al. 1980; Fiil et al. 1980), and in at least one case the kinetics of mRNA degradation are changed in an RNase-III$^-$ mutant compared with wild type (Cannistraro and Kennell 1979).

It is only with bacteriophage mRNAs that solid evidence for RNA-processing steps have so far been found. In the case of bacteriophage T7 early mRNA synthesis, Dunn and Studier were able to show that, when a transcript of the entire 7000-base early region of this bacteriophage DNA was synthesized in vitro, it could be cleaved precisely into segments corresponding in size and map position to the in vivo mRNAs by the action of an enzyme present in *E. coli* subcellular extracts (Dunn and Studier 1973a). This enzyme was shown to be RNase III, by genetic means (Kindler et al. 1973) and by comparison with authentic RNase III (Dunn and Studier 1973b).

The action of RNase III on phage T7 early mRNA precursor has been studied extensively. The cleavage sites have been sequenced (Dunn and Studier 1981) and the relation of this in vitro pathway to the situation in vivo explored. Two rather compelling conclusions have emerged from these studies. First, the cleavage sites differ from the 18 – 23 bp RNA-RNA duplexes found in the 30S rRNA precursor. They are shorter, less perfect helices capable of folding into regular stemlike structures only through additional tertiary interactions. Second, the processing of T7 mRNAs by RNase III—required in some cases to achieve full translational expression (Dunn and Studier 1975)—may not mimic a host mRNA-processing pathway. Nevertheless, this pathway does provide a prototype for studying enzymatic recognition of RNA-processing signals.

Bacteriophage λ is another virus that makes use of RNase III. Apparently certain posttranscriptional events during λ early gene expression are controlled by the presence or absence of potential RNase-III cleavage sites within particular λ gene transcripts (Lozeron et al. 1976, 1977). These latter molecules behave more like typical host mRNAs than do the T7 mRNA precursors, and the involvement of RNase III in their metabolism is more as a regulatory agent than as a maturation enzyme.

In summary, extensive processing of prokaryotic mRNAs has not been observed. The elusive nature of the high-molecular-weight precursor RNAs, along with the tangential and rather subtle effects of the RNase-III$^-$ mutation on cellular and viral growth, suggest that prokaryotic mRNA is not extensively processed. The enzymes that initiate mRNA breakdown, however, remain unknown. When they are identified and characterized, we may find that the last step of RNA processing and the first step of RNA breakdown are one and the same.

EUKARYOTIC mRNA PROCESSING

The metabolism of eukaryotic cellular and viral mRNA is now one of the most intensively investigated fields in all of molecular biology. While few enzymes have been isolated and characterized, it is likely that eukaryotic mRNA processing will eventually represent the field with the most varied group of processing enzymes and signals. From a knowledge of the DNA structure of a number of mRNA coding regions, the existence of various in vivo intermediates, and finally from the structure of mature mRNA itself, it is possible to identify several important ways in which RNA processing must be involved in the maturation of these transcripts. First, most eukaryotic mRNAs contain poly(A) tails on their 3′ termini (Perry 1976), and recent work suggests that, in several cases, the site of addition for poly(A) is determined by endonucleolytic cleavage which creates the site for polyadenylation. Second, there are characteristic regions of double-stranded structure within the heterogeneous nuclear mRNA precursors of higher mammalian cells (Robertson et al. 1977) that contain representative segments of a highly reiterated DNA sequence called the *Alu* family sequence (Rubin et al. 1980). These regions, which represent 5–10% of the nuclear mRNA precursors' mass, are virtually absent in mature mRNA. Third, numerous studies (Abelson 1979) make it abundantly clear that most viral and cellular mRNA precursors contain intervening sequences that must be removed following transcription, allowing the mRNA fragments to be rejoined in their final spliced form for transport to the cytoplasm. Thus, creation of polyadenylation sites, removal of sequences present in double-stranded regions, and the splicing out of intervening sequences are the three areas where progress is likely to be made with respect to the isolation and characterization of specific nucleases.

POLYADENYLATION

With regard to creation of polyadenylation sites in mRNA transcripts, it seems likely that the enzymes responsible might be straightforward RNA-processing enzymes such as RNase P or RNase III of *E. coli*. In the case of adenoviral late mRNA, it has been known for some time that cleavage is a necessary prelude to polyadenylation. Here a single primary transcript, which may cover as much as 80% of the total genome, is precursor to five groups of coterminal mRNAs. Only one mature mRNA is derived from each late mRNA precursor (Nevins and Darnell 1978) and internal cleavage within the precursor must therefore occur to define the particular 3′ terminus to be polyadenylated.

Recent work on SV40 (Fitzgerald and Shenk 1981) has suggested that, in this system as well, alternative 3′ termini can arise depending upon alternative cleavage events within the mRNA precursor. In both of these viral cases, the 6-base nucleotide sequence AAUAAA is present within a reasonable distance of the alternative polyadenylation sites (Fiers et al. 1978; Fraser and Ziff 1978). It had been previously thought that this signal might correspond to a transcription termination signal in the DNA, just as it is now thought that, in many cases, the site for addition of 5′-terminal "cap" structure represents a transcription initiation signal in the DNA (Salditt-Georgieff et al. 1980). However, it is equally likely that this 6-base region represents part of a signal for RNA processing. Cases where mRNA transcription can be shown to proceed well past the point where polyadenylation occurs have recently been identified in cellular mRNA transcripts also. Work of Hofer and Darnell (1981) has suggested that the mouse globin genes are transcribed in vivo at least 1000 bases beyond the site for 3′-terminal polyadenylation. In addition, studies by Young et al. (1981) have suggested that mouse α-amylase mRNAs can differ in their 3′-terminal polyadenylation sites in different tissues, and that the most likely explanation for these results is alternative cleavage sites in a transcript whose synthesis proceeds well beyond the 3′ ends of the mature mRNAs. It will be important to understand the factors governing the use of these alternative cleavage sites.

So far there is no in vitro system that is capable of cleaving a eukaryotic mRNA precursor at the site for polyadenylation. The need for these enzymes has only recently been defined, however, and it is likely that their discovery awaits only the isolation and proper study of the correct RNA substrate. Most RNA precursors, especially that for adenoviral late mRNAs, contain a bewildering variety of potential cleavage sites, some of which may lead to poly(A) addition and others to splicing. Simpler subsets of these transcripts would seem to represent more appropriate substrates. For example, a local region of adenovirus containing the precursor to polypeptide IX (Ziff 1980), which contains a polyadenylation site but no intervening sequence, would seem to be ideal. Alternatively, cellular genes whose mRNA transcripts have been shown to contain polyadenylation sites but not intervening sequences could be used. If transcripts from such genes—e.g. those for certain interferons—could be isolated, they would represent powerful probes for the cleavage required to precede polyadenylation. In vitro systems already in existence (Sugden and Keller 1973; Wilhem et al. 1976; Ferrari et al. 1980; Manley et al. 1980; Baglioni et al. 1981; Wolgemuth 1981; Nilsen et al. 1981) may allow detection of these elusive endonucleases.

With regard to the extensive double-stranded regions containing *Alu* family transcripts within hnRNA precursors of mammalian cells, one possible model for their removal during the maturation of mRNAs would invoke a double-stranded RNase which cleaves within the sequences to bring about their destabilization and subsequent destruction. However, when Ferrari et al. (1980) used isolated, nuclear, double-stranded RNA from HeLa cells as a probe for such an RNase-III-like activity in nuclear and cytoplasmic fractions, they were unable to detect even minor nicking with the *Alu* family regions. Perhaps the double-stranded RNA regions of hnRNA may be removed by splicing out larger segments containing them.

In addition to considering an RNase III-like activity or splicing for the removal of *Alu* sequences, recent studies on ATP-dependent cleavage of HeLa-cell hnRNA by HeLa-cell extracts (Nilsen et al. 1981) have suggested a third candidate for the enzyme that removes double-stranded RNA regions. This enzyme, RNase L, is activated by the group of exotic oligonucleotides 2′, 5′ oligo(A) that contain adenosine residues in a 2′→5′ linkage and were originally identified during studies of interferon action (Kerr et al. 1977). RNase L is present in a number of cell types. Double-stranded RNA (including that in hnRNA) activates an enzyme also present in normal cells that makes 2′, 5′ oligo(A) thus causing hnRNA to "activate" its own cleavage. In vitro, this cleavage of hnRNA occurs adjacent to double-stranded regions (Nilsen et al. 1981). The possible relevance of these in vitro events to those in the cell is of great interest.

RNA SPLICING MECHANISMS

RNA splicing was discovered in studies of adenoviral mRNA transcripts and was subsequently extended to a variety of other viral and cellular systems. A typical eukaryotic mRNA precursor containing intervening sequences (that for globin mRNA for example) can be represented as an aggregate of alternating extra and mature sequences (Abelson 1979). This structure bears a striking similarity to the prokaryotic rRNA precursor, as shown in Figure 4A. The information is similarly laid out in the DNA—mature and extra sequences are interspersed—and the separation of the two types of sequences by highly precise endonucleolytic cleavage appears to be the same in both cases. Subsequent ligation by an enzymatic activity unique to eukaryotic cells provides the crucial distinction. From this analogy, it is clear that, as in the case of tRNA splicing discussed above, the processing of eukaryotic mRNA could be carried out by sequential action of two rather familiar enzymatic activities. One of these would be analogous to prokaryotic enzymes and would carry out the cleavages required. The second would be an RNA ligase like that found in yeast, which carries out the rejoining of cleaved tRNA segments or, alternatively, the wheat germ RNA ligase activity described by Konarska et al. (1981).

The kinds of RNA signals recognized by the cleavage and ligation enzymes (or the splicing complex) have been sought both by attempting to define local or long-distance structures involving precise duplex regions (Trapnell et al. 1980) or by attempting to find "consensus" sequences in the vicinity of cleavage sites (sometimes called splice junctions) that could be recognized by the RNA-processing machinery (Sharp 1981). So far these approaches have yielded only circumstantial evidence for the involvement of RNA:RNA structure in recognition, and limited nonrandom sequence composition in the vicinity of the cleavage sites (Lamb 1980). Neither of these properties appears sufficient to define adequately the signal for cleavage with respect to the many thousands of bases in the surrounding sequence that remain uncleaved. As with the so-called "Shine-Dalgarno" sequence invoked to explain specificity of ribosome binding (particularly in prokaryotic mRNAs) in 1975–1978 (Shine and Dalgarno 1975), it is clear that a limited region of nonrandom sequence near a functional site of an RNA molecule is not adequate to explain recognition events as complex as ribosome binding or RNA processing.

What enzymatic activities, then, have been found that might be involved in the splicing of eukaryotic mRNA? None have been positively identified, even in crude extracts. Candidates for RNA cleavage activities include the cleavage enzyme specific for HeLa cell hnRNA reported by Ferrari et al. (1980). This activity, which causes a size shift of hnRNA but cleaves no other prokaryotic or eukaryotic RNA or RNA precursor tested, can also cleave adenoviral late mRNA to give precise and reproducible bands. It is clear that more suitable substrates must be sought than those containing multiple polyadenylation and splicing sites, and one candidate is influenza mRNA. Two of the eight influenza mRNA transcripts are converted in HeLa-cell nuclei by splicing into smaller mRNA species (Lamb 1980; Lamb et al. 1981). These transcripts must therefore utilize host splicing enzymes. Furthermore they are not polyadenylated by the host machinery and do not transcribe beyond the 3′ terminus (Plotch and Krug 1977). Thus, the only reproducible endonucleolytic cleavages expected in these influenza mRNA transcripts are those that relate to their splicing in vivo. Subcellular fractions of HeLa cells are capable of cleaving these transcripts also, with a sequence specificity that is under investigation.

In marked contrast to the situation with rRNA, where the coupling of transcription and processing has yielded promising results, the two most widely used RNA polymerase-II systems that transcribe eukaryotic DNA to give correct initiation of mRNA transcripts seem incapable of correct processing (Manley et al. 1980; Weil et al. 1980). One of the reasons for this may be that so far no complete transcripts have been synthesized in vitro, and there could be a requirement for polyadenylation to precede splicing.

Weingartner and Keller (1981) have presented evidence for the processing of adenovirus RNA in an in vitro transcription processing system made from HeLa cells. The DNA template used was from the early region III of the adenoviral genome, encoding eight different mRNAs. These mRNAs have a common 5′-leader sequence of 372 nucleotides and differ from each other by the position of their 3′ ends. Moreover, five out of the eight mRNAs carry a second splice junction ending at four alternative sites. An analysis of the RNAs generated in vitro by hybridization and nuclease S1 mapping techniques allowed the authors to conclude that RNA size reductions consistent with RNA splicing and the formation of specific 3′ termini had indeed occurred in some of the in vitro products.

Goldenberg and Raskas (1981) have also used an in vitro system to seek splicing of adenoviral RNA precursors to yield products similar to those found in vivo. Whole-cell extracts were prepared from mouse MOPC-35 myeloma cells and incubated with a purified 28S precursor from early region II of adenovirus. The mature mRNA from this region, which encodes a 72K DNA binding protein, has a gel mobility equivalent to 20S. These authors demonstrated that an RNA of about 1700 bases and with a mobility equivalent to 20S is formed in vitro from the 28S precursor. Furthermore, the hybridization properties of this RNA suggest that it has lost sequences from the two introns present in the precursor. The degree of precision of this reaction was not demonstrated, nor were the biochemical requirements precisely proven. Furthermore, the conclusions of both Weingartner and Keller (1981) and Goldenberg and Raskas (1981) rely in part or completely on the nuclease S1 assay of radioactive DNA probes following hybridization to treated RNA precursors. Since the relative S1 resistance or sensitivity of these probes themselves has not been thoroughly characterized, unequivocal proof that accurate splicing has occurred must await RNA sequence determination of the putative spliced products. However, these systems appear to have great potential for identifying the molecules and mechanisms involved in the splicing of mRNA.

Taking advantage of the fact that the sequence in the vicinity of splice junctions in adenovirus is nonrandom, Murray and Holliday (1979) proposed that a small nuclear RNA present in adenoviral infection, VA-RNA, could form structures between its 5′ and 3′ termini and the consensus sequences in the vicinity of adenoviral splice junctions. This RNA:RNA structure was thought to provide the opportunity for the splicing enzyme(s) to work with a spatially constrained substrate in a manner possibly analogous to that of the RNA subunit of RNase P (see above). A more general version of this same idea was subsequently proposed for various eukaryotic mRNA precursors (Lerner et al. 1980; Rogers and Wall 1980). The RNA subunit analogous to that in RNase P, which was proposed in this latter case, is one or more of the small nuclear RNAs, U1–U6, present in various mammalian tissue culture cells (Ro-Choi and Busch 1974; Ramachandra and Busch 1981). It is possible to draw hydrogen-bonded structures relating these RNAs to splice-junction sequences. So far, although RNase P of eukaryotic cells seems to have an RNA subunit (Koski et al. 1976), there is no evidence that any activity involved in mRNA splicing contains such an RNA subunit.

A final consideration in mRNA splicing is the fact that mRNAs and their precursors are seldom found in vivo as free RNAs—they are bound to proteins in mRNP or hnRNP complexes. It is not yet possible to argue that the proteins of these complexes have a particular role in RNA metabolism, and in particular whether they protect any parts of the RNA precursor from random cleavage. It would be relevant to know whether the proteins that bind mRNA precursors are phased (by analogy to the recently defined "phasing" of nucleosomes on the DNA; see Calvet and Pederson 1977; Wu et al. 1979) so that certain regions are always bound to the same proteins. One clue to this latter question comes from studies carried out by Pederson and his colleagues (Calvet and Pederson 1977; 1978) who have shown that the characteristic double-stranded RNA regions of HeLa-cell hnRNA are protein-free. Perhaps both the enzyme and the substrate for eukaryotic mRNA processing are RNA:protein complexes, in which case three potential macromolecular interactions—protein:protein, protein:RNA, and RNA:RNA—could be involved.

SPECIFICITY DETERMINANTS FOR RNA PROCESSING

There are four major types of specificity determinants to be considered: substrate conformation and/or sequence, cofactors, modifications of the processing enzymes, and subcellular compartmentalization. Substrate conformation and sequence were discussed above. There is no case of an RNA-processing enzyme which, like a DNA restriction enzyme, merely recognizes a limited sequence and cleaves at or near it. While sequence homologies can be found to a limited extent in prokaryotic rRNA and mRNA precursors, there is no simple way to relate these homologies to processing specificity. In like manner, consensus sequences near splice junctions of eukaryotic mRNA precursors have no defined role in the processing. Equally impressive is the extraordinary specificity of cleavage (and ligation in the case of yeast tRNA) obtainable in the absence of any strict RNA sequence homology requirement. *E. coli* RNase P has no sequence requirement at all, relying primarily on tertiary structure. In a similar manner, extensive genetic engineering of the precursors for prokaryotic tRNA and *B. subtilis* 5S RNA have shown that the yeast RNA splicing machinery and the *B. subtilis* RNase m5 achieve extraordinary precision with no sequence requirement. Substrate conformation can be crucial, particularly in the case of RNase-III processing of various RNA precursors. Relatively small variations in monovalent cation concentration can have profound effects on the stability of RNA structures, and hence on the specificity of RNase-III processing.

With regard to cofactors, the most important ones so far identified are the RNA subunits of prokaryotic and eukaryotic RNase P activities. In addition to these cases, Martin and Underbrink-Lyon (1981) have recently suggested that yeast mitochondria code for an RNA species that is essential for mitochondrial tRNA maturation, possibly through interaction with RNase P. The ways in which different subunits could alter the specificity of processing by allowing alternative RNA:RNA interactions also make this a potentially important regulatory mechanism. It is therefore not surprising that the search for RNA subunits of other RNA-processing activities—and in particular those involved in eukaryotic mRNA maturation—has begun enthusiastically.

Other cofactors include the salt conditions of the environment. In the case of RNase III, there is some evidence for a salt-dependent monomer-dimer transition, which could mediate alterations in the specificity of RNase-III cleavage. In addition to the processing of rRNA and bacteriophage T7 early mRNA precursor, there are two other well-known reactions of RNase III: (1) solubilization of double-stranded RNA, which has an identical salt optimum to that for the processing reaction; and (2) secondary cleavage of normally resistant single-stranded RNA molecules, which has a much lower ionic strength optimum (for a summary of these data, see Robertson 1977). Experiments in which the RNase-III enzymatic activity was sized on Sephadex G-100 revealed two peaks of activity, with the major one running in the position expected for a dimer of two 25K subunits (Dunn 1976; Robertson 1977). The other peak runs with the expected position of an RNase-III monomer and has a much higher rate of secondary cleavage under the high-salt conditions optimal for RNA processing than does the dimer peak. Since the two in vitro reactions of RNase III (solubilization of double-stranded RNA and secondary cleavage) may reflect two different aspects of processing signal recognition, it is interesting to speculate that the enzyme monomer may contain information to recognize nucleotide sequence or non-Watson-Crick structure, while the dimer may be required to recognize and cleave within RNA-RNA duplex regions.

There is as yet no concrete evidence for modification of RNA-processing enzymes (e.g., protease cleavage, phosphorylation) in vivo except for the case of RNase D in T4-infected cells (see above). However, enzyme modification remains another important potential pathway for regulation of processing. Several cases exist where RNA-processing patterns of the same transcript change during the course of viral infection. The most striking such case is the adenoviral mRNA encoding the 52,55K polypeptide, which changes its processing between the early and late phases of infection (Akusjarvi and Persson 1981). This effect could be caused either by a modification of a host enzyme or synthesis of a new subunit (or enzyme) by the virus.

Regarding compartmentalization, an mRNA could remain sensitive to nuclear enzymes, but rapid transport to the cytoplasm could terminate processing at its proper time. The existence of transport pathways complicates the study of RNA cleavage because it imposes a requirement for an additional set of

RNA signals, protein-binding sites, and proteins that must be carefully identified in order not to be confused with signals for the RNA-processing events.

WHY HAS RNA PROCESSING EVOLVED?

The embedding of mature RNA sequences in surrounding extra sequences could be accidental. If this is the case, then RNA processing would rescue the needed sequences from their surroundings. The biosynthesis of any particular RNA species could then be regarded as a metabolic pathway with various intermediates ultimately leading to a final product. The function of RNA-processing enzymes in this context would be analogous to that of enzymes in any metabolic pathway—a "simple" function.

Many examples of potential benefits derived from RNA processing can be listed.

1. There is a great need, in certain cases, to separate functional segments of RNA from each other. One striking case of this is in human mtDNA (Anderson et al. 1981), where protein-coding sequences and functional RNA sequences are compactly arranged. RNA processing provides a convenient means to generate the final products, which are tRNAs, rRNAs, or mRNAs. Just as impressive is the case of adenoviral and SV40 mRNA production, where the flexibility provided by RNA processing enables the virus to program different mRNAs from the same DNA segment (Akusjarvi and Persson 1981). Similarly, the mode of expression of both immunoglobulin (Davis et al. 1980) and amylase (Young et al. 1981) genes is affected by changes in the availability of certain RNA molecules and their splicing.
2. RNA processing can ensure the coordinate appearance of molecules—e.g., rRNAs (Darnell 1968; Dunn and Studier 1973b; Nikolaev et al. 1973)—needed at the same time in stoichiometric amounts.
3. In contrast, through cotranscription of different classes of RNA via the same precursor, RNA processing could serve to influence the relative levels of products (Lee et al. 1981).
4. RNA processing enables the cell to alter or replace a single protein (the processing enzyme or its cofactors) and thereby change the nature of a large number of RNAs.
5. RNA processing can define correct ends of RNA molecules, a specificity that would otherwise be exclusively provided by RNA polymerase.
6. The ligation step in RNA splicing allows formation of both mature RNAs and circular RNAs (Konarska et al. 1981), which may play a role in gene regulation. For example, such regulation might involve RNA molecules as diffusible elements interacting with the genome (Martin and Underbrink-Lyon 1981). Alternatively, the spliced products could lead to autoregulatory features as seen in the yeast mitochondrial "box" locus (Lazowska et al. 1981).

7. RNA-processing pathways may confer the advantage of having a "storage form" of RNA available, by analogy with protein zymogens. Specifically, precursors not yet activated (e.g., in embryonic development) by the correct processing enzyme may be stored in the cell for future use.

Although our understanding of the functional significance of RNA processing is in its infancy, its results are dramatic. A classic example is the globin "gene," depicted in textbooks for years as an uninterrupted coding entity, which in fact only exists as such at the RNA level. Globin DNA is instead a functional mosaic, the properties of which would have been unrecognizable a few years ago.

PROSPECTS

Progress in the detailed characterization of RNA-processing nucleases has been painfully slow. There are several reasons for this, including technical problems of purification and a state of apathy among nucleic acid biochemists.

Over the past several years it has become apparent that the best way to study RNA processing is to use a natural substrate throughout the purification. However, natural substrates have proved troublesome to isolate and are unstable. Since many of the enzymes are endonucleases, which put one or a few cuts into large RNA molecules, rather slow assays and rigorous characterization of products are needed. Assays using synthetic substrates and based on acid solubility or acid precipitability may be useful, but such assays must always be validated with natural substrates.

Recombinant DNA technology offers an alternative route for the preparation of suitable substrates with which to assay these enzymes so that we can expect considerable progress in this exciting area.

ACKNOWLEDGMENTS

We thank colleagues, in particular the contributors to the Nuclease Meeting at Cold Spring Harbor Laboratory, for providing us with unpublished material. Work reported here which was performed in the authors' laboratories was supported by grants from the USPHS, NSF, and ACS to S.A. and H.D.R.

REFERENCES

Abelson, J.N. 1979. RNA processing. *Annu. Rev. Biochem.* **481:**1053.

Akusjarvi, G. and H. Persson. 1981. Controls of RNA splicing and termination in the major late adenovirus transcription unit. *Nature* **292:**420.

Altman, S. 1971. Isolation of tyrosine tRNA precursor molecules. *Nature New Biol.* **229:**19.

———. 1978. Transfer RNA biosynthesis. *Int. Rev. Biochem.* **17:**19.

———. 1981. Transfer RNA processing enzymes. *Cell* **23:**3.

Altman, S., P. Model, G.H. Dixon, and M.A. Wosnick. 1981. An *E. coli* gene coding for a protamine-like protein. *Cell* **26:**299.

Altman, S., E.J. Bowman, R.L. Garber, R. Kole, R.A. Koski, and B.C. Stark. 1980. Aspects of RNase P structure and function. In *Transfer RNA: Biological aspects* (ed. D. Söll, J.N. Abelson, and P.R. Schimmel), p. 71. Cold Spring Harbor Laboratory, Cold Spring Harbor, New York.

An, G. and J.D. Friesen. 1980. The nucleotide sequence of *tufB* and four nearby tRNA structural genes of *Escherichia coli*. *Gene* **12:**33.

Anderson, S., T. Bankier, B.G. Barrell, M.H.L. deBruyn, A.R. Coulson, J. Drouin, I.C. Eperon, D.P. Nierlich, B.A. Roe, F. Sanger, P.H. Schrier, A.J.H. Smith, R. Staden, and I.G. Young. 1981. Sequence and organization of the human mitochondrial genome. *Nature* **290:**457.

Apirion, D., B.K. Ghora, G. Plautz, T.K. Misra, and P. Gegenheimer. 1980. Processing of ribosomal RNA and transfer RNA in *Escherichia coli* cooperation between processing enzymes. In *Transfer RNA: Biological aspects* (ed. D. Söll, J.N. Abelson, and P.R. Schimmel), p. 139. Cold Spring Harbor Laboratory, Cold Spring Harbor, New York.

Baglioni, C., P.A. Maroney and T.W. Nilsen. 1981. Assay for the 2′-5′-oligo (A)-dependent endonuclease. *Methods Enzymol.* **79:**249.

Barry, G., C. Squires, and C.L. Squires. 1980. Attenuation and processing of RNA from the *rpl JL-rpo BC* transcription unit of *Escherichia coli*. *Proc. Natl. Acad. Sci.* **77:**3331.

Bikoff, E.K., B.F. LaRue, and M.L. Gefter. 1975. *In vitro* synthesis of transfer RNA. II. Identification of required enzymatic activities. *J. Biol. Chem.* **250:**6248.

Calvet, J.P. and T. Pederson. 1977. Ribonucleoprotein organization of eukaryotic RNA. 7. Secondary structure of heterogeneous nuclear RNA: Two classes of double-stranded RNA in native ribonucleoprotein. *Proc. Natl. Acad. Sci.* **74:**3705.

__________. 1978. Ribonucleoprotein organization of eukaryotic RNA. 8. Nucleoprotein organization of inverted repeat DNA transcripts in heterogeneous nuclear RNA-ribonucleoprotein particles from HeLa cells. *J. Mol. Biol.* **122:**361.

Cannistraro, V.J. and D. Kennell. 1979. *Escherichia coli lac* operator messenger RNA affects translation initiation of beta-galactosidase messenger RNA. *Nature* **277:**407.

Carin, M., B.F. Jensen, K.D. Jentsch, J.C. Leer, O.F. Nielsen, and O. Westergaard. 1980. *In vitro* splicing of the ribosomal RNA precursor in isolated nucleoli from *Tetrahymena*. *Nucleic Acids Res.* **8:**5551.

Colby, D., P.S. Leboy, and C. Guthrie. 1981. Yeast tRNA precursor mutated at a splice junction is correctly processed *in vivo*. *Proc. Natl. Acad. Sci.* **78:**415.

Cudny, H. and M.P. Deutscher. 1980. Reactions at the 3′ terminus of transfer-RNA. 27. Apparent involvement of ribonuclease D in the 3′ processing of transfer RNA precursors. *Proc. Natl. Acad. Sci.* **77:**837.

Cudny, H., P. Roy, and M.P. Deutscher. 1981a. Alteration of *E. coli* RNase D by infection with bacteriophage T4. *Biochem. Biophys. Res. Commun.* **98:**337.

Cudny, H., R. Zaniewski, and M.P. Deutscher. 1981b. *Escherichia coli* RNase D: Purification and structural characterization of a putative processing nuclease. *J. Biol. Chem.* **256:**5627.

Dahlberg, A., J.E. Dahlberg, E. Lund, H. Tokimatsu, A.B. Rabson, P.C. Calvert, F. Reynolds, and M. Zahalak. 1978. Processing of 5′-end of *Escherichia coli* 16S ribosomal RNA. *Proc. Natl. Acad. Sci.* **75:**3598.

Darnell, J.E. 1968. Ribonucleic acids from animal cells. *Microbiol. Rev.* **32:**262.

Davis, N.N., S.K. Kim, and L. Hood. 1980. Immunoglobulin class switching: Developmentally regulated DNA rearrangements during differentiation. *Cell* **22:**1.

Dunn, J.J. 1976. RNaseIII cleavage of single-stranded RNA. *J. Biol. Chem.* **267:**3807.

Dunn, J.J. and F.J. Studier. 1973a. T7 early RNAs are generated by site-specific cleavages. *Proc. Natl. Acad. Sci.* **70:**1559.

__________. 1973b. T7 early RNAs and *Escherichia coli* ribosomal RNAs are cut from large precursor RNAs *in vivo* by ribonuclease III. *Proc. Natl. Acad. Sci.* **70:**3296.

__________. 1975. Effect of RNase III cleavage on translation of bacteriophage T7 messenger RNAs. *J. Mol. Biol.* **99:**487.

———. 1981. Nucleotide sequence from the genetic left end of bacteriophage T7 DNA of the beginning of gene 4. *J. Mol. Biol.* **143:**303.

Feldman, H. 1976. Arrangement of transfer RNA genes. *Nucleic Acids Res.* **3:**2379.

Ferrari, S., C.O. Yehle, H.D. Robertson, and E. Dickson. 1980. Specific RNA cleaving activities from HeLa cells. *Proc. Natl. Acad. Sci.* **77:**2395.

Fiers, W., R. Contreras, G. Haegeman, R. Rogiers, A. Van de Voorde, H. Heuverswyn, J.V. Herreweghe, G. Volckaert, and M. Ysebaert. 1978. Complete nucleotide sequence of SV40 DNA. *Nature* **273:**113.

Fiil, N.P., J.D. Friesen, W.L. Downing, and P.P. Dennis. 1980. Post-transcriptional regulatory mutants in a ribosomal protein-RNA polymerase operon of *E. coli. Cell* **19:**837.

Fitzgerald, M. and T. Shenk. 1981. The sequence 5′-AAUAAA-3′ forms part of the recognition site for polyadenylation of late SV40 mRNAs. *Cell* **24:**251.

Fraser, N. and E. Ziff. 1978. RNA structures near poly(A) of adenovirus-2 late messenger RNAs. *J. Mol. Biol.* **124:**27.

Garber, R.L. and L.P. Gage. 1979. Transcription of a cloned *Bombyx mori* $tRNA_2Ala$ gene: nucleotide sequence of the tRNA precursor and its processing *in vitro. Cell* **18:**817.

Gegenheimer, P. and D. Apirion. 1978. Processing of ribosomal RNA by RNAse P: Spacer transfer RNAs are linked to 16S ribosomal RNA in a RNase P: RNAse III mutant strain of *Escherichia coli. Cell* **15:**527.

Gegenheimer, P., N. Watson, and D. Apirion. 1977. Multiple pathways for primary processing of ribosomal RNA in *Escherichia coli. J. Biol. Chem.* **252:**3064.

Goldfarb, A. and V. Daniel. 1980. An *E. coli* endonuclease responsible for primary cleavage of *in vitro* transcripts of bacteriophage T4 tRNA gene cluster. *Nucleic Acids Res.* **8:**4501.

Goldenberg, C.J. and H.J. Raskas. 1981. *In vitro* splicing of purified precursor RNA as specified by early region 2 of the adenovirus 2 genome. *Proc Natl. Acad. Sci.* **78:**5430.

Grabowski, P.J., A.J. Zuag, and T.R. Cech. 1981. The intervening sequence of the ribosomal RNA precursor is converted to a circular RNA in isolated nuclei of *Tetrahymena. Cell* **23:**467.

Grimm, M.F., M.D. Cole, and A.M. Lambowitz. 1981. Ribonucleic acid splicing in *Neurospora* mitochondria: Secondary structure of the 35S ribosomal precursor ribonucleic acid investigated by digestion with ribonuclease III and by electron microscopy. *Biochemistry* **20:**2836.

Grummt, I., S.H. Hall, and R.T. Crouch. 1979. Localization of an endonuclease specific for double-stranded RNA within the nucleolus and its implication in processing ribosomal transcripts. *Eur. J. Biochem.* **94:**437.

Gumport, R.I. and O.C. Uhlenbeck. 1982. T4 RNA ligase as a nucleic acid synthesis and modification agent. In *Gene amplification and analysis* (ed. J.G. Clirikijian and T.S. Papas), vol. 2. Elsevier-North Holland, New York. (In press)

Hofer, E. and J.E. Darnell. 1981. The primary transcription unit of the mouse beta-major globin gene. *Cell* **23:**585.

Hudson, L., J. Rossi, and A. Landy. 1981. Dual function transcripts specifying tRNA and mRNA. *Nature* **294:**422.

Johnson, J. and J. Abelson. 1980. Transcription and processing of a yeast tRNA gene containing a modified intervening sequence. *Proc. Natl. Acad. Sci.* **77:**2564.

Kerr, I.M., R.E. Brown, and A.G. Hovanessian. 1977. Nature of inhibitor of cell free protein synthesis formed in response to interferon and double-stranded RNA. *Nature* **268:**540.

Kindler, P., T.U. Keil, and P.H. Hofschneider. 1973. Isolation and characterization of a ribonuclease-III deficient mutant of *E. coli. Mol. Gen. Genet.* **126:**53.

Kline, L., S. Nishikawa, and D. Soll. 1981. Partial purification of RNase P from *S. pombe. J. Biol. Chem.* **256:**5058.

Kole, R. and S. Altman. 1979. Reconstitution of RNase P activity from inactive RNA and protein. *Proc. Natl. Acad. Sci.* **76:**3795.

Kole, R. and S. Altman. 1981. Properties of purified ribonuclease P from *Escherichia coli. Biochemistry* **20:**1902.

Kole, R., M.F. Baer, B.C. Stark, and S. Altman. 1980. *E. coli* RNase P has a required RNA component *in vivo*. *Cell* **19:**881.

Konarska, M., W. Filipowicz, H. Domdey, and H.J. Gross. 1981. Formation of a 2′-phosphomonoester, 3′, 5′-phosphodiester linkage by a novel RNA ligase in wheat germ. *Nature* **293:**112.

Koski, R.A., E. Bothwell, and S. Altman. 1976. Identification of a ribonuclease P-like activity from human KB cells. *Cell* **9:**101.

Koski, R.A., S.G. Clarkson, J. Kurjan, B.D. Hall, and M. Smith. 1980. Mutations of the yeast sup4 $tRNA^{tyr}$ locus: Transcription of the mutant genes *in vitro*. *Cell* **22:**415.

Lamb, R.A. 1980. Sequence of interrupted and uninterrupted messenger RNAs and cloned DNA coding for the two overlapping nonstructural proteins of influenza virus. *Cell* **21:**475.

Lamb, R.A., C.-J. Lai, and P.N. Choppin. 1981. Sequences of messenger RNAs derived from genomic RNA segment 7 of influenza virus: Coding and interrupted messenger RNAs code for overlapping proteins. *Proc. Natl. Acad. Sci.* **78:**4170.

Lazowska, J., C. Jacq, and P.P. Slonimski. 1981. Sequence of introns and flanking exons in wild-type and box3 mutants of cytochrome b reveals an interlaced splicing protein coded by an intron. *Cell* **22:**333.

Lee, J.S., G. An, J.D. Friesen, and N.P. Fiil. 1981. Location of the *tuf B* promoter of *E. coli*: Co-transcription of *tuf B* with four transfer RNA genes. *Cell* **25:**251.

Lerner, M.R., J.A. Boyle, S.M. Mount, S.L. Wolin, and J.A. Steitz. 1980. Are snRNPs involved in splicing? *Nature* **283:**220.

Lozeron, H.A., P.J. Anevski, and D. Apirion. 1977. Anti-termination and absence of processing of leftward transcript of coliphage lambda in RNase III deficient host. *J. Mol. Biol.* **109:**359.

Lozeron, H.A., J.E. Dahlberg, and W. Szybalski. 1976. Processing of major leftward messenger RNA of coliphage lambda. *Virology* **71:**262.

Lund, E., J.E. Dahlberg, and C. Guthrie. 1980. Processing of spacer transfer RNAs from ribosomal RNA transcripts of *Escherichia coli*. In *Transfer RNA: Biological aspects* (ed. D. Söll, J.N. Abelson, and P.R. Schimmel), p. 123. Cold Spring Harbor Laboratory, Cold Spring Harbor, New York.

Manley, J.L., A. Fire, A. Cano, P.A. Sharp, and M. Gefter. 1980. DNA-dependent transcription of adenovirus genes in a soluble whole-cell extract. *Proc. Natl. Acad. Sci.* **77:**3855.

Mao, J., O. Schmidt, and D.G. Soll. 1980. Dimeric tRNA precursors in *S. pombe*. *Cell* **21:**509.

Martin, N. and K. Underbrink-Lyon. 1981. A mitochondrial locus is necessary for the synthesis of mitochondrial tRNA in the yeast *Saccharomyces cerevisiae*. *Proc. Natl. Acad. Sci.* **78:**4743.

McClain, W.H. 1977. Seven terminal steps in a biosynthetic pathway leading from DNA to transfer RNA. *Accts. Chem. Res.* **10:**418.

McCorkle, G.M. and S. Altman. 1982. Large deletion mutants of *E. coli* $tRNA_1^{tyr}$. *J. Mol. Biol.* **155:**83.

Melton, D.A., E.M. deRobertis, and R. Cortese. 1980. Order and intracellular location of the events involved in the maturation of spliced tRNA. *Nature* **284:**143.

Misra, T.K. and D. Apirion. 1979. RNase E, an RNA processing enzyme from *Escherichia coli*. *J. Biol. Chem.* **254:**11154.

Murray, V. and R. Holliday. 1979. Mechanism for RNA splicing of gene transcripts. *FEBS Lett.* **106:**5.

Nakajima, N., H. Ozeki, and Y. Shimura. 1981. Organization and structure of an *E. coli* tRNA operon containing seven tRNA genes. *Cell* **23:**239.

Nevins, J.R. and J.E. Darnell. 1978. Groups of adenovirus type 2 mRNAs derived from a large primary transcript: Probable nuclear origin and possible common 3′ ends. *J. Virol.* **25:**811.

Nikolaev, N., L. Silengo, and D. Schlessinger. 1973. Synthesis of a large precursor to ribosomal RNA in a mutant of *Escherichia coli*. *Proc. Natl. Acad. Sci.* **70:**3361.

Nilsen, T.W., P.A. Maroney, H.D. Robertson, and C. Baglioni. 1982. Heterogeneous nuclear RNA promotes synthesis of 2′, 5′ oligo(A) and is cleaved by the 2′, 5′ oligo(A)- activated endoribonuclease. *Mol. Cell Biol.* **2:**154-160.

Nomura, M. and L.E. Post. 1980. Organization of ribosomal genes and regulation of their expression in *E. coli*. In *Ribosomes* (ed. G. Chambliss, G.R. Craven, J. Davies, K. Kavis, C. Kehen, and M. Nomura), p. 671. University Park Press, Baltimore.

Ozeki, H. 1980. The organization of transfer RNA genes in *Escherichia coli*. In *RNA-polymerase, tRNA and ribosomes: Their genetics and evolution* (ed. S. Asawa, H. Ozeki, H. Uchida, and T. Yura), p. 173. Tokyo University Press, Tokyo.

Pace, N.R., B. Meyhack, B. Pace, and M.L. Sogin. 1980. The interaction of RNase m5 with a 5S ribosomal RNA precursor. In *Transfer RNA: Biological aspects* (ed. D. Söll, J.N. Abelson, and P. Schimmel), p. 155. Cold Spring Harbor Laboratory, Cold Spring Harbor, New York.

Perry, R.P. 1976. Processing of RNA. *Annu. Rev. Biochem.* **45:**605.

———. 1982. RNA processing comes of age. *J. Cell Biol. Suppl.* (in press).

Plotch, S.J. and R.M. Krug. 1977. Influenza virion transcriptase: Synthesis *in vitro* of large, polyadenylate-containing complementary RNA. *J. Virol.* **21:**24.

Ramachandra, R. and H. Busch. 1981. snRNAs of nuclear snRNPs. In *The cell nucleus* (ed. H. Busch), vol. VIII, p. 261. Academic Press, New York.

Ray, B.K. and D. Apirion. 1981. RNase P is dependent on RNase E action in processing monomeric RNA precursors that accumulate in an RNase E^- mutant of *Escherichia coli*. *J. Mol. Biol.* **149:**599.

Robertson, H.D. 1977. Structure and function of RNA processing signals. In *Protein-nucleic acid interactions* (ed. H. Vogel), p. 549. Academic Press, New York.

Robertson, H.D. and F. Barany. 1978. Enzymes and mechanisms in RNA processing. *FEBS Proc. Meet.* **12:**285.

Robertson, H.D., E. Dickson, and W. Jelinek. 1977. Determination of nucleotide sequences from double-stranded regions of HeLa cell nuclear RNA. *J. Mol. Biol.* **115:**571.

Robertson, H.D., E.G. Pelle, and W.H. McClain. 1980. RNA processing in an *Escherichia coli* strain deficient in both RNase P and RNase III. In *Transfer RNA: Biological aspects* (ed. D. Söll, J.N. Abelson, and P. Schimmel), p. 107. Cold Spring Harbor Laboratory, Cold Spring Harbor, New York.

Robertson, H.D., R.E. Webster, and N.D. Zinder. 1968. Purification and properties of ribonuclease III from *Escherichia coli*. *J. Biol. Chem.* **243:**82.

Ro-Choi, T.S. and H. Busch. 1974. Low molecular weight nuclear RNAs. In *The cell nucleus* (ed. H. Busch), vol. III, p. 157. Academic Press, New York.

Rogers, J. and R. Wall. 1980. A mechanism for RNA splicing. *Proc. Natl. Acad. Sci.* **77:**1877.

Rubin, C.M. and D.S. Hogness. 1975. Effect of heat shock on synthesis of low molecular weight RNAs in *Drosophila:* Accumulation of a novel form of 5S RNA. *Cell* **6:**207.

Rubin, C.M., C.M. Houck, P.L. Deininger, T. Friedmann, and C.W. Schmid. 1980. Partial nucleotide sequence of the 300 nucleotide interspersed repeated human DNA sequences. *Nature* **284:**372.

Sakano, H. and Y. Shimura. 1975. Sequential processing of precursor tRNA molecules in *Escherichia coli*. *Proc. Natl. Acad. Sci.* **72:**3369.

Sakano, H., S. Yamada, T. Ikemura, Y. Shimura, and H. Ozeki. 1974. Temperature sensitive mutants of *Escherichia coli* for tRNA synthesis. *Nucleic Acids Res.* **1:**355.

Salditt-Georgieff, M., M. Harpold, S. Chen-Kiang, and J.E. Darnell. 1980. Addition of the 5′ cap structures occurs early in hnRNA synthesis and prematurely terminated molecules are capped. *Cell* **19:**69.

Sanger, F. and A. Coulson. 1978. The use of thin acrylamide gels for DNA sequencing. *FEBS Lett.* **87:**107.

Schedl, P. and P. Primakoff. 1973. Mutants of *Escherichia coli* temperature sensitive for the biosynthesis of transfer RNA. *Proc. Natl. Acad. Sci.* **70:**2091.

Schedl, P., P. Primakoff, and J. Roberts. 1974. Processing of *E. coli* tRNA precursors. *Brookhaven Symp. Biol.* **26:**53.

Schmidt, O., J. Mao, A. Ogden, J. Beckman, H. Sakano, J. Abelson, and D. Soll. 1980. Dimeric tRNA precursors in yeast. *Nature* **287:**750.

Sekiya, T., R. Contreras, H. Kupper, A. Landy, and H.G. Khorana. 1976. *Escherichia coli* tyrosine transfer ribonucleic acid genes. *J. Biol. Chem.* **251:**5124.

Sharp, P. 1981. Speculations in RNA splicing. *Cell* **23:**643.

Shine, J. and L. Dalgarno. 1975. Determination of cistron specificity in bacterial ribosomes. *Nature* **254:**34.

Sogin, M.L., B. Pace, and N.R. Pace. 1977. Partial purification and properties of a ribosomal-RNA maturation endonuclease from *Bacillus subtilis*. *J. Biol. Chem.* **252:**1350.

Stark, B.C. 1977. "Further purification and properties of ribonuclease P from *Escherichia coli*." Ph.D. thesis, Yale University, New Haven, Connecticut.

Sugden, B. and W. Keller. 1973. Mammalian deoxyribonucleic acid-dependent ribonucleic acid polymerases. 1. Purification and properties of an alpha-amanitin sensitive ribonucleic acid polymerase and stimulatory factors from HeLa and KB cells. *J. Biol. Chem.* **218:**3777.

Tekamp, P., R.L. Garcea, and W.J. Rutter. 1980. Secondary structures for splice junctions in eukaryotic and viral messenger RNA precursors. *Nucleic Acids Res.* **8:**3659.

Trapnell, B.C., P. Tolstoshev, and R.G. Crystal. 1980. Secondary structures for splice junctions in eukaryotic and viral messenger RNA precursors. *Nucleic Acids Res.* **8:**3659.

Weil, P.A., D.S. Luce, J. Segall, and R.G. Roeder. 1980. Selective and accurate initiation of transcription at the Ad2 major late promoter in a soluble system dependent on purified RNA polymerase II and DNA. *Cell* **18:**469.

Weingartner, B. and W. Keller. 1981. Transcription and processing of adenoviral RNA by extracts from HeLa cells. *Proc. Natl. Acad. Sci.***78:**4092.

Wilhem, J., O. Brison, C. Kedinger, and P. Chambon. 1976. Characterization of adenovirus type 2 transcriptional complexes isolated from infected HeLa cell nuclei. *J. Virol.* **19:**61.

Wolgemuth, D.J. and M.T. Hsu. 1981. Visualization of nascent RNA transcripts and simultaneous transcription and replication in viral nucleoprotein complexes from adenovirus 2-infected HeLa cells. *J. Mol. Biol.* **147:**247.

Wu, M. and N. Davidson. 1975. Use of gene 32 protein staining of single-strand polynucleotides for gene mapping by electron-microscopy: Application to $\phi 80d_3 ilvsu^+7$ system. *Proc. Natl. Acad. Sci.* **72:**4506.

Young, R.A. and J.A. Steitz. 1978. Complementary sequences 1700 nucleotides apart from a ribonuclease III cleavage site in *Escherichia coli* ribosomal precursor RNA. *Proc. Natl. Acad. Sci.* **75:**3593.

Young, R.A., O. Hagenbuchle, and U. Schibler. 1981. A single mouse alpha-amylase gene specifies two different tissue specific messenger RNAs. *Cell* **23:**451.

Ziff, E. 1980. Transcription and RNA processing by the DNA tumor-viruses. *Nature* **287:**491.

Use of Nucleases in RNA Sequence and Structural Analyses

Uttam L. RajBhandary, Raymond E. Lockard,* and Regina M. Reilly
Department of Biology
Massachusetts Institute of Technology
Cambridge, Massachusetts 02139

*Medical School Department of Biochemistry
George Washington University
Washington, DC 20037

As might be expected, nucleases have played crucial roles in the sequence and structural analyses of RNAs. Along with other evidence, studies on the effect of pancreatic RNase and spleen phosphodiesterase on chemically synthesized nucleoside 2′- and 3′-phosphodiesters helped to establish that the phosphodiester linkages in RNA were 3′−5′ rather than 2′−5′ (Brown and Todd 1953). The availability of base-specific endoribonucleases such as pancreatic RNase and T1-RNase made possible the specific cleavage of RNAs to short oligonucleotides, a key step in sequence analysis of RNAs. Exonucleases such as snake venom phosphodiesterase and spleen phosphodiesterase have found extensive use in the sequence analysis of these oligonucleotides (Laskowski, this volume). Nuclease P1, a random endonuclease, has proved useful not only for sequencing short oligonucleotides, but also for the terminal analysis of large RNAs that are labeled at either their 5′ or 3′ ends (Silberklang et al. 1977). Base-specific nucleases are also being used for direct sequence analysis by partial digestion of end-labeled RNAs, followed by gel-electrophoretic analysis of the labeled fragments. This has provided a rapid method for sequencing up to 400 nucleotides from the labeled end of an RNA (Donis-Keller et al. 1977; Simoncsits et al. 1977; Lockard et al. 1978; 1982).

Many of the nucleases are sensitive to the secondary and tertiary structure of the RNAs. S1 nuclease from *Aspergillus oryzae* cleaves only single-stranded regions of RNA or DNA (Ando 1966), whereas a nuclease from cobra venom cleaves only within double-stranded regions of an RNA. These and other enzymes are currently used widely as probes for RNA structure.

This paper reviews briefly the use of various nucleases both in RNA sequence and in RNA structural analyses. A more detailed description of the use of nucleases in RNA sequencing can be found elsewhere (Silberklang et al. 1979).

NUCLEASES IN RNA SEQUENCE ANALYSIS

Base-specific Endonucleases in RNA Sequence Analysis

Total Digests of RNAs

In earlier work, a key step in RNA sequence analysis was the specific fragmentation of the RNA to small oligonucleotides. Enzymes commonly used for this were pancreatic RNase which cleaves RNA on the 3′ side of pyrimidines, T1-RNase which cleaves on the 3′ side of G residues, and U2-RNase which cleaves on the 3′ side of purine residues. Except for oligonucleotides originating from the termini, the products of such digestion all contain a 5′-hydroxyl and a 3′-phosphomonoester group. Depending on the scale of the digestion, the oligonucleotides may be separated either by ion-exchange column chromatography (Tomlinson and Tener 1962) or, if the RNA used is uniformly labeled with ^{32}P, by two-dimensional electrophoresis (Sanger et al. 1965). Alternatively, oligonucleotides present in total enzymatic digests of nonradioactive RNA can, subsequent to removal of phosphomonoester residues with alkaline phosphatase, be labeled with ^{32}P at their 5′-ends using polynucleotide kinase (Simsek et al. 1973). These 5′-^{32}P-labeled oligonucleotides, all of which contain a free 3′-hydroxyl group, are then separated by two-dimensional electrophoresis (Fig. 1) or by two-dimensional homochromatography (Silberklang et al. 1979). These oligonucleotides can then be sequenced using one of several procedures described below.

Partial Digestion of RNA and Isolation of Large Fragments

To complete an RNA sequence, oligonucleotides present in total enzymatic digests of an RNA must be ordered to yield a unique sequence. One approach to obtain this overlap information is to isolate large fragments of the RNA and determine within them the arrangement of oligonucleotides whose sequences are already known. Partial digestion with base-specific nucleases is also useful for the isolation of large oligonucleotide fragments of an RNA. Because base-specific nucleases are also sensitive to RNA structure, partial digestion of an

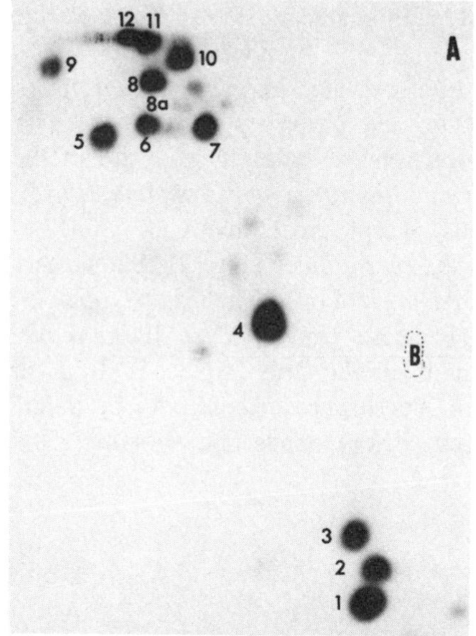

Figure 1 Autoradiogram of 5′-^{32}P oligonucleotides obtained by labeling of a T1-RNase digest of *N. crassa* mitochondrial tRNA^{Tyr}. Electrophoresis in the first dimension was on a strip of cellulose acetate at pH 3.5, and, in the second dimension, on DEAE-cellulose paper using 7% formic acid. The position of the xylene cyanole FF blue dye marker is indicated B.

RNA under conditions where RNA structure is preserved yields primarily a limited number of large fragments in which the nuclease has cleaved the most susceptible bonds in the RNA. Depending on the size of the fragment and the scale of digestion, the large oligonucleotide fragments may be separated either by ion-exchange column chromatography (Penswick and Holley 1965) or by one- or two-dimensional polyacrylamide gel electrophoresis (DeWachter and Fiers 1972). As in the case of smaller oligonucleotides present in total enzymatic digests of an RNA, the large fragments can also be end-labeled with ^{32}P at their 5′ ends prior to their fractionation (Silberklang et al. 1979).

Direct Sequence Analysis of RNA Using Partial Digestion of End-labeled RNA with Base-specific Nucleases

Partial digestion of end-labeled RNAs with base-specific nucleases followed by mapping of the base-specific cleavage sites relative to the labeled end by polyacrylamide gel electrophoresis under denaturing conditions provides a rapid method for sequencing up to 400 nucleotides from the labeled end of an RNA. Conditions used for the partial digestion should be such as to disrupt RNA secondary structure. The enzymes commonly used are T1-RNase for cleavage at G residues, U2-RNase for cleavage at A residues, and an extracellular RNase from *B. cereus*, or in some cases pancreatic RNase for cleavage at pyrimidine residues (Donis-Keller et al. 1977; Simoncsits et al. 1977; Lockard et al. 1978). Unlike pancreatic RNase, which under the conditions of partial digestion is more sequence specific and cleaves pyr-pyr bonds with difficulty, the endonuclease from *B. cereus* cleaves after virtually all of the pyrimidines. An extracellular enzyme from *Physarum polycephalum* (*Phy*I), which cleaves all phosphodiester bonds except those involving CpN sequences, is used to distinguish between the pyrimidines (Fig. 2). Should any ambiguities arise in distinguishing C from U, several other enzymes may be used. These include a C-specific endonuclease from chicken liver, which under most conditions cleaves only bonds involving CpN sequences (Boguski et al. 1980), and an endonuclease from *Neurospora crassa*, which, in the presence of 7 M urea, preferentially cleaves those bonds not involving CpN residues (Krupp and Gross 1979).

Use of RNaseH in RNA Sequencing

Since RNAs can be labeled at either their 5′ or 3′ ends (RajBhandary 1980) and sequences of up to 400 nucleotides can be determined from either end, a key problem in the sequencing of large RNAs is the isolation of specific long fragments of the RNA. One possible approach to this is sequence-specific cleavage of an RNA in a manner somewhat analogous to restriction enzyme cleavage on DNA (Donis-Keller 1979). A potentially useful enzyme is RNaseH (see Chapter 9), which can be used to cleave RNA at specific sites in the presence of chemically synthesized oligodeoxyribonucleotides. The oligodeoxyribonucleotides hybridize to specific regions of the RNA and thereby direct cleavage at these regions (Fig. 3). Different oligodeoxyribonucleotides will direct cleavage of the RNA at different locations and thereby provide overlaps among the large fragments. The RNA fragments produced all contain a 3′-hydroxyl and a 5′-phosphoryl group and can either be labeled at the 3′ end with ^{32}P-Cp using RNA ligase (England and Uhlenbeck 1978) or at the 5′ end with ^{32}P with polynucleotide kinase following removal of the 5′-phosphate group.

Use of Exonucleases in RNA Sequence Analysis

As noted above, an important step in RNA sequencing is the sequence determination of oligonucleotides which are present in complete enzymatic digests of an RNA. Exonucleases such as spleen phosphodiesterase and snake venom phosphodiesterase are the two most commonly used enzymes for this purpose. Spleen phosphodiesterase cleaves an oligonucleotide exonucleolytically from the 5′-hydroxyl end releasing nucleoside 3′-phosphates (Razzell and Khorana 1959a). The action of snake venom phosphodiesterase is complementary to that of the spleen enzyme. It cleaves an oligonucleotide exonucleolyti-

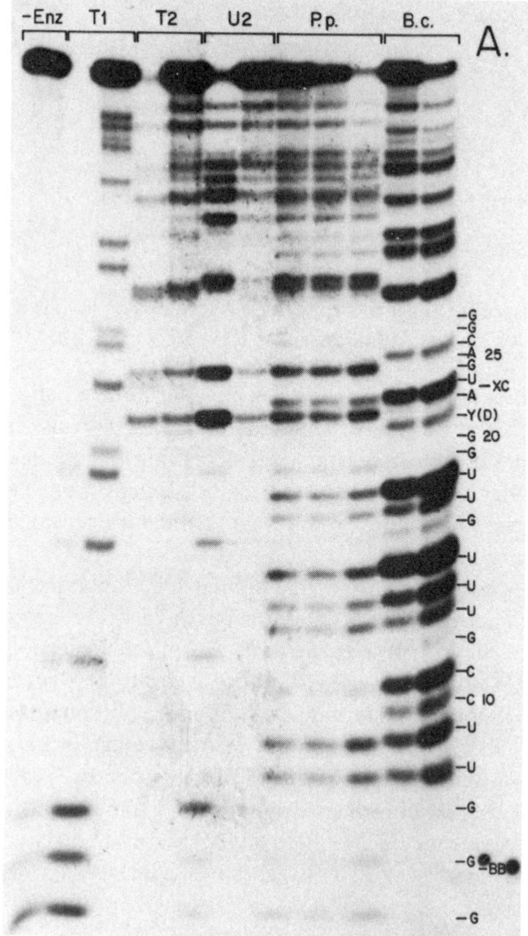

Figure 2 Autoradiogram of partial enzymatic digests on 5′-^{32}P-labeled *N. crassa* mitochondrial tRNA^{Tyr}. *Left* to *right*: (-Enz) incubated in absence of enzyme; (T1, T2, U2) incubation of end-labeled tRNA at two different concentrations of these RNases; (P.p.) incubation of the RNA with three different concentrations of *Phy*I RNase (see text); and (B.c.) incubation with two concentrations of a pyrimidine-specific RNase from *B. cereus*.

cally from the 3′-hydroxyl terminus and releases nucleoside 5′-phosphates in a sequential manner (Razzell and Khorana 1959b).

The use of spleen phosphodiesterase for rapid sequencing of uniformly ^{32}P-labeled oligonucleotides was a key factor in the procedures developed by Sanger and coworkers for sequencing uniformly ^{32}P-labeled RNA. Partial digestion of an oligonucleotide with spleen phosphodiesterase produces, be-

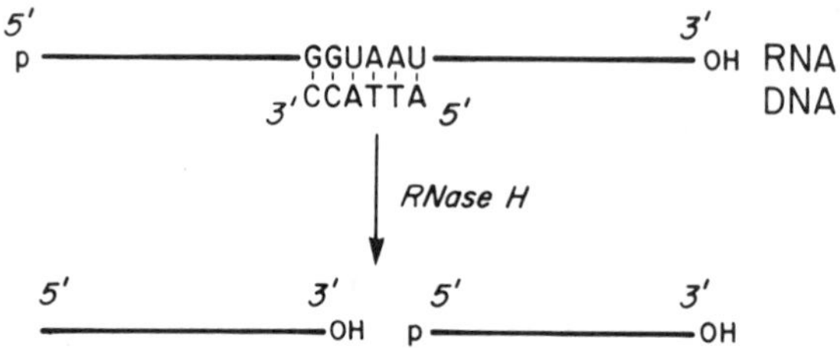

Figure 3 Scheme for sequence-specific enzymatic cleavage of RNA using RNaseH.

sides nucleoside 3′-phosphates, a homologous series of oligonucleotides, all of which contain the original 3′ end of the oligonucleotide. Ionophoresis on DEAE-cellulose paper separates these oligonucleotides according to size, and the sequence of the original oligonucleotide can be deduced directly from the relative mobilities of the partial digestion products in the one-dimensional electrophoresis system (Sanger et al. 1965).

Snake venom phosphodiesterase may be used in a similar way for sequencing oligonucleotides which contain a free 3′-hydroxyl group and are hence susceptible to the action of this enzyme. It is particularly useful for sequencing oligonucleotides which are ^{32}P-labeled at their 5′ ends (see above). Partial digestion of a 5′-^{32}P-labeled oligonucleotide with snake venom phosphodiesterase yields a series of 5′-^{32}P products ranging in size from the original oligonucleotide down to the 5′-terminal nucleotide (Fig. 4). These 5′-^{32}P-oligonucleotides are separated by two-dimensional homochromatography (Silberklang et al. 1979), and the sequence of the oligonucleotide can be deduced directly from the characteristic mobility shifts between the successive intermediates in the partial digest (Fig. 5).

Use of Nuclease P1 in RNA Sequence Analysis

Nuclease P1, a random endonuclease isolated from *Penicillium citrinum*, cleaves both RNA and DNA to yield oligonucleotides containing 3′-hydroxyl and 5′-phosphoryl groups (Fujimoto et al. 1974). The products of complete digestion on RNA or DNA are nucleoside 5′-phosphates. Although this enzyme acts differently from snake venom phosphodiesterase (the snake venom enzyme is an exonuclease, whereas P1 is an endonuclease), the radioactively labeled products of partial digestion of most 5′-^{32}P-oligonucleotides with either enzyme are the same. Consequently, nuclease P1 can also be used for sequencing 5′-^{32}P-labeled oligonucleotides and can be used along with snake venom phosphodiesterase in special cases (Silberklang et al. 1977, 1979). Furthermore, since P1 is an endonuclease, it can also be used for sequencing

1. PARTIAL DIGESTION WITH SNAKE VENOM PHOSPHODIESTERASE AND/OR NUCLEASE P1.

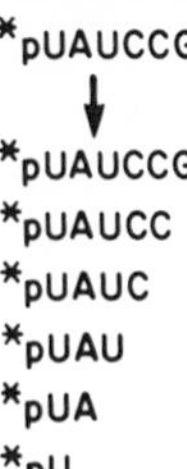

2. ANALYSIS OF PARTIAL DIGESTS BY TWO DIMENSIONAL HOMOCHROMATOGRAPHY.

Figure 4 Scheme for sequence analysis of 5′-^{32}P-labeled oligonucleotides.

the termini of large RNAs labeled at their 5′ or 3′ ends. As in the case of short oligonucleotides, the products are analyzed by two-dimensional homochromatography and the sequence of over 20 nucleotides can usually be determined from analysis of a single partial digest (Fig. 6).

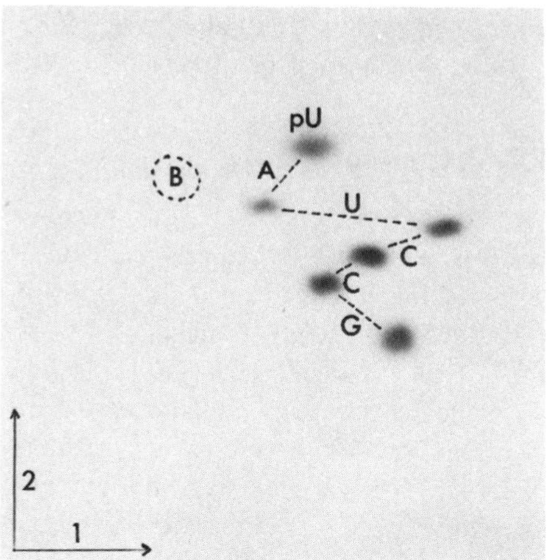

Figure 5 Autoradiogram of a partial snake venom phosphodiesterase digest on 5′-^{32}P-UAUCCG as analyzed by two-dimensional homochromatography. *First dimension*, electrophoresis on cellulose acetate at pH3.5; *second dimension*, homochromatography on a 20 × 20 cm DEAE-cellulose thin layer plate. Circled B indicates position of xylene cyanole FF dye.

NUCLEASES AS PROBES FOR HIGHER-ORDER STRUCTURE OF RNAs

A number of approaches are available for probing RNA structure in solution. These include the use of spectroscopic methods (Reid 1981), chemical modifications (Rich and RajBhandary 1976; Peattie and Gilbert 1980), oligonucleotide binding (Uhlenbeck 1972), and nucleases (Wurst et al. 1978; Vournakis et al. 1981). The enzymatic approach allows one to identify exposed single-stranded regions (nonbase-paired) and structured regions (either base-paired or involved in tertiary interactions) within an RNA. S1 nuclease (Rushizky et al. 1975), *N. crassa* nuclease (Linn and Lehman 1965), mung bean nuclease (Kowalski et al. 1976), T1-RNase, and pancreatic RNase are the most common enzymes used to probe for single-stranded regions, whereas the more recently discovered enzyme(s) from cobra venom (Favarova et al. 1981; Lockard and Kumar 1981) is useful as a probe for base-paired regions. In general, the accessibility of single-stranded regions of an RNA towards nucleases depends not only on whether or not this region is single-stranded and exposed on the surface of the molecule, but also on the size and specificity of the enzyme. Thus, although lack of cleavage by a single-strand-specific enzyme does not conclusively identify a region of RNA as being base-paired, residues that are cleaved can be concluded to be nonbase-paired and hence exposed. Similarly, although not all phosphodiester bonds in base-paired regions of an RNA are cleaved by cobra venom nuclease, those that are can be identified as being in duplex regions or in regions involved in tertiary interactions. Although this review will focus mostly on tRNAs, nucleases are being used widely in the structural analysis of rRNAs (Ross and Brimacombe 1979; Steigler et al. 1981), mRNAs (Pavlakis et al. 1980; Vournakis et al. 1981), and other small RNAs (Krol et al. 1981; Machatt et al. 1981).

Single-strand-specific Nucleases As Probes for tRNA Structure

First indications that certain sites in RNA were more susceptible to nucleases than others came during sequence analysis of tRNAs. Under conditions of limited digestion with T1-RNase, yeast $tRNA^{Ala}$ was cleaved predominantly in the anticodon (Penswick and Holley 1965). Subsequent analysis of partial digests on a number of tRNAs with T1-RNase, pancreatic RNase, and other structure-specific nucleases such as S1 nuclease and *N. crassa* nuclease led to the general conclusion (Chang and RajBhandary 1968; Harada and Dahlberg 1975; Tal 1975; Rushizky and Mozejko 1977) that the most exposed regions in tRNAs are the anticodon loop followed by the CCA sequence at the 3′ end. The D and TψC loops are much less accessible towards single-strand nucleases. In general, results obtained with these nucleases are in good agreement with the predictions from X-ray structure analysis (Jack et al. 1976; Quigley and Rich 1976) and from studies of the sites on tRNAs most reactive towards chemical modifications.

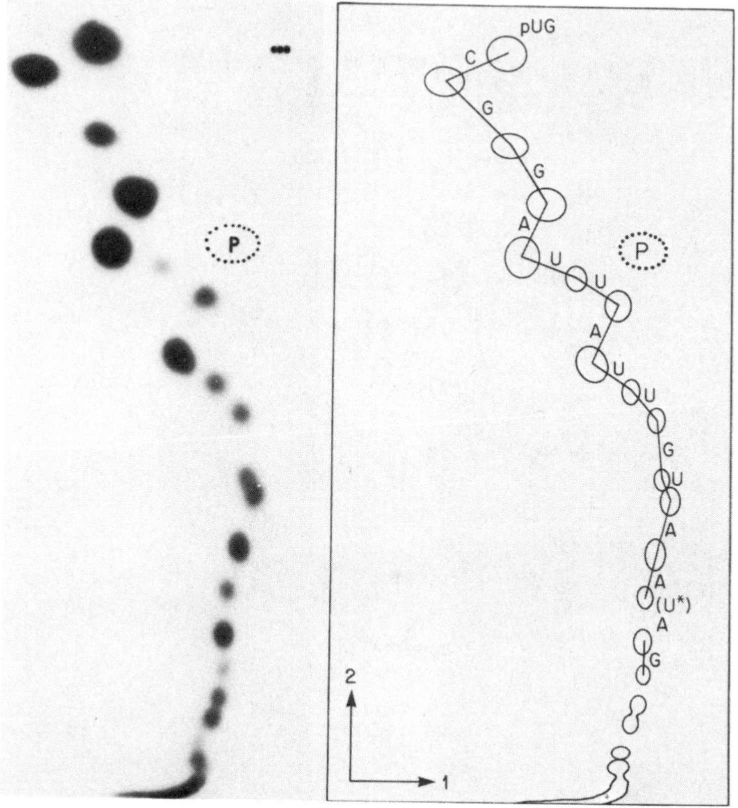

Figure 6 Autoradiogram of a partial nuclease P1 digest of 5′-^{32}P *N. crassa* mitochondrial initiator tRNA as analyzed by two-dimensional homochromatography. *First dimension*, electrophoresis on cellulose acetate strip at pH 3.5; *second dimension*, homochromatography on a 20 × 40 cm DEAE-cellulose thin layer plate. Circled P indicates position of the pink dye marker.

The most commonly used procedure for structure mapping of RNA is based upon the rapid RNA sequencing method discussed above. It utilizes ^{32}P-end-labeled RNA, which is partially digested with single-strand-specific nucleases under nondenaturing conditions and subjected to polyacrylamide gel electrophoresis to resolve oligonucleotides (Wurst et al. 1978). S1 nuclease and T1-RNase are among enzymes that are used most often. Mung bean nuclease, which has slightly different properties from S1 nuclease, can also be used (Vournakis et al. 1981). Because of its relatively large size (mw 32,000), S1 nuclease is useful as a probe for single-stranded regions containing several contiguous nonbase-paired residues, whereas the G-specific T1-RNase, being a smaller enzyme (mw 11,000), appears effective in cleaving all single-stranded regions containing G residues.

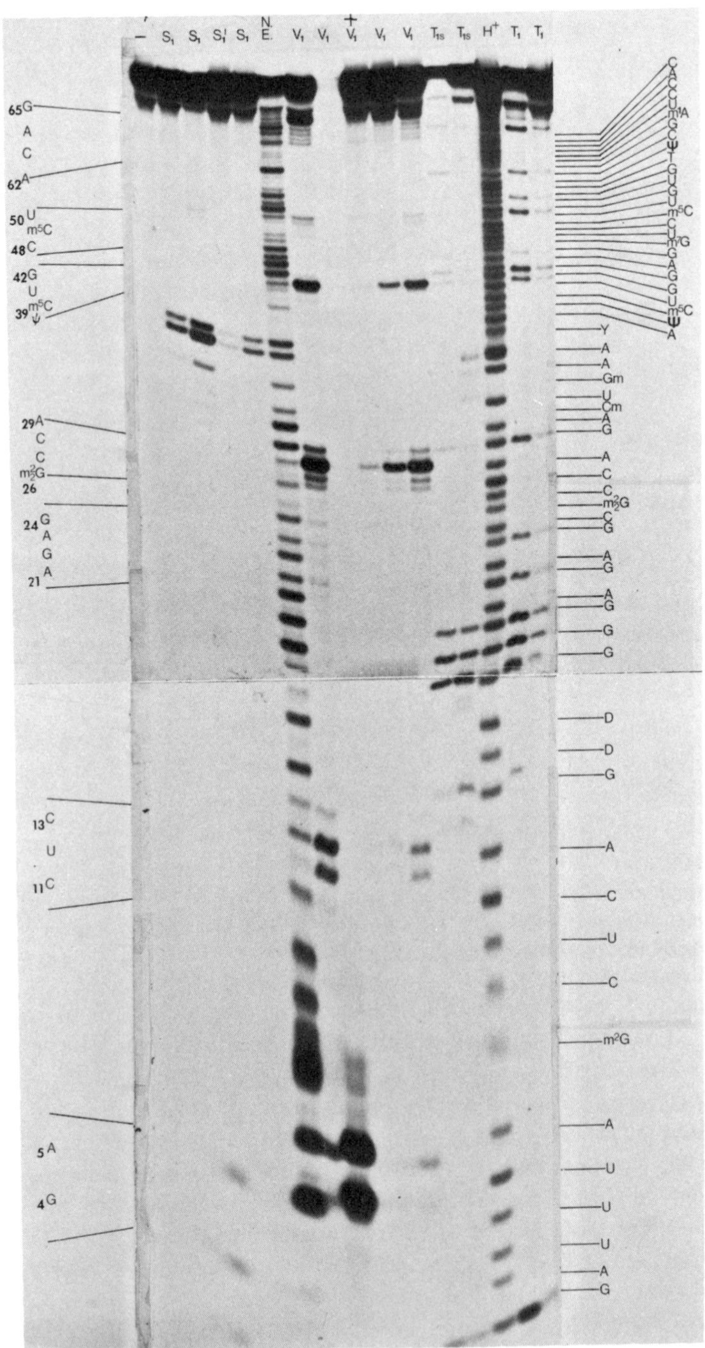

Figure 7 (See facing page for legend.)

The sites of cleavage of ^{32}P-end-labeled RNA by single-strand-specific enzymes are identified by comparing the locations of radioactive bands with bands of known lengths produced by other nucleases. For 5′-^{32}P-labeled RNA fragments, the electrophoretic mobility on polyacrylamide gels is sensitive to the presence or absence of a phosphoryl group at the 3′ end. For the analysis of partial digests with S1 nuclease, which produces fragments with 3′-hydroxyl groups at the sites of cleavage, a ladder generated by random cleavage of the same end-labeled tRNA with *N. crassa* nuclease under denaturing conditions (Krupp and Gross 1979) provides the set of necessary markers (Fig. 7). For T1-RNase, which produces fragments with 3′-phosphoryl groups at the sites of cleavage, a uniform ladder generated by either partial acid or alkaline hydrolysis of the same ^{32}P-end-labeled RNA or partial digestion with T1-RNase under denaturing conditions (which cuts at every G residue present) helps identify the most accessible G residues in the RNA.

Cobra Venom Nuclease(s) As Probe for tRNA Structure

The analysis of sites susceptible to single-strand-specific nucleases allows identification of only nonbase-paired nucleotides within an RNA. Care must be exercised before inferring that base-paired regions exist merely because they are not cleaved by these enzymes. Two recently described nucleases from cobra venom provide more direct probes for such regions (Favarova et al. 1981; Lockard and Kumar 1981). Although these nucleases cleave within base-paired regions of RNA, there are certain differences in their properties. Whether these enzymes (one of which is designated V1) represent distinct enzymatic activities is not yet known. Since V1 nuclease, like S1 nuclease, generates fragments with 3′-hydroxyl groups at the sites of cleavage, a ladder generated by *N. crassa* nuclease under denaturing conditions when electrophoresed on a lane between S1 and V1 partial digests allows simultaneous identification of both single- and double-strand-specific cleavages (Fig. 7).

Figure 7 Autoradiogram of partial digests of 5′-^{32}P-end-labeled yeast $tRNA^{Phe}$ electrophoresed on a 15% polyacrylamide slab gel in 90% formamide. Partial digests contained the following amounts of enzyme per μg of RNA. *Left* to *right*: (−) minus enzyme; (S_1) S1 nuclease, 0.06 U for 1 minute, 0.06 U for 10 minutes, 0.006 U for 1 minute, 0.006 U for 10 minutes; (N.E.) *Neurospora crassa* endonuclease, 5×10^{-2} U; (V_1) V1 nuclease, 1×10^{-5} U for 1 minute, 1×10^{-5} U for 10 minutes, 1×10^{-5} U for 10 minutes + 20 mM EDTA, 1×10^{-6} U for 1 minute, 1×10^{-6} U for 10 minutes; (T_1s) T1-RNase under nondenaturing conditions, 3×10^{-4} U, and 3×5^{-5} U; (H^+) controlled acid hydrolysis; (T_1) T1-RNase under denaturing conditions, 0.005 U and 0.0005 U. Bracketed nucleotide sequences at left indicate regions digested by nuclease V1.

V1 nuclease does not cleave all base-paired nucleotides even upon prolonged incubation (Fig. 8). Also even within base-paired regions, some phosphodiester bonds are cleaved well whereas others are cleaved only weakly. A weak cleavage after G_4 and a strong cleavage after U_{69} in the acceptor stem of yeast $tRNA^{Phe}$ (Fig. 8) imply that V1 nuclease can cleave nucleotides that are not base-paired in the standard Watson-Crick form. In addition, cleavages after $m_2^2G_{26}$ and C_{48} suggest that V1 nuclease can also cleave after some nucleotides that are involved in tertiary interactions in a tRNA.

Correct interpretation of RNA structure based on susceptibility towards nucleases requires a careful consideration of several factors.

1. The end-labeled RNA must be intact and conditions used for structural analysis be such as to retain the overall conformation of the RNA.
2. Kinetic analysis of the time course of digestion and variation of enzyme-to-substrate ratio should be performed to rule out the possibility that some of the cleavages observed are due to secondary cleavages on an RNA which has become unfolded as a result of a primary cleavage elsewhere.
3. Most nucleases do not cleave phosphodiester bonds involving modified nucleotides. For tRNAs in particular that contain many modified nucleotides, lack of cleavage by a single-strand-specific enzyme next to a modified nucleotide cannot be used as evidence that this region of the RNA is not single-stranded. For example, yeast $tRNA^{Phe}$ contains 2′-*0*-methyl G in the anticodon, the most exposed region in the tRNA, yet T1-RNase does not cleave this tRNA in the anticodon because of its absolute requirement for a 2′-hydroxyl group on the G residue (Fig. 7).

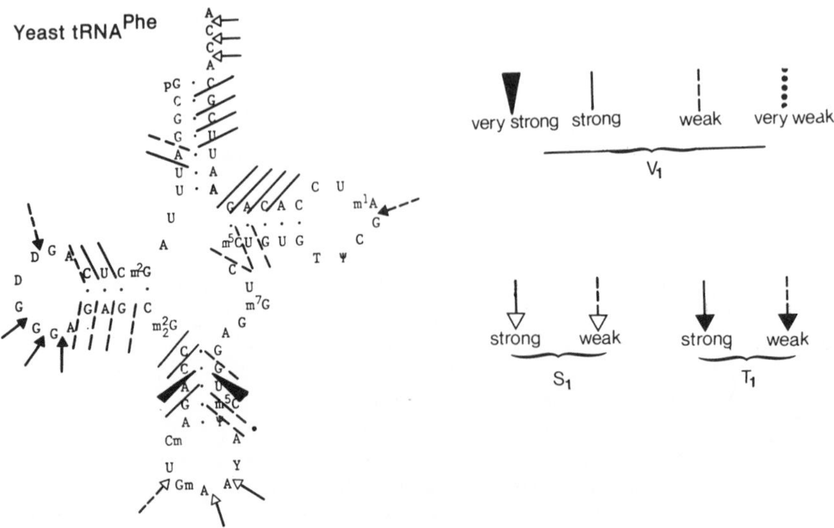

Figure 8 Structural map of yeast $tRNA^{Phe}$.

4. Nucleases of different size, base specificity, and structural requirements should be used to obtain maximal information on RNA structure.
5. If an RNA contains a readily accessible region proximal to the labeled end, this will often obscure weak cleavages that occur further away. Thus, although the 3′ end of the tRNA is accessible to S1 nuclease, the strong cleavage in the anticodon region by this enzyme obscures cleavage at the 3′ end when only 5′-end-labeled RNA is used for structural analysis. This problem is avoided if 3′-end-labeled RNA is also used. Consequently, for small RNA molecules such as tRNAs, 5S RNAs, or for fragments of large RNAs, both 5′ and 3′-end-labeled RNAs should be used for structural analysis.
6. Finally, comparative structural analysis on intact RNA and fragments of the RNA can provide valuable information on whether lack of cleavage of certain residues in the intact RNA is due to their involvement in tertiary structure.

ACKNOWLEDGMENTS

Work from the authors' laboratories was supported by grants GM-17151 from the National Institutes of Health and NP-114 from the American Cancer Society (ULR), and grants GM-27692 and GM-28985 from the National Institutes of Health to REL.

REFERENCES

Ando, S. 1966. A nuclease specific for heat-denatured DNA isolated from a product of *Aspergillus oryzae*. *Biochim. Biophys. Acta* **114:**158.

Boguski, M.S., P.A. Hieter, and C.C. Levy. 1980. Identification of a cytidine-specific ribonuclease from chicken liver. *J. Biol. Chem.* **255:**2160.

Brown, D.M. and A.R. Todd. 1953. Evidence on the nature of the chemical bonds in nucleic acids. In *The nucleic acids* (ed. E. Chargaff and J.N. Davidson), vol. I, pp. 409. Academic Press, New York.

Chang, S.H. and U.L. RajBhandary. 1968. Studies on polynucleotides. LXXXI. Yeast phenylalanine transfer ribonucleic acid. *J. Biol. Chem.* **243:**592.

DeWachter, R. and W. Fiers. 1972. Preparative two-dimensional polyacrylamide gel electrophoresis of ^{32}P-labeled RNA. *Anal. Biochem.* **49:**184.

Donis-Keller, H. 1979. Site specific enzymatic cleavage of RNA. *Nucleic Acids Res.* **7:**179.

Donis-Keller, H., A.M. Maxam, and W. Gilbert. 1977. Mapping adenines, guanines, and pyrimidines in RNA. *Nucleic Acids Res.* **4:**2527.

England, T.E. and O.C. Uhlenbeck. 1978. 3′-Terminal labelling of RNA with T4 RNA ligase. *Nature* **275:**560.

Favarova, O.O., F. Fasiolo, G. Keith, S.K. Vasilenko, and J.P. Ebel. 1981. Partial digestion of tRNA-aminoacyl-tRNA synthetase complexes with cobra venom ribonuclease. *Biochemistry* **20:**1006.

Fujimoto, M., A. Kuninaka, and H. Yoshino. 1974. Nuclease from *Penicillium citrinum*. Substrate specificity of nuclease P1. *Agric. Biol. Chem.* **38:**1555.

Harada, F. and J.E. Dahlberg. 1975. Specific cleavage of tRNA by nuclease S_1. *Nucleic Acids Res.* **2:**865.

Jack, A., J.E. Ladner, and A. Klug. 1976. Crystallographic refinement of yeast phenylalanine transfer RNA at 2·5Å resolution. *J. Mol. Biol.* **108:**619.

Kowalski, D., W.D. Kroeker, and M. Laskowski. 1976. Mung bean nuclease I. Physical, chemical, and catalytic properties. *Biochemistry* **15:**4457.

Krol, A., C. Branlant, E. Lazar, J. Gallinaro, and M. Jacob. 1981. Primary and secondary structures of chicken, rat and man nuclear U4 RNAs. Homologies with U1 and U5 RNAs. *Nucleic Acids Res.* **9:**2699.

Krupp, G. and H.J. Gross. 1979. Rapid RNA sequencing: Nucleases from *Staphylococcus aureus and Neurospora crassa* discriminate between uridine and cytidine. *Nucleic Acids Res.* **6:**3481.

Linn, S. and I.R. Lehman. 1965. An endonuclease from *Neurospora crassa* specific for nucleotides lacking an ordered structure. *J. Biol. Chem.* **240:**1287.

Lockard, R.E. and A. Kumar. 1981. Mapping tRNA structure in solution using double-strand-specific ribonuclease V_1 from cobra venom. *Nucleic Acids Res.* **9:**5125.

Lockard, R.E., J.F. Connaughton, and A. Kumar. 1982. Nucleotide sequence of the 5'- and 3'-domains for rabbit 18S ribosomal RNA. *Nucleic Acids Res.* **10:**3445.

Lockard, R.E., B. Alzner-DeWeerd, J.E. Heckman, J. MacGee, M.W. Tabor, and U.L. RajBhandary. 1978. Sequence analysis of 5'[^{32}P] labeled mRNA and tRNA using polyacrylamide gel electrophoresis. *Nucleic Acids Res.* **5:**37.

Machatt, M.A., J.P. Ebel, and C. Branlant. 1981. The 3'-terminal region of bacterial 23S ribosomal RNA: Structure and homology with the 3'-terminal region of eukaryotic 28S rRNA and with chloroplast 4.5S rRNA. *Nucleic Acids Res.* **9:**1533.

Pavlakis, G.N., R.E. Lockard, N. Vamvakopoulos, L. Rieser, U.L. RajBhandary, and J.N. Vournakis. 1980. Secondary structure of mouse and rabbit α- and β-globin mRNAs: Differential accessibility of α and β initiator AUG codons towards nucleases. *Cell* **19:**91.

Peattie, D.A. and W. Gilbert. 1980. Chemical probes for higher-order structure in RNA. *Proc. Natl. Acad. Sci.* **77:**4679.

Penswick, J.R. and R.W. Holley. 1965. Specific cleavage of the yeast alanine RNA into two large fragments. *Proc. Natl. Acad. Sci.* **53:**543.

Quigley, G.J. and A. Rich. 1976. Structural domains of transfer RNA molecules. *Science* **194:**796.

RajBhandary, U.L. 1980. Recent developments in methods for RNA sequencing using in vitro ^{32}P-labeling. *Fed. Proc.* **39:**2815.

Razzell, W.E. and H.G. Khorana. 1959a. Studies on polynucleotides. III. Enzymic degradation. Substrate specificity and properties of snake venom phosphodiesterase. *J. Biol. Chem.* **234:**2105.

———. 1959b. Studies on polynucleotides. IV. Enzymic degradation. The stepwise action of venom phosphodiesterase on deoxyribo-oligonucleotides. *J. Biol. Chem.* **234:**2114.

Reid, B.R. 1981. NMR studies on RNA structure and dynamics. *Annu. Rev. Biochem.* **50:**969.

Rich, A. and U.L. RajBhandary. 1976. Transfer RNA: Molecular structure, sequence, and properties. *Annu. Rev. Biochem.* **45:**805.

Ross, A. and R. Brimacombe. 1979. Experimental determination of interacting sequences in ribosomal RNA. *Nature* **281:**271.

Rushizky, G.W. and J.H. Mozejko. 1977. Optimization of conditions for cleavage of tRNA at the anticodon loop by S_1 nuclease. *Anal. Biochem.* **77:**562.

Rushizky, G.W., V.A. Shaternikov, J.H. Mozejko, and H.A. Sober. 1975. S_1 nuclease hydrolysis of single-stranded nucleic acids with partial double-stranded configuration. *Biochemistry* **14:**4221.

Sanger, F., G.G. Brownlee, and B.G. Barrell. 1965. A two-dimensional fractionation procedure for radioactive nucleotides. *J. Mol. Biol.* **13:**373.

Silberklang, M., A.M. Gillum, and U.L. RajBhandary. 1977. The use of nuclease P_1 in sequence analysis of end group labeled RNA. *Nucleic Acids Res.* **4:**4091.

———. 1979. Use of *in vitro* ^{32}P labeling in the sequence analysis of nonradioactive tRNAs. *Methods Enzymol.* **59:**58.

Simoncsits, A., G.G. Brownlee, R.S. Brown, J.R. Rubin, and H. Guilley. 1977. New rapid gel sequencing method for RNA. *Nature* **269:**833.

Simsek, M., J. Ziegenmeyer, J.E. Heckman, and U.L. RajBhandary. 1973. Absence of the sequence G-T-ψ-C-G(A)- in several eukaryotic cytoplasmic initiator transfer RNAs. *Proc. Natl. Acad. Sci.* **70:**1041.

Steigler, P., P. Carbon, M. Zucker, J.P. Ebel, and C. Ehresman. 1981. Structural organization of the 16S ribosomal RNA from *E. coli*. Topography and secondary structure. *Nucleic Acids Res.* **9:**2153.

Tal, J. 1975. The cleavage of transfer RNA by a single strand specific endonuclease from *Neurospora crassa*. *Nucleic Acids Res.* **2:**1073.

Tomlinson, R.V. and G.M. Tener. 1963. The effect of urea, formamide, and glycols on the secondary binding forces in the ion-exchange chromatography of polynucleotides on DEAE-cellulose. *Biochemistry* **2:**697.

Uhlenbeck, O.C. 1972. Complementary oligonucleotide binding to transfer RNA. *J. Mol. Biol.* **65:**25.

Vournakis, J.N., J. Celentano, M. Finn, R.E. Lockard, T. Mitra, A. Troutt, G. Pavlakis, M. van den Berg, and R. Wurst. 1981. Sequence and structure analysis of end-labeled RNA with nucleases. In *Gene amplification and analysis* (ed. J.G. Chirikjian and T.S. Papas), vol. II, p. 268. Elsevier-North Holland.

Wurst, R.M., J.J. Vournakis, and A.M. Maxam. 1978. Structure mapping of 5′-^{32}P-labeled RNA with S1 nuclease. *Biochemistry* **17:**4493.

The Deoxyribonucleases of *Escherichia coli*

Stuart Linn
Department of Biochemistry
University of California
Berkeley, California 94720

INTRODUCTION

The organism that has been most extensively characterized with regard to its DNases is undoubtedly *Escherichia coli*. The number and complexity of the enzymes that have been identified to date are impressive and were certainly unexpected. To obtain an overview of the dynamics of DNA metabolism in this or other organisms, a familiarity with the general features of these enzymes is necessary. It is the purpose of this section to list these enzymes and to summarize some highlights of their properties. More in-depth discussions of particular enzymes can be found elsewhere in this volume, in the specific references cited below and in several other reviews and monographs (Lehman

1963, 1971; Cantoni and Davies 1966; Privat de Garilhe 1967), and various chapters in Boyer (1981).

EXONUCLEASES

The exonucleases of *E. coli* have been very recently reviewed by Weiss (1981) with emphasis upon exonucleases I, III, IV, VII, and VIII.

Exonuclease I

Exonuclease I is a monomeric protein of M_r about 72,000 (Mackay and Linn 1974), though active oligomeric forms have been described. The enzyme hydrolyzes single-strand substrates about 40,000-fold faster than duplex DNA. It acts in a 3′→5′ direction, releasing nucleoside 5′-phosphates in a processive manner, but leaves the 5′-terminal dinucleotide intact. A free 3′-hydroxyl terminus is required for activity (Lehman and Nussbaum 1964).

The *sbcB* gene at 44 minutes on the *E. coli* genetic map specifies exonuclease I (Bachmann and Low 1980). Mutants lacking the enzyme appear to be perfectly normal, but in the presence of *recB* and/or *recC* mutations, the *sbcB* mutation suppresses the rec^- and repair-defective phenotypes of the former mutants provided that the multiple mutants remain $recF^+$ (Clark 1973). Hence, its biological function remains unclear, although the catalytic specificities of the enzyme have been exploited extensively (Weiss 1981).

Exonucleases Associated with the DNA Polymerases

DNA Polymerase I

The 3′→5′ exonuclease activity associated with the monomeric, 109K DNA polymerase I has historically been called exonuclease II. The activity degrades DNA to nucleoside 5′-phosphates and prefers single-stranded or oligomeric substrates; it requires 3′-hydroxyl termini (Lehman and Richardson 1964). The activity is able to remove mismatched nucleotides from 3′ termini of primers and is thought to serve in a "proofreading" capacity during DNA synthesis (Brutlag and Kornberg 1972).

The 5′→3′ exonuclease activity of polymerase I has historically been called exonuclease VI. It can be separated from the other activities of polymerase I by proteolysis, which yields a 35K fragment with 5′→3′ exonuclease and a 76K fragment which has the polymerase and 3′→5′ exonuclease activities (Setlow and Kornberg 1972). Alternatively, it can be separated by certain mutations of the *polA* gene at 83 minutes on the *E. coli* map (Lehman 1981).

The 5′→3′ activity utilizes duplex substrates. It can remove pyrimidine dimers and other damage from the 5′ side of an incision nick and can act, in concert with polymerase action, to "translate" a nick along a duplex so as to repair DNA lesions (see Chapter 4). The activity also removes RNA from DNA-RNA hybrids (see Chapters 8 and 9). The activity produces mono- and oligonucleotides with 5′-phosphomonester termini. By using the linear portion of a D loop in negatively twisted circular DNA as a primer, the 5′→3′ activity appears able to cleave, endonucleolytically, the displaced strand of the duplex. This endonucleolytic cleavage may be due to resemblance of this distorted structure to DNA damage (Lehman 1981).

As discussed in Chapters 4, 8, and 9, the 3′→5′ exonuclease activity of DNA polymerase I is probably used to assure fidelity during DNA synthesis by polymerase I. Since mutants that lack this exonuclease activity as well as the polymerase activity are viable, it is presumably dispensable for the cell's normal DNA metabolism, though such cells are more sensitive to DNA-damaging agents. Conversely, mutant studies imply that the 5′→3′ exonuclease of DNA polymerase I is indispensable (Lehman 1981). It appears to be necessary for DNA replication—specifically for the joining of Okazaki fragments, probably by removal of RNA primers. It also plays a major role in DNA excision-repair.

DNA Polymerase II

Polymerase II has a 3′→5′ exonuclease, but lacks a 5′→3′ exonuclease activity (Wickner et al. 1972). The 120K enzyme is coded by the *polB* locus, which is at 2 minutes on the genetic map (Bachmann and Low 1980). The activity prefers single-stranded DNA. To date genetic studies have not uncovered a role for this activity (Campbell et al. 1972).

DNA Polymerase III

Polymerase III has both 3′→5′ and 5′→3′ exonuclease activities (Livingston and Richardson 1975). The 3′→5′ exonuclease is single-strand-specific and hydrolyzes DNA to nucleoside 5′-phosphates and a 5′-terminal dinucleotide. The activity will not degrade DNA lacking terminal 3′-hydroxyl groups. It will remove nonbase-paired 3′-termini from duplex DNA, presumably reflecting an editing function.

The 5′→3′ exonuclease differs significantly from that of polymerase I in that it requires single-stranded DNA or duplexes with 5′-single-stranded termini from which it can proceed into a duplex. Degradation is processive to form oligonucleotides and the enzyme can evidently release pyrimidine dimers from the product of the *Micrococcus luteus* UV endonuclease.

The nuclease activities of polymerase III are associated with the 140K α peptide, the product of the *polC* (= *dnaE*) locus at 4 minutes on the *E. coli* genetic map (Bachman and Low 1980). Effects of the other subunits of the polymerase-III holoenzyme on the nuclease activities remain to be studied (McHenry and Kornberg 1981). Presumably the nuclease activities of polymerase III have roles analogous to those of polymerase I.

Exonuclease III (= Endonuciease VI)

This monomeric 28K protein is the product of the *xthA* gene at 38 minutes on the *E. coli* chromosome. It is remarkable in that it has four activities: exodeoxyribonuclease, RNaseH, DNA-3′-phosphatase, and AP endonuclease. The enzyme has recently been reviewed in detail by Weiss et al. (1978) and by Weiss (1981).

The exodeoxyribonuclease is duplex specific, hydrolyzing DNA in a non-processive fashion from the 3′ terminus to form nucleoside 5′-phosphates and residual single strands (Richardson et al. 1964; Weiss 1981). The enzyme can initiate hydrolysis from single-strand breaks so as to produce gaps. The exonuclease has been widely applied as a reagent in DNA sequencing and DNA sequence modification protocols.

When presented with a DNA-RNA hybrid, the enzyme degrades the RNA strand from the 3′-terminus to release nucleoside 5′-phosphates (Keller and Crouch 1972; see Chapter 9). Though the enzyme demonstrates this RNaseH activity, it does not degrade single-stranded or duplex RNA.

The enzyme also releases inorganic phosphate specifically from 3′-phosphomonoester groups on DNA, but not from RNA or short oligonucleotides (Richardson and Kornberg 1964). Removal of these groups converts the terminus into a primer for DNA polymerases.

The AP endonuclease activity of exonuclease III is discussed below as "endonuclease VI" and in Chapter 4.

The biological role of exonuclease III is unclear. Genetic studies have failed to reveal a function for either the exonuclease, phosphatase, or RNaseH activities. The relatively weak phenotypic characteristics of *xthA* mutants might be best ascribed to the apurinic endonuclease activity (see below).

Exonucleases IVA and IVB

These enzymes were described by Jorgensen and Koerner (1966) as enzymes specific for oligonucleotides, cleaving them entirely to nucleoside-5′-phosphates. The two fractions are distinguishable only by chromatography upon DEAE-cellulose. Unfortunately, further purification and study of these activities in the context of our more extensive knowledge of the *E. coli* DNases has not been reported.

Exonuclease V (The *recBC* Enzyme)

This enzyme and its analogs from other bacteria have been extensively discussed in Chapters 2 and 4 and reviewed recently by Muskavitch and Linn (1981). The *E. coli* enzyme contains two subunits of roughly 140K and 130K (Goldmark and Linn 1972), which are the products of the *recB* and *recC* genes, respectively (Hickson and Emmerson 1981). These genes map at 60 minutes on the *E. coli* map. A separate 60K subunit is often found associated with the enzyme activity, but it has no apparent effect upon the enzyme other than to aid in reassociation of the separated subunits to an active form (Lieberman and Oishi 1974; K.M.T. Muskavitch and S. Linn, unpubl.).

The enzyme has a battery of catalytic properties. It is an ATP-dependent double-strand exonuclease that works processively from the 3′ or 5′ strand to yield oligonucleotides; it is an ATP-dependent single DNA endonuclease that forms large oligonucleotides; it is a DNA-dependent ATPase; and finally, it is a duplex DNA-unwinding enzyme.

As summarized by Muskavitch and Linn (1981) the relative efficiency of each activity is quite strongly influenced by reaction conditions. For example, low ATP levels favor duplex exonuclease (Goldmark and Linn 1972); Ca^{2+} inhibits the DNase activities and allows the enzyme to unwind DNA without degrading it (Rosamond et al. 1979); SSB (MacKay and Linn 1976) and *recA* protein (Williams et al. 1981) protect single-stranded DNA from degradation, but in a complex manner, particularly when both SSB and *recA* protein are present (K.T. Muskavitch and S. Linn, unpubl.).

Several models for the mechanics of degradation of duplex DNA have been presented, the most recent being those of Taylor and Smith (1980) and Muskavitch and Linn (1982). They attempt to integrate the unwinding and nuclease activities with various looped and tailed intermediate structures which are seen in the electron microscope. It remains to be seen how these models apply to cellular function.

The *recBC* enzyme appears to have several roles in the cell. Because mutants in the *recB* or *recC* loci are defective in recombination and sensitive to DNA-damaging agents (Clark 1973), the enzyme is thought to take part in DNA recombination by the "recBC pathway" (see Chapter 2) and to be involved in DNA repair (see Chapter 4). It also appears to be implicated in the breakdown of restricted DNA (Simmon and Lederberg 1972) and, since many phage produce inhibitors of the enzyme, it may also be able to "restrict" infecting DNA itself in some instances. *recB* and *recC* mutants have a slower growth rate and reduced viability (Capaldo and Barbour 1973). Hence, the enzyme might be involved in an additional cellular function, perhaps DNA replication.

Exonuclease VII

Exonuclease VII is an 88K protein that is the product of the *xseA* gene which lies at 53 minutes on the genetic map of *E. coli* (Chase and Richardson 1974a,b; Chase et al. 1979). It is discussed in Chapter 4 and reviewed by Weiss (1981). The enzyme is specific for single-stranded DNA or single-stranded DNA termini. It forms oligonucleotides, acting in a 3′ → 5′ and a 5′ → 3′ direction. It is processive and not inhibited by EDTA. Since it acts from nicks produced at pyrimidine dimers by the *M. luteus* UV endonuclease (see Chapter 4), it presumably can act at distorted regions adjacent to nicks.

xseA mutants are somewhat "hyper-rec" and slightly sensitive to DNA-damaging agents. The enzyme appears to function in DNA repair, though it appears to be largely replaceable by other enzymes (see Chapter 4).

Exonuclease VIII

Exonuclease VIII is a dimer of a 140K protein that is the product of the *recF* gene (Kushner et al. 1974; Gillen et al. 1977). This gene is at 30 minutes on the genetic map and is derepressed in mutants of the *sbcA* gene to which it is closely linked (Gillen et al. 1981). These two genes are present in strains that contain the integrated cryptic Rac prophage. Expression of exonuclease VIII occurs only in *sbcA* mutants and results in the suppression of the phenotypes of bacterial mutants, which lack the *recBC* enzyme, and of λ phage mutants which are defective in λ exonuclease. Evidently exonuclease VIII can functionally substitute for either of these enzymes in spite of their lack of physical similarities. (λ Exonuclease is a 52K protein and does not cross-react antigenically with exonuclease VIII [Gillen et al. 1977]; see exonuclease V discussion above.)

The enzyme shows a 40-fold preference for duplex DNA. It is an exonuclease that appears to lack endonuclease and DNA phosphatase activities and has no ATP requirement. The enzyme has a 5′→3′ polarity and requires a 5′-hydroxyl group. It does not act from gaps or nicks.

The enzyme is likely to function in DNA recombination and is discussed in Chapter 2.

ENDONUCLEASES

Endonuclease I

Endonuclease I is a 12K protein that is the product of the *endA* gene which maps at 63 minutes on the *E. coli* chromosome (Lehman et al. 1962a,b; Wright 1971). It appears to exist in a series of oligomeric forms. The enzyme is evidently in the periplasmic space and is easily released from the cells during the formation of spheroplasts with osmotic shock or simply by treatment with EDTA (Nossal and Heppel 1966). The enzyme is present at highest levels during very early log phase, then drops off at least 100-fold during growth up to stationary phase, though the extent of this effect depends upon the growth medium (Shortman and Lehman 1964).

Endonuclease I prefers duplex DNA and degrades it to oligonucleotides of an average chain length of 7 that contain 5′-phosphomonoester groups (Lehman et al. 1962a). The enzyme cleaves both strands of a duplex simultaneously and during each such cleavage it releases roughly 400 nucleotides as oligonucleotides in a "shattering" event (Lehman 1971).

Endonuclease I is strongly inhibited by double-stranded RNA and is generally found in an inhibited form which can be activated by pancreatic RNase

(Lehman et al. 1962b). tRNA is a competitive inhibitor with a $K_i = 10^{-8}$ M versus a K_m of 2.5×10^{-5} M for DNA substrate (both expressed as nucleotide residues). Goebel and Helinski (1970) made the very interesting observation that in the "inhibited" state, a tRNA-endonuclease-I complex isolated from cells, or formed in vitro, can put exactly one random nick into a supercoiled circular DNA molecule.

The function of endonuclease I has remained totally undefined—no phenotype has been observed in *endA* mutants. While the ability to digest DNA so completely, together with its periplasmic location, might suggest some degradative (e.g., scavenger or restrictive) role, the presence of high activity only during very early log phase and the very specific activity in the presence of tRNA suggest a more noble function.

"Endonuclease II"

Endonuclease II was an activity first described as able to degrade DNA treated with methylmethane sulfonate, apparently being able to act both at alkylated and baseless sites (Friedberg et al. 1969; Hadi and Goldthwait 1971). Subsequently, Verly and Rassart (1975) isolated the major AP endonuclease activity of *E. coli,* which apparently had been previously described as one of the activities of "endonuclease II," in a form that did not act upon alkylated lesions. They considered this activity to be a new enzyme, which they called endonuclease VI (though it was subsequently shown to be identical physically with exonuclease III). At a meeting on DNA repair mechanisms, it was decided that endonuclease II be reserved for an endonuclease degrading alkylated DNA (should one exist), whereas endonuclease VI should be reserved for the enzyme described by Verly and Rassart (Linn 1978). One can legitimately argue, however, that the term "endonuclease VI" should be dropped, as it is merely an alternate designation for exonuclease III (Weiss 1981). At any rate, in view of the recent discovery of the DNA glycosylases, it is not unlikely that "endonuclease II" was a mixture of several glycosylases and AP endonuclease and that as time progresses it will slowly fade away.

Endonuclease III

Endonuclease III is also described in Chapter 4. The enzyme was first described by Radman (1976) as an endonuclease specific for duplex, UV-irradiated DNA. Further characterization (Gates and Linn 1977b) showed it to act on duplex DNA only at a minor UV-photoproduct, as well as on DNA exposed to X-rays, OsO_4, or acid. It also acts upon DNA exposed to hydrogen peroxide (B. Demple and S. Linn, unpubl.). The enzyme is inhibited by tRNA but is active in EDTA.

The diverse specificity was reconciled by Demple and Linn (1980) when the enzyme was shown to be a combination of a DNA-glycosylase which acts upon saturated thymines of the dihydroxy-dihydrothymine type and an AP endonuclease. Endonuclease III thus recognizes these damaged thymine residues in DNA and generates AP (baseless) sites. These, or other apyrimidinic and apurinic sites in DNA, can then serve as substrates for the AP endonuclease activity of the enzyme. Thus, the enzyme is similar to the *M. luteus* and T4 UV endonucleases which act upon DNA containing pyrimidine dimers (see Chapter 4).

Endonuclease III cleaves AP DNA as a "class-I" AP endonuclease (Chapter 4; Warner et al. 1980; Mosbaugh and Linn 1980):

B B B B
...P P P P... ⟶ ...P P OH + P P...

Given a substrate previously cleaved by a class-II AP endonuclease so as to leave a 5′-terminal sugar phosphate residue, it will form a free sugar phosphate (Mosbaugh and Linn 1980):

B B B B
...P P P P... —class II / APendo⟶ ...P OH + P P P...

B B
P P P... —Endonuclease / III⟶ ...P OH + P P...

Endonuclease III sediments near 2.7S (Radman 1976; Gates and Linn 1977b) corresponding to a spherical protein of roughly 27,000. A second form at 2.0S is also occasionally observed. The subunit structure is not known, nor is the genetic basis of the activity, and no mutants with altered enzyme have been reported. The enzyme is almost certainly involved in DNA repair.

Endonuclease IV

Endonuclease IV is also described in Chapter 4. The enzyme specifically forms single-strand breaks only at AP sites in DNA and represents perhaps 10% of the AP endonuclease activity of *E. coli* (Lindahl 1979). It has no other known substrate and is active in the presence of EDTA or tRNA (Ljunquist 1977).

Endonuclease IV is a class-II AP endonuclease (Chapter 4; Warner et al. 1980; Mosbaugh and Linn 1980):

...P–(B)–P–(B)–P–(B)–P... ⟶ ...P–(B)–OH + P–P–(B)–P–(B)–P...

Moreover, it will remove the 3′-AP termini generated by class-I AP endonucleases such as *E. coli* endonuclease III:

...P–(B)–P–OH —Endonuclease IV⟶ ...P–(B)–OH + P–OH .

Endonuclease IV sediments at 3.4S, corresponding to a molecular weight of roughly 33,000 ± 3000. Neither the subunit structure nor mutants with altered endonuclease-IV activity have been reported. The enzyme is presumably directly involved in repair of AP sites in DNA.

Endonuclease V

This enzyme is also described in Chapter 4. It is a protein of roughly 27,000 whose genetic basis is unknown (Gates and Linn 1977a). It is most active upon single-stranded DNA but hydrolyzes duplex DNA, particularly if damaged by UV (it does not act at pyrimidine dimers), acid, O_sO_4, or 7-bromomethylbenz[*a*]anthracene. It is most active upon phage PBS-2 DNA which contains uracil in place of thymine and it easily degrades this substrate to acid-soluble material. It does not act upon RNA or RNA-DNA hybrids. It requires Mg^{2+} and has a pH optimum of 9.5. It forms 3′-hydroxyl and 5′-phosphoryl termini.

With duplex DNA, endonuclease V generates roughly one double-strand break per eight nicks regardless of substrate (Demple and Linn 1982). It acts processively, completing its endonucleolytic action upon one DNA molecule before proceeding to the next. It also shows cooperative kinetics.

The enzyme seems suited for involvement in DNA repair, surveying the DNA for structural distortions generated by damage for which specific repair systems might not exist. However, simple disruption of base-pairing is not recognized, as the enzyme does not act at pyrimidine dimers, methylmethane sulfonate-generated damage, or psoralen adducts. Perhaps it recognizes disruptions in base stacking.

"Endonuclease VI" (=Exonuclease III)

As discussed above, "endonuclease VI" is a name given to the AP endonuclease activity associated with exonuclease III. It is discussed in Chapter 4 and reviewed by Weiss et al. (1978) and Weiss (1981).

The activity functions in the absence of a divalent cation, but is inhibited by EDTA. It is a class-II enzyme (Weiss et al. 1978; Warner et al. 1980) cleaving AP sites as follows:

B B B B

...P P P P... ⟶ ...P OH + P P P...

Like endonuclease IV, it will also release a sugar phosphate residue from 3′-AP termini. It is not clear whether the exonuclease activity of the enzyme plays a role biologically after the cleavage of an AP site, as it would remove normal nucleotides from the 3′ end of the incision, rather than removing the baseless site from the 5′ end. Instead, Weiss (1981) has proposed a model for the enzyme that consolidates the exonuclease, phosphatase, and AP endonuclease activities associated with the protein into a single active site and the same three DNA substrate-binding sites.

The AP endonuclease activity is certainly involved in DNA repair, though *xth* mutants remain viable, presumably because of ‘‘back-up’’ by endonuclease IV and other enzymes. The *xth* mutation does become lethal in the presence of nonlethal *dut* mutations (dUTPase negative). Presumably this is because the AP sites, generated by uracil DNA-glycosylase at dUMP residues, which accumulate in DNA in the absence of dUTPase, are not repaired in the *xth* mutants (Weiss et al. 1978; Taylor and Weiss 1982). Indeed the lethality is reversed by the addition of any *ung* (uracil DNA-glycosylase) mutation to the double mutant. *xth* mutants are also very sensitive to H_2O_2 (B. Demple and S. Linn, unpubl.), presumably because of the need to handle AP sites generated by endonuclease III at sites in DNA that were formed by hydroxyl radicals generated by the peroxide (see above).

In summary, under normal growth conditions endonuclease-VI activity is probably dispensable due to ‘‘back-up’’ functions; but, under conditions that result in a high frequency of AP sites in DNA, it appears to become necessary for survival.

Endonuclease VII

Recent reports of a peculiar activity, ‘‘endonuclease VII,’’ have appeared. This activity cleaves uracil-containing, single-strand synthetic polymers or DNAs that have been treated with uracil DNA-glycosylase to generate depyrimidinated sites (Friedberg et al. 1981; Bonura et al. 1982). Duplex DNAs with baseless sites are not attacked. The activity sediments at 4.2S and has an apparent molecular weight of 56,000 by gel filtration. It is active in EDTA but inhibited by tRNA.

Eco Restriction Endonucleases

The type-I restriction endonucleases, *Eco*B and *Eco*K, are apparently true *E. coli* enzymes rather than being enzymes associated with prophage, cryptic prophage, or plasmids, as is the case for the types-II and -III restriction enzymes associated with certain *E. coli* strains. These type-I enzymes (and a possibly nonallelic type-I enzyme, *Eco*A) are reviewed in detail in Chapter 5 and by Endlich and Linn (1981) and Yuan (1981).

The *Eco*B and *Eco*K enzymes are controlled by three linked genes, *hsdR*, *hsdM*, and *hsdS*, which map in this order at 98 minutes on the genetic map and code for the three polypeptides α (135K), β (60K), and γ (55K), respectively (Sain and Murray 1980). The enzymes are made up of various aggregate forms, all of which contain at least one of each subunit. The *K* and *B* genes are functionally allelic.

All three wild-type gene products are required for restriction, whereas only β and γ are necessary for modification. Restriction requires ATP and S-adenosylmethionine (SAM); α appears to contain the ATP-binding site, β, the SAM-binding site, and γ, the DNA recognition sequence-binding site (Endlich and Linn 1981; Yuan 1981; Chapter 5).

As summarized in Chapter 5, the hyphenated recognition sequences for *Eco*B and *Eco*K have been determined to be, respectively, 5′-TGA(N_8)TGCT-3′ and 5′-AAC(N_6)GTGC-3′. It is noteworthy that the trimeric and tetrameric domains are separated by spacings of 8 and 6 for the B and K specificities, respectively, in spite of apparent close relatedness of the γ_K and γ_B recognition subunits (they can both functionally act with either K or B and β subunits in vivo).

Summaries of schemes for the complex mode of action of *Eco*B and *Eco*K are given in Chapter 5 and in recent reviews (Endlich and Linn 1981; Yuan 1981), which should be referred to for details. In essence, the enzyme binds to a nonmethylated recognition site, travels roughly 1–5 kb (specifically, to the side of the trimeric domain of the recognition sequence), at least for the case of the *Eco*B sites on fd RF DNA, then creates a gap of about 75 nucleotides. This action requires prior activation of the enzyme by binding to *S*-adenosylmethionine and ATP. During the translocation, the enzyme remains bound to the recognition site, thus forming a DNA loop between the recognition site and the enzyme-binding site. The loop is generally supercoiled in the case of *Eco*K (Yuan 1981) and also for *Eco*B when closed circular, but not linear, substrates are utilized (B. Endlich and S. Linn, unpubl.).

Upon nicking (gapping) the DNA, the enzyme remains bound to the DNA loop and hydrolyzes ATP. It does not induce additional DNA cleavages; instead, a second enzyme molecule, finding the recognition and cleavage sites juxtaposed, completes the double-strand break. At this point the enzyme still maintains a terminal looped structure and continues to hydrolyze ATP. Disruption of these structures by salt, phenol, heat, etc., can break up the complex, but even with the enzyme removed, the 5′ ends are refractory to phosphorylation by polynucleotide kinase, although not to degradation by exonuclease. This block is not caused by intact enzyme subunits, or parts of the ATP or *S*-adenosylmethionine molecules and remains unexplained.

In summary, these complex enzymes have endonuclease, ATPase, translocation, DNA-winding, and SAM-binding activities. They also can specifically modify (methylate) DNA under proper conditions (see Chapter 5). Such a complex enzyme might a priori appear to have a function other than restriction, but *E. coli* C strains naturally lack the enzyme and have no DNA that hybridizes to the *hsd* genes (Sain and Murray 1980); yet the lack of restriction and modification is the only obvious phenotype of this strain attributable to this gene difference.

uvrABC Endonuclease

This enzyme is discussed in Chapter 4 and reviewed by Seeberg (1981) and by Friedberg et al. (1981). The enzyme activity requires three gene products: *uvrA* (114K), *uvrB* (84K), and *uvrC* (68K), the genes of which map at 92, 17, and 42 minutes on the *E. coli* chromosome, respectively (Bachmann and Low 1980). Each gene has been cloned and large quantities of each product are available (Sancar et al. 1981a,b; Yoakum and Grossman 1981).

All three peptides are required for cleavage of damaged DNA, but no evidence exists for any polypeptide complex either in extracts or during the reaction. The enzyme attacks at sites of pyrimidine dimers, bulky adducts, psoralen cross-links, and other types of damage (Hanawalt et al. 1979). Mutant extracts complement one another (Seeberg 1978), allowing purification of active peptides. Generally, the mechanism of cleavage is unknown as are the site of cleavage relative to damage, the type of phosphomonoester group formed, and the role of ATP. However, W.D. Rupp and A. Sancar (unpubl.) have recently observed incision of UV-irradiated DNA to occur between the seventh and eighth bases on the 5′ side of a pyrimidine dimer and between the third and fourth bases on the 3′ side. Conceivably, the complex acts both to incise damaged DNA and to excise the damage. The *uvrA* and *uvrC* proteins bind to single-stranded DNA, while *uvrA* protein binds to irradiated duplex DNA and has ATPase activity. The *uvrB* peptide is apparently present in large excess over the *uvrA* or *uvrC* proteins (Sancar et al. 1981a,b; Yoakum and Grossman 1981).

The *uvr* enzyme complex is clearly involved in DNA excision-repair, though the basis of enzyme specificity is not clear. It differs from endonuclease V, which has somewhat complementary specificities (Demple and Linn 1982). The excess of *uvrB* protein and the particular nonviability of a *uvrB polA* mutant (Morimyo and Shimazo 1976), but not a *uvrA polA* mutant, suggest that this polypeptide may have some other function.

Eco Topoisomerases

This class of enzymes is the basis of Chapter 3 and is also reviewed by Cozzarelli (1980), Wang (1981), and Gellert (1981). Though not usually considered to be "DNases," these enzymes do in fact cleave DNA, and the cleaved intermediates can be isolated. The cleavage mechanism is often complex and is reviewed in detail in Chapter 4.

The type-I topoisomerase, the 110K ω protein, is the product of the *top* gene which maps at 28 minutes on the *E. coli* map. The enzyme acts by putting transient single-strand nicks into duplex DNA to relax supercoiling or to intertwine or decatenate various rings. The transient nicked, protein-bound intermediate isolated after disruption has a free 3′-hydroxyl group and a 5′-phosphoryl group linked to the protein through the O^4-position of a tyrosine (Wang 1981).

The type-II topoisomerase, DNA gyrase, contains two each of the gene products for the *gyrA* gene (105K) and *gyrB* gene (95K); the genes map at 48 minutes and 82 minutes on the *E. coli* genetic map, respectively. (A second, 50K protein, which might be a proteolytic cleavage product of *gyrB* protein, can replace the 95K protein.) The enzyme requires ATP to introduce supertwists into DNA, but can relax DNA in the absence of ATP. Cleavage occurs on both strands simultaneously. When cleaved DNA is isolated in the presence of the inhibitor oxolinic acid (Cozzarelli 1980; Gellert 1981), the strand breaks are 4 bp apart with single-stranded 5′-termini. Although the breaks have no obvious sequence specificity, they are not randomly distributed in the DNA (Cozzarelli 1980).

CONCLUSIONS

There Are Many DNases in *E. coli*

The highlights of each DNase are listed in Table 1. The first obvious conclusion from the table is that there are many DNases in *E. coli*, more than had been anticipated. Preliminary reports of other DNases (see, for example, Blackmore and Linn 1974) promise that yet more DNases remain to be characterized. Even this impressive number of DNases is not always sufficient—it is increased upon phage infection or plasmid introduction, during which new DNases involved in recombination, replication, repair, restriction, salvaging of host nucleotides, etc., are often expressed.

Perusal of the list of DNases and the properties of the enzymes and the mutants lacking them leads to the conclusion that under normal growth conditions many DNases are not necessary; they can be "backed-up" by enzymes with similar activities and functions. Only in times of stress—for example, excess damage or recombination, high levels of replication of cellular, viral, or plasmid DNAs, infection by foreign DNA, or in the absence of

Table 1 The deoxyribonucleases of *E. coli*[a]

Enzyme	Gene	(map position)[a]	Subunit M_r (kilodaltons)	Substrate[b]	Major characteristics and products
Exonuclease I	*sbcB*	(44)	72	SS DNA	3′→5′; mononucleotides, 5′-terminal dinucleotide, processive
DNA polymerase I	*polA*	(86)	109	mismatched DS 3′ termini, SS DNA, oligonucleotides,	3′→5′; mononucleotides
				DS DNA, DNA-RNA hybrid	5′→3′; oligonucleotides
DNA polymerase II	*polB*	(2)	120	SS DNA	3′→5′; mononucleotides
DNA polymerase III	*polC*	(4)	140	mismatched DS 3′ termini, SS DNA	3′→5′; mononucleotides; 5′-terminal dinucleotide, nonprocessive
Exonuclease III	*xthA*	(38)	28	DS DNA	3′→5′; mononucleotides; residual single strands
				AP sites	3′-hydroxynucleotide termini; 5′-deoxyribose phosphate termini
				DNA-3′-phosphates	P_i
				RNA-DNA hybrid	deoxyribo- and ribonucleoside-5′-phosphates; residual single strands
Exonucleases IVA, IVB	?		?	oligonucleotides	mononucleotides
Exonuclease V	*recB*	(60)	140	DS DNA	ATP required; duplex unwound during reaction; 3′→5′ and 5′→3′; oligonucleotides[c]; processive; ATPase
	recC	(60)	130	SS DNA	ATP required; 3′→5′ and 5′→3′ exonuclease; oligonucleotides[c]; ATPase ATP-stimulated endonuclease; large fragments[c]
Exonuclease VII	*xseA*	(53)	88	SS DNA, SS termini	3′→5′ and 5′→3′; oligonucleotides; processive
Exonuclease VIII	(*recE*)	(30)	140	DS DNA	5′→3′; mononucleotides
Endonuclease I	(*endA*)	(63)	12	DS DNA	DS breaks; "shatters DNA"; oligonucleotides periplasmic; inhibited by DS RNA

Endonuclease III	?		27(native)[d]	AP sites	3′-deoxyribose termini; 5′-phosphomonoester nucleotide termini
				thymine glycol sites	free thymine glycol via glylosylase action, resulting AP site cleaved; active in EDTA; inhibited by tRNA
Endonuclease IV	?		33(native)[d]	AP sites	3′-hydroxynucleotide termini; 5′-deoxyribose phosphate termini; active in EDTA
Endonuclease V	?		27(native)[d]	SS DNA; damaged duplex DNA	recognizes some, but not all damage in duplex DNA; forms 1 DS break per 8 SS nicks; processive and cooperative
Endonuclease VII	?		55(native)[d]	SS DNA with AP sites	active in EDTA (preliminary reports)
*Eco*K, *Eco*B	*hsdR* *hsdM* *hsdS*	(98) (98) (98)	135 60 55	B: 5′-TGA(N_8)TGCT K: 5′-AAC(N_6)GTGC (unmodified duplex DNA)	require SAM, ATP; do not turnover as nuclease; translocate to form loops between recognition and cleavage sites; ATPase activity; leave blocked 5′ termini
uvrABC	*uvrA* *uvrB*	(92) (17)	114 84	damaged duplex DNA	requires ATP, ATPase; excises at least some forms of damage
	uvrC	(42)	68		
Eco topoisomerase I	*top*	(128)	110	duplex, supercoiled DNA	SS nicks with 5′-phosphate termini bound to enzyme; "ω" protein
Eco topoisomerase II	*gyrA* *gyrB*	(48) (82)	105 95	duplex DNA	DS breaks with 5′-phosphate termini bound to enzyme with 4-nucleotide 5′-extended termini. ATP required if introducing "supertwists," "DNA gyrase"

References are cited in the text. No *E. coli* DNases are known that form 3′-phosphoryl termini.

[a]The map position on the *E. coli* chromosome is expressed in minutes according to Bachmann and Low (1980).

[b]Abbreviations: SS, single stranded; DS, double stranded; AP sites, apurinic/apyrimidinic sites in duplex DNA.

[c]The extent of reaction, specificity, and products are affected by the presence of *recA* proteins, SSB, and Ca^{2+}.

[d]The subunit structure of these enzymes has not been determined, but their native molecular weights have been estimated.

other DNases due to mutation or inhibitors—do some of these activities appear to be necessary. Even then, some enzymes appear not to be absolutely required. Clearly, DNA metabolic functions are important to the cell, and as with DNA polymerases, ligases, helicases, and topoisomerases, DNases are present in excess to allow for the availability of alternative enzymes to achieve the same goal. The alternatives might simply replace the enzyme in a particular function (for example, exonuclease III for the 3′-5′ polymerase-I exonuclease), or they might shunt the substrate into another pathway (for example, from excision-repair to recombinational repair).

DNases Are Elegant and Complex Enzymes

A final generalization about the DNases of *E. coli* is their complex nature. They do not simply hydrolyze DNA phosphodiester bonds at random. They are often quite specific and can be associated with such activities as DNA polymerase, ATPase, duplex DNA unwinding, DNA phosphatase, DNA translocase, and topoisomerase. Each enzyme is an interesting object for study, both in its own right as an enzyme and as part of an important physiological function.

ACKNOWLEDGMENTS

The work cited from our laboratory was made possible by funds from the National Institutes of Health, the Department of Energy, the American Cancer Society, and the University of California.

REFERENCES

Bachmann, B.J. and K.B. Low. 1980. Linkage map of *Escherichia coli* K-12. Edition 6. *Microbiol. Rev.* **44:**1.

Blackmore, R.V. and S. Linn. 1974. Partial purification and characterization of four endodeoxyribonucleases from *Escherichia coli* K-12. *Nucleic Acids Res.* **1:**1.

Bonura, T., R. Schultz, and E.C. Friedberg. 1982. An enzyme activity from *Escherichia coli* that attacks single-stranded deoxyribopolymers and single-stranded deoxyribonucleic acid containing apyrimidinic sites. *Biochemistry* **21:**2548.

Boyer, P.D., ed. 1981. *The enzymes*, vol. 14A. Academic Press, New York.

Brutlag, D. and A. Kornberg. 1972. Enzymatic synthesis of deoxyribonucleic acid. XXXVI. A proofreading function for the 3′ ⟶ 5′ exonuclease activity of deoxyribonucleic acid polymerase. *J. Biol. Chem.* **247:**241.

Campbell, J.L., L. Soll, and C.C. Richardson. 1972. Isolation and partial characterization of a mutant deficient in DNA polymerase II. *Proc. Natl. Acad. Sci.* **69:**2090.

Cantoni, G.L. and D.R. Davies. 1966. *Procedures in nucleic acid research*. Harper and Row, New York.

Capaldo, F.N. and S.P. Barbour. 1973. Isolation of the nonviable cells produced during normal growth of recombination-deficient strains of *Escherichia coli* K-12. *J. Bactiol.* **115:**928.

Chase, J.W. and C.C. Richardson. 1974a. Exonuclease VII of *Escherichia coli*. Purification and properties. *J. Biol. Chem.* **249:**4545.

———— . 1974b. Exonuclease VII of *Escherichia coli*. Mechanism of action. *J. Biol. Chem.* **249:**4553.

Chase, J.W., W.E. Masker, and J.B. Murphy. 1979. Pyrimidine dimer excision in *Escherichia coli* strains deficient in exonuclease V and VII and in the 5′⟶3′ exonuclease of DNA polymerase I. *J. Bacteriol.* **137:**234.

Clark, A.J. 1973. Recombination deficient mutants of *E. coli* and other bacteria. *Annu. Rev. Genetics* **7:**67.

Cozzarelli, N.R. 1980. DNA gyrase and the supercoiling of DNA. *Science* **207:**953.

Demple, B. and S. Linn. 1980. DNA *N*-glycosylases and UV repair. *Nature* **287:**203.

———— . 1982. On the recognition and cleavage mechanism of *Escherichia coli* endonuclease V, a possible DNA repair enzyme. *J. Biol. Chem.* **257:**2848.

Endlich, B. and S. Linn. 1981. Type I restriction enzymes. In *The enzymes* (ed. P.D. Boyer), vol. 14, p. 137. Academic Press, New York.

Friedberg, E.C., S.-M. Hadi, and D.A. Goldthwait. 1969. Endonuclease II of *Escherichia coli*. II. Enzyme properties and studies on the degradation of alkylated and native deoxyribonucleic acid. *J. Biol. Chem.* **244:**5879.

Friedberg, E.C., T. Bonura, E.H. Radany, and J.D. Love. 1981. Enzymes that incise damaged DNA. In *The enzymes* (ed. P.D. Boyer), vol. 14, p. 251. Academic Press, New York.

Gates, F.T. III and S. Linn. 1977a. Endonuclease V of *Escherichia coli*. *J. Biol. Chem.* **252:**1647.

———— . 1977b. An endonuclease from *E. coli* that acts specifically upon duplex DNA damaged by ultraviolet light, osmium tetroxide, acid, or X-rays. *J. Biol. Chem.* **252:**2802.

Gellert, M. 1981. DNA gyrase and other type II topoisomerases. In *The enzymes* (ed. P.D. Boyer), vol. 14, p. 345. Academic Press, New York.

Gillen, J.R., A.E. Karu, H. Nagaishi, and A.J. Clark. 1977. Characterization of the deoxyribonuclease determined by lambda reverse as exonuclease VIII of *Escherichia coli*. *J. Mol. Biol.* **113:**27.

Gillen, J.R., D.K. Willis, and A.J. Clark. 1981. Genetic analysis of the *recE* pathway of genetic recombination in *Escherichia coli* K-12. *J. Bacteriol.* **145:**521.

Goebel, W. and D.R. Helinski. 1970. Nicking activity of an endonuclease I-transfer ribonucleic acid complex of *Escherichia coli*. *Biochemistry* **9:**4793.

Goldmark, P.J. and S. Linn. 1972. Purification and properties of the *recBC* DNase of *E. coli* K-12. *J. Biol. Chem.* **247:**1849.

Hadi, S.M. and D.A. Goldthwait. 1971. Endonuclease II of *Escherichia coli*. Degradation of partially depurinated deoxyribonucleic acid. *Biochemistry* **10:**4986.

Hanawalt, P.C., P.K. Cooper, A.K. Ganesan, and C.A. Smith. 1979. DNA repair in bacteria and mammalian cells. *Annu. Rev. Biochem.* **48:**783.

Hickson, I.D. and P.T. Emmerson. 1981. Identification of the *Escherichia coli recB* and *recC* gene products. *Nature* **294:**578.

Jorgensen, S.E. and J.F. Koerner. 1966. Separation and characterization of deoxyribonucleases of *Escherichia coli*-B. I. Chromatographic separation and properties of two deoxyribooligonucleotidases. *J. Biol. Chem.* **241:**3090.

Keller, W. and R. Crouch. 1972. Degradation of DNA RNA hybrids by ribonuclease H and DNA polymerases of cellular and viral origin. *Proc. Natl. Acad. Sci.* **69:**3360.

Kushner, S.R., H. Nagaishi, and A.J. Clark. 1974. Isolation of exonuclease VIII: The enzyme associated with *sbcA* indirect supressor. *Proc. Natl. Acad. Sci.* **71:**3593.

Lehman, I.R. 1963. The nucleases of *Escherichia coli*. *Prog. Nucleic Acid Res.* **2:**83.

———— . 1971. Bacterial deoxyribonucleases. In *The enzymes* (ed. P.D. Boyer), vol. 14, p. 251. Academic Press, New York.

———— . 1981. DNA polymerase I of *Escherichia coli*. In *The enzymes* (ed. P.D. Boyer), vol. 14, p. 15. Academic Press, New York.

Lehman, I.R. and A.L. Nussbaum. 1964. The deoxyribonucleases of *Escherichia coli*. V. On the specificity of exonuclease I (phosphodiesterase). *J. Biol. Chem.* **239:**2628.

Lehman, I.R. and C.C. Richardson. 1964. The deoxyribonucleases of *Escherichia coli*. IV. An exonuclease activity present in purified preparations of deoxyribonucleic acid polymerase. *J. Biol. Chem.* **239:**233.

Lehman, I.R., G.G. Roussos, and E.A. Pratt. 1962a. The deoxyribonucleases of *Escherichia coli*. II. Purification and properties of a ribonucleic acid-inhibitable endonuclease. *J. Biol. Chem.* **237:**819.

_________ . 1962b. The deoxyribonucleases of *Escherichia coli*. III. Studies of the nature of the inhibition of endonuclease by ribonucleic acid. *J. Biol. Chem.* **237:**829.

Lieberman, R.P. and M. Oishi. 1974. The *recBC* deoxyribonuclease of *Escherichia coli*: Isolation and characterization of the subunit proteins and reconstitution of the enzyme. *Proc. Natl. Acad. Sci.* **71:**4816.

Lindahl, T. 1979. DNA glycosylases, endonucleases for apurinic/apyrimidinic sites, and base excision repair. *Prog. Nucleic Acid Res.* **22:**135.

Linn, S. 1978. Workshop summary: Enzymology of base excision repair. In *DNA repair mechanisms* (ed. P.C. Hanawalt, E.C. Friedberg, and C.F. Fox), p. 175. Academic Press, New York.

Livingston, D.M. and C.C. Richardson. 1975. Deoxyribonucleic acid polymerase III of *Escherichia coli*. Characterization of associated exonuclease activities. *J. Biol. Chem.* **250:**470.

Ljungquist, S. 1977. A new endonuclease from *Escherichia coli* acting at apurinic sites in DNA. *J. Biol. Chem.* **252:**2808.

McHenry, C. and A. Kornberg. 1981. DNA polymerase III holoenzyme. In *The enzymes* (ed. P.D. Boyer), vol. 14, p. 39. Academic Press, New York.

Mackay, V. and S. Linn. 1974. Molecular structure of exonuclease I from *Escherichia coli* B. *Biochim. Biophys. Acta* **349:**131.

Morimyo, M. and Y. Shimazu. 1976. Evidence that gene *uvrB* is indispensable for a polymerase I deficient strain of *Escherichia coli* K-12. *Molec. Gen. Genet.* **147:**243.

Mosbaugh, D.W. and S. Linn. 1980. Further characterization of human fibroblast apurinic/apyrimidinic DNA endonucleases: The definition of two mechanistic classes of enzyme. *J. Biol. Chem.* **255:**11743.

Muskavitch, K.M.T. and S. Linn. 1981. *recBC*-like enzymes: Exonuclease V deoxyribonucleases. In *The enzymes* (ed. P.D. Boyer), vol. 14, p. 233. Academic Press, New York.

_________ . 1982. A unified mechanism for the nuclease and unwinding activities of the *recBC* enzyme of *Escherichia coli*. *J. Biol. Chem.* **257:**2641.

Nossal, N.G. and L.A. Heppel. 1966. The release of enzymes by osmotic shock from *Escherichia coli* in exponential phase. *J. Biol. Chem.* **241:**3055.

Privat deGarilhe, M. 1967. *Enzymes in nucleic acid research*. Holden-Day, San Francisco.

Radman, M. 1976. An endonuclease from *Escherichia coli* that introduces single polynucleotide chain scissions in ultraviolet-irradiated DNA. *J. Biol. Chem.* **251:**1438.

Richardson, C.C. and A. Kornberg. 1964. A deoxyribonucleic acid phosphatase-exonuclease from *Escherichia coli*. I. Purification of the enzyme and characterization of the phosphatase activity. *J. Biol. Chem.* **239:**242.

Richardson, C.C., I.R. Lehman, and A. Kornberg. 1964. A deoxyribonucleic acid phosphatase-exonuclease from *Escherichia coli*. II. Characterization of the exonuclease activity. *J. Biol. Chem.* **239:**251.

Rosamond, J., K.M. Telander, and S. Linn. 1979. Modulation of the action of the *recBC* enzymes of *Escherichia coli* by Ca^{2+}. *J. Biol. Chem.* **254:**8642.

Sain, B. and N.E. Murray. 1980. The *hsd* (host specificity) genes of *Escherichia coli* K12. *Mol. Gen. Genet.* **180:**35.

Sancar, A., N.D. Clarke, J. Griswold, W.J. Kennedy, and W.D. Rupp. 1981a. Identification of the *uvrB* gene product. *J. Mol. Biol.* **148:**63.

Sancar, A., R.P. Wharton, S. Selzer, B.M. Kacinski, N.D. Clarke, and W.D. Rupp. 1981b. Identification of the *uvrA* gene product. *J. Mol. Biol.* **148:**45.

Seeberg, E. 1978. Reconstitution of an *Escherichia coli* repair endonuclease activity from the separated *uvrA*$^+$ and *uvrB*$^+$/*uvrC*$^+$ gene products. *Proc. Natl. Acad. Sci.* **75:**2569.

———. 1981. Multiprotein interactions in strand cleavage of DNA damaged by UV and chemicals. *Prog. Nucleic Acid Res. Mol. Biol.* **26:**217.

Setlow, P. and A. Kornberg. 1972. Deoxyribonucleic acid polymerase: Two distinct enzymes in one polypeptide II. A proteolytic fragment containing 5′ ⟶ 3′ exonuclease function. Restoration of intact enzyme functions from the two proteolytic fragments. *J. Biol. Chem.* **247:**232.

Shortman, K. and I.R. Lehman. 1964. The deoxyribonucleases of *Escherichia coli.* VI. Changes in enzyme levels in response to alterations in physiological state. *J. Biol. Chem.* **239:**2964.

Simmon, V.F. and S. Lederberg. 1972. Degradation of bacteriophage lambda deoxyribonucleic acid after restriction by *Escherichia coli* K-12. *J. Bacteriol.* **112:**161.

Taylor, A. and G.R. Smith. 1980. Unwinding and rewinding of DNA by the recBC enzyme. *Cell* **22:**447.

Taylor, A.F. and B. Weiss. 1982. Role of exonuclease III in the base excision repair of uracil-containing DNA. *J. Bacteriol.* **151:**351.

Verly, W.G. and E. Rassart. 1975. Purification of *Escherichia coli* endonuclease specific for apurinic sites in DNA. *J. Biol. Chem.* **250:**8214.

Wang, J. 1981. Type I DNA topoisomerases. In *The enzymes* (ed. P.D. Boyer), vol. 14, p. 332. Academic Press, New York.

Warner, H.R., B.F. Demple, W.A. Deutsch, C.M. Kane, and S. Linn. 1980. Apurinic/apyrimidinic endonucleases in the repair of pyrimidine dimers and other lesions in DNA. *Proc. Natl. Acad. Sci.* **77:**4602.

Weiss, B. 1981. Exodeoxyribonucleases of *Escherichia coli*. In *The enzymes* (ed. P.D. Boyer), vol. 14, p. 203. Academic Press, New York.

Weiss, B., S.G. Rogers, and A.F. Taylor. 1978. The endonuclease activity of exonuclease III and the repair of uracil-containing DNA in *E. coli*. In *DNA repair mechanisms* (eds. P.C. Hanawalt, E.C. Friedberg, and C.F. Fox), p. 191. Academic Press, New York.

Wickner, R.B., B. Ginsberg, I. Berkower, and J. Horwitz. 1972. Deoxyribonucleic acid polymerase II of *Escherichia coli*. I. The purification and characterization of the enzyme. *J. Biol. Chem.* **247:**489.

Williams, J.G.K., T. Shibata, and C.M. Radding. 1981. *Escherichia coli* recA protein protects single-stranded DNA or gapped duplex DNA from degradation by *recBC* DNase. *J. Biol. Chem.* **256:**7573.

Wright, M. 1971. Mutants of *Escherichia coli* lacking endonuclease I, ribonuclease I or ribonuclease II. *J. Bacteriol.* **107:**87.

Yoakum, G.H. and L. Grossman. 1981. Identification of the *Escherichia coli uvrC* gene product. *Nature* **292:**171.

Yuan, R. 1981. Structure and mechanism of multifunctional restriction endonucleases. *Annu. Rev. Biochem.* **50:**285.

APPENDIX A
Restriction Endonucleases

Richard J. Roberts
Cold Spring Harbor Laboratory
Cold Spring Harbor, New York 11724

Microorganism	Source	Enzyme[a]	Sequence[b]	Number of cleavage sites[c]					References
				λ	Ad2	SV40	ϕX	pBR	
Acetobacter aceti sub. liquefaciens	IFO 12388	AacI (BamHI)	GGATCC	5	3	1	0	1	180
Acetobacter aceti sub. liquefaciens	M. Van Montagu	AaeI (BamHI)	GGATCC	5	3	1	0	1	180
Acetobacter aceti sub. orleanensis	NCIB 8622	AorI (EcoRII)	CC↑$\binom{A}{T}$GG	>35	>35	16	2	6	180
Acetobacter pasteurianus sub. pasteurianus	NCIB 7215	ApaI	GGGCC↑C	1	>10	1	0	0	180
Achromobacter immobilis	ATCC 15934	AimI	?	?	?	?	?	?	48
Acinetobacter calcoaceticus	R.J. Roberts	AccI	GT↑$\binom{A}{C}\binom{G}{T}$AC	7	8	1	2	2	236
		AccII (FnuDII)	CGCG	>50	>50	0	14	23	236
		AccIII	?	>10	>6	?	?	?	149
Agmenellum quadruplicatum	W.F. Doolittle	AquI (AvaI)	CPyCGPuG	8	15	0	1	1	114
Agrobacterium tumefaciens	ATCC 15955	AtuAI	?	>30	>30	?	?	?	179
Agrobacterium tumefaciens B6806	E. Nester	AtuBI (EcoRII)	CC$\binom{A}{T}$GG	>35[d]	>35	16	2	6	170
Agrobacterium tumefaciens IIBV7	G. Roizes	AtuBVI	?	>14	?	1	0	?	169
Agrobacterium tumefaciens ID 135	C. Kado	AtuII (EcoRII)	CC$\binom{A}{T}$GG	>35[d]	>35	16	2	6	119
Agrobacterium tumefaciens C58	E. Nester	AtuCI (BclI)	TGATCA	7[d]	5	1	0	0	179
Alcaligenes species	N. Brown	AspAI (BstEII)	G↑GTNACC	11	8	0	0	0	19
Anabaena catanula	CCAP 1403/1	AcaI	?	?	?	?	?	?	89
Anabaena cylindrica	CCAP 1403/2a	AcyI	GPu↑CGPyC	>14	>14	0	7	6	39

Microorganism	Source	Enzyme[a]	Sequence[b]	Number of cleavage sites[c]					References
				λ	Ad2	SV40	ϕX	pBR	
Anabaena flos-aquae	A.E. Walsby	AflI (AvaII)	G↑G($^{A}_{T}$)CC	17	30	6	1	8	23
		AflII	C↑TTAAG	3	?	1	2	0	23
		AflIII	?	?	?	?	?	?	23
Anabaena oscillarioides	CCAP 1403/11	AosI (MstI)	TGC↑GCA	>10	>15	0	1	4	40
		AosII (AcyI)	GPu↑CGPyC	>14	>14	0	7	6	40
Anabaena strain Waterbury	ATCC 29208	AstWI (AcyI)	GPu↑CGPyC	>14	>14	0	7	6	38
Anabaena subcylindrica	CCAP 1403/4b	AsuI	G↑GNCC	>30	>30	11	2	15	90
		AsuII	TT↑CGAA	7	1	0	0	0	149,38;38
		AsuIII (AcyI)	GPu↑CGPyC	>14	>14	0	7	6	38
Anabaena variabilis	ATCC 27892	AvaI	C↑PyCGPuG	8	15	0	1	1	146;91
		AvaII	G↑G($^{A}_{T}$)CC	>17	>30	6	1	8	146;206,91, 54
		AvaIII	ATGCAT	?	?	3	0	0	168;168,186
Anabaena variabilis uw	E.C. Rosenvold	AvrI (AvaI)	CPyCGPuG	8	15	0	1	1	171
		AvrII	CCTAGG	2	2	2	0	0	171
Aphanothece halophytica	ATCC 29534	AhaI (CauII)	CC($^{C}_{G}$)GG	>30	>30	0	1	10	23
		AhaII	?	?	?	?	?	?	23
		AhaIII	TTT↑AAA	13	>16	11	2	3	23
Arthrobacter luteus	ATCC 21606	AluI	AG↑CT	>50	>50	35	24	16	165
Arthrobacter pyridinolis	R. DiLauro	ApyI (EcoRII)	CC↑($^{A}_{T}$)GG	>35[d]	>35	16	2	6	41
Bacillus acidocaldarius	ATCC 27009	BacI (SacII)	CCGCGG	4	>25	0	1	0	136,149
Bacillus amyloliquefaciens F	ATCC 23350	BamFI (BamHI)	GGATCC	5	3	1	0	1	185

Microorganism	Source	Enzyme[a]	Sequence[b]	Number of cleavage sites[c]					References
				λ	Ad2	SV40	ΦX	pBR	
Bacillus amyloliquefaciens H	F.E. Young	BamHI	G↑GATC*C	5	3	1	0	1	229;167; 80;80
Bacillus amyloliquefaciens K	T. Kaneko	BamKI (BamHI)	GGATCC	5	3	1	0	1	185
Bacillus amyloliquefaciens N	T. Ando	BamNI (BamHI)	GGATCC	5	3	1	0	1	184
		$BamN_x$ (AvaII)	G↑G($^{A}_{T}$)CC	>17	>30	6	1	8	183,184; 94
Bacillus brevis S	A.P. Zarubina	BbvSI	GC*($^{A}_{T}$)GC	specific methylase					221
Bacillus brevis	ATCC 9999	BbvI	GCAGC(8/12)	>30	>30	23	14	21	63;62,177; 80;80
Bacillus caldolyticus	A. Atkinson	BclI	T↑GATCA	7[d]	5	1	0	0	10
Bacillus centrosporus	A.A. Janulaitis	BcnI (NciI)	CC↑CGG	>15	>15	0	1	10	96
Bacillus cereus	ATCC 14579	Bce14579	?	>10	?	?	?	?	185
Bacillus cereus	IAM 1229	Bce1229	?	>10	?	?	?	?	185
Bacillus cereus	T. Ando	Bce170 (PstI)	CTGCAG	18	25	2	1	1	185
Bacillus cereus Rf sm st	T. Ando	BceR (FnuDII)	CGCG	>50	>50	0	14	23	185
Bacillus globigii	G.A. Wilson	BglI	GCCNNNN↑NGGC	22	12	1	0	3	43,230;6, 217
		BglII	A↑GATCT	6	12	0	0	0	43,230; 157
Bacillus megaterium 899	B899	Bme899	?	>5	?	?	?	?	185
Bacillus megaterium B205-3	T. Kaneko	Bme205	?	>10	?	?	?	?	185
Bacillus megaterium	J. Upcroft	BmeI	?	>10	>20	4	?	?	59
Bacillus pumilus AHU1387A	T. Ando	BpuI	?	6	>30	2	?	?	93
Bacillus sphaericus	IAM 1286	Bsp1286	?	?	?	?	?	?	185

Microorganism	Source	Enzyme[a]	Sequence[b]	Number of cleavage sites[c]					References
				λ	Ad2	SV40	ΦX	pBR	
Bacillus sphaericus R	P. Venetianer	BspRI (HaeIII)	GG↑CC	>50	>50	19	11	22	104;223; 107
Bacillus stearo-thermophilus C1	N. Welker	BstCI (HaeIII)	GGCC	>50	>50	19	11	22	113
Bacillus stearo-thermophilus C11	N. Welker	BssCI (HaeIII)	GGCC	>50	>50	19	11	22	113
Bacillus stearo-thermophilus G3	N. Welker	BstGI (BclI)	TGATCA	7[d]	5	1	0	0	113
		BstGII (EcoRII)	CC($^{A}_{T}$)GG	>35[d]	>35	16	2	6	113
Bacillus stearo-thermophilus G6	N. Welker	BssGI (BstXI)	?	10	9	0	3	0	113
		BssGII (MboI)	GATC	>50[d]	>50	8	0	22	113
Bacillus stearo-thermophilus H1	N. Welker	BstHI (XhoI)	CTCGAG	1	6	0	1	0	113
Bacillus stearo-thermophilus H3	N. Welker	BssHI (XhoI)	CTCGAG	1	6	0	1	0	113
		BssHII (BsePI)	GCGCGC	6	>18	0	1	0	113
Bacillus stearo-thermophilus H4	N. Welker	BsrHI (BsePI)	GCGCGC	6	>18	0	1	0	113
Bacillus stearo-thermophilus P1	N. Welker	BssPI	?	>8	?	?	?	?	113
Bacillus stearo-thermophilus P5	N. Welker	BsrPI	?	11	>20	?	0	0	113
		BsrPII	?	>50	?	?	?	?	113
Bacillus stearo-thermophilus P6	N. Welker	BsePI	GCGCGC	6	>18	0	1	0	113
Bacillus stearo-thermophilus P8	N. Welker	BsaPI	?	>20	?	?	?	?	113
Bacillus stearo-thermophilus P9	N. Welker	BsoPI (BsrPI)	?	11	>20	?	0	0	113

Microorganism	Source	Enzyme[a]	Sequence[b]	Number of cleavage sites[c] λ	Ad2	SV40	ϕX	pBR	References
Bacillus stearo-thermophilus T12	N. Welker	BstTI (BstXI)	?	10	9	0	3	0	113
Bacillus stearo-thermophilus X1	N. Welker	BstXI	?	10	9	0	3	0	113
		BstXII (MboI)	GATC	>50[d]	>50	8	0	22	113
Bacillus stearo-thermophilus 1503-4R	N. Welker	BstI (BamHI)	G↑GATCC	5	3	1	0	1	26;29
Bacillus stearo-thermophilus 240	A. Atkinson	BstAI	?	?	?	?	?	?	12
Bacillus stearo-thermophilus ET	N. Welker	BstEI	?	?	?	?	?	?	138
		BstEII	G↑GTNACC	11	8	0	0	0	138;115
		BstEIII (MboI)	GATC	>50[d]	>50	>8	0	22	138;68; 149
Bacillus stearo-thermophilus	ATCC 12980	BstPI (BstEII)	G↑GTNACC	11	8	0	0	0	159
Bacillus stearo-thermophilus	D. Comb	BstNI (EcoRII)	CC↑$\binom{A}{T}$GG	>35[d]	>35	16	2	6	178
Bacillus stearo-thermophilus strain 822	T. Oshima	BseI (HaeIII)	GGCC	>50	>50	19	11	22	189
		BseII (HpaI)	GTTAAC	13	6	4	3	0	189
Bacillus subtilis strain R	T. Trautner	BsuRI (HaeIII)	GG↑C*C	>50	>50	19	11	22	17;18;73
Bacillus subtilis Marburg 168	T. Ando	BsuM	?	>10	?	?	?	?	185
Bacillus subtilis	ATCC 6633	Bsu6633	?	>20	?	?	?	?	185
Bacillus subtilis	IAM 1076	Bsu1076 (HaeIII)	GGCC	>50	>50	19	11	22	185
Bacillus subtilis	IAM 1114	Bsu1114 (HaeIII)	GGCC	>50	>50	19	11	22	185
Bacillus subtilis	IAM 1247	Bsu1247 (PstI)	CTGCAG	18	25	2	1	1	185;86

Microorganism	Source	Enzyme[a]	Sequence[b]	Number of cleavage sites[c] λ	Ad2	SV40	ϕX	pBR	References
Bacillus subtilis	ATCC 14593	Bsu1145	?	>20	?	?	?	?	185
Bacillus subtilis	IAM 1192	Bsu1192	?	>10	?	?	?	?	185
Bacillus subtilis	IAM 1193	Bsu1193	?	>30	?	?	?	?	185
Bacillus subtilis	IAM 1231	Bsu1231	?	>20	?	?	?	?	185
Bacillus subtilis	IAM 1259	Bsu1259	?	>8	?	?	?	?	185
Bifidobacterium breve	H. Takahashi	BdeI (NarI)	GGCGC↑C	>2	18	0	2	4	103
Bordetella bronchiseptica	ATCC 19395	BbrI (HindIII)	AAGCTT	6	11	6	0	1	149
Bordetella pertussis	P. Novotny	BpeI (HindIII)	AAGCTT	6	11	6	0	1	69
Brevibacterium albidum	ATCC 15831	BalI	TGG↑C*CA	15	17	0	0	1	60;60; 214
Brevibacterium luteum	ATCC 15830	BluI (XhoI)	C↑TCGAG	1	6	0	1	0	64
		BluII (HaeIII)	GGCC	>50	>50	19	11	22	218
Calothrix scopulorum	CCAP 1410/5	CscI (SacII)	CCGC↑GG	4	>25	0	1	0	45
Caryophanon latum L	H. Mayer	ClaI	AT↑CGAT	15	2	0	0	1	134
Caryophanon latum	ATCC 15219	ClmI (HaeIII)	GGCC	>50	>50	19	11	22	196
		ClmII (AvaII)	GG($^{A}_{T}$)CC	>17	>30	6	1	8	196
Caryophanon latum	DSM 484	CltI (HaeIII)	GG↑CC	>50	>50	19	11	22	136
Caryophanon latum RII	H. Mayer	CluI	?	>20	?	?	?	?	136
Caryophanon latum H7	W.C. Trentini	CalI	?	14	?	?	?	?	136
Caulobacter crescentus CB-13	R.J. Syddall	CcrI	?	1	>10	0	1	1	207
		CcrII (XhoI)	CTCGAG	1	6	0	1	0	207

Microorganism	Source	Enzyme[a]	Sequence[b]	Number of cleavage sites[c]					References
				λ	Ad2	SV40	ϕX	pBR	
Chloroflexus aurantiacus	A. Bingham	CauI (AvaII)	GG($^{A}_{T}$)CC	>30	>30	6	1	8	11
		CauII	CC↑($^{G}_{C}$)GG	>30	>30	0	1	10	11,122
Chromatium vinosum	G.C. Grosveld	CvnI (SauI)	CC↑TNAGG	>10	>15	0	1	0	72
Chromobacterium violaceum	ATCC 12472	CviI	?	?	?	?	?	?	48
Citrobacter freundii	A.A. Janulaitis	CfrI	Py↑GGCPu	>25	>35	0	2	6	97
Clostridium formico-aceticum	ATCC 23439	CfoI (HhaI)	GCGC	>50	>50	2	18	31	130
Clostridium pasteurianum	NRCC 33011	CpaI (MboI)	GATC	>50[d]	>50	8	0	22	223
Corynebacterium humiferum	ATCC 21108	ChuI (HindIII)	AAGCTT	6	11	6	0	1	48
		ChuII (HindII)	GTPyPuAC	34	>20	7	13	2	48
Corynebacterium petrophilum	ATCC 19080	CpeI (BclI)	TGATCA	7[d]	5	1	0	0	51
Cystobacter velatus Plv9	H. Reichenbach	CveI	?	?	?	?	?	?	136
Desulfovibrio desulfuricans Norway strain	H. Peck	DdeI	C↑TNAG	>50	>50	19	14	8	130;61
		DdeII (XhoI)	CTCGAG	1	6	0	1	0	149
Desulfovibrio desulfuricans	ATCC 27774	DdsI (BamHI)	GGATCC	5	3	1	0	1	129
Diplococcus pneumoniae	S. Lacks	DpnI	G$\overset{*}{A}$↑TC	only cleaves methylated DNA					110;57,111
Diplococcus pneumoniae	S. Lacks	DpnII (MboI)	GATC	>50[d]	>50	8	0	22	110;111
Enterobacter aerogenes	P.R. Whitehead	EaeI (CfrI)	Py↑GGCCPu	>25	>35	0	2	6	23
Enterobacter cloacae	H. Hartmann	EclI	?	14	?	?	?	?	78
		EclII (EcoRII)	CC($^{A}_{T}$)GG	>35[d]	>35	16	2	6	78

Microorganism	Source	Enzyme[a]	Sequence[b]	Number of cleavage sites[c] λ	Ad2	SV40	ϕX	pBR	References
Enterobacter cloacae	DSM 30056	EcaI (BstEII)	G↑GTNACC	11	8	0	0	0	85
		EcaII (EcoRII)	CC(A/T)GG	>35[d]	>35	16	2	6	149
Enterobacter cloacae	DSM 30060	EccI (SacII)	CCGCGG	4	>25	0	1	0	135;149
Escherichia coli pDX1	A. Piekarowicz	EcoDXI	ATCA(N)$_7$ATTC	?	?	1	1	0	154
Escherichia coli J62 pLG74	L.I. Glatman	EcoRV	GATAT↑C	14	8	1	0	1	102
Escherichia coli RY13	R.N. Yoshimori	EcoRI	G↑AA*TTC	5	5	1	0	1	71;81;71;42
		EcoRI'	PuPuA↑TPyPy	>10	>10	24	16	15	145
Escherichia coli R245	R.N. Yoshimori	EcoRII	↑CC*(A/T)GG	>35[d]	>35	16	2	6	235;8,14; 235;14
Escherichia coli B	W. Arber	EcoB	TGA*(N)$_8$TGCT	Type I		1	0	0	50;116,161; 117;220
Escherichia coli K	M. Meselson	EcoK	AAC(N)$_6$GTGC	Type I		0	0	2	139;7,99;75
Escherichia coli (PI)	K. Murray	EcoPI	AGA*CC	Type III[e]		4	7	4	74;4;15,16; 4,79
Escherichia coli P15	W. Arber	EcoP15	CAGCAG	Type III[e]		12	5	7	162;76
Flavobacterium okeanokoites	IFO 12536	FokI	GGATG(9/13)	>50	>50	11	8	6	202
Fremyella diplosiphon	PCC 7601	FdiI (AvaII)	G↑G(A/T)CC	>17	>30	6	1	8	216,200
		FdiII (MstI)	TGC↑GCA	>10	>15	0	1	4	216,200
Fusobacterium nucleatum A	M. Smith	FnuAI (HinfI)	G↑ANTC	>50	>50	10	21	10	125
		FnuAII (MboI)	GATC	>50[d]	>50	8	0	22	125;149
Fusobacterium nucleatum C	M. Smith	FnuCI (MboI)	↑GATC	>50[d]	>50	8	0	22	125

Microorganism	Source	Enzyme[a]	Sequence[b]	Number of cleavage sites[c] λ	Ad2	SV40	ϕX	pBR	References
Fusobacterium nucleatum D	M. Smith	FnuDI (HaeIII)	GG↑CC	>50	>50	19	11	22	125
		FnuDII	CG↑CG	>50	>50	0	14	23	125
		FnuDIII (HhaI)	GCG↑C	>50	>50	2	18	31	125
Fusobacterium nucleatum E	M. Smith	FnuEI (MboI)	↑GATC	>50[d]	>50	8	0	22	125
Fusobacterium nucleatum 48	M. Smith	Fnu48I	?	>50	?	?	>10	?	124
Fusobacterium nucleatum 4H	M. Smith	Fnu4HI	GC↑NGC	>50	>50	25	31	42	121
Gluconobacter dioxy-acetonicus	IAM 1814	GdiI (StuI)	AGG↑CCT	5	12	7	1	0	218
		GdiII	Py↑GGCCG	>10	>20	0	2	5	218
Gluconobacter dioxy-acetonicus	IAM 1840	GdoI (BamHI)	GGATCC	5	3	1	0	1	180
Gluconobacter oxydans sub. melonogenes	IAM 1836	GoxI (BamHI)	GGATCC	5	3	1	0	1	180
Haemophilus aegyptius	ATCC 11116	HaeI	(A/T)GG↑CC(A/T)	?	?	11	6	7	148
		HaeII	PuGCGC↑Py	>30	>30	1	8	11	164;215
		HaeIII	GG↑C*C	>50	>50	19	11	22	140;18;132; 132
Haemophilus aphrophilus	ATCC 19415	HapI	?	>30	?	?	?	?	149
		HapII (HpaII)	C↑CGG	>50	>50	1	5	26	210;203
Haemophilus gallinarum	ATCC 14385	HgaI	GACGC(5/10)	>50	>50	0	14	11	210;21,201
Haemophilus haemo-globinophilus	ATCC 19416	HhgI (HaeIII)	GGCC	>50	>50	19	11	22	149
Haemophilus haemolyticus	ATCC 10014	HhaI	GC*G↑C	>50	>50	2	18	31	166;166; 133
		HhaII (HinfI)	GANTC	>50	>50	10	21	10	131

Microorganism	Source	Enzyme[a]	Sequence[b]	Number of cleavage sites[c] λ	Ad2	SV40	φX	pBR	References
Haemophilus influenzae GU	J. Chirikjian	HinGUI (HhaI)	GCGC	>50	>50	2	18	31	197 and 28
		HinGUII (FokI)	GGATG	>50	>50	11	8	6	197;150
Haemophilus influenzae 173	J. Chirikjian	Hin173 (HindIII)	AAGCTT	6	11	6	0	1	197
Haemophilus influenzae 1056	J. Stuy	Hin1056I (FnuDII)	CGCG	>50	>50	0	14	22	154
		Hin1056II	?	>30	>30	0	5	?	154
Haemophilus influenzae serotype b, 1076	J. Stuy	HinbIII (HindIII)	AAGCTT	6	11	6	0	1	154
Haemophilus influenzae serotype c, 1160	J. Stuy	HincII (HindII)	GTPyPuAC	34	>20	7	13	2	154
Haemophilus influenzae serotype c, 1161	J. Stuy	HincII (HindII)	GTPyPuAC	34	>20	7	13	2	154
Haemophilus influenzae serotype e	A.Piekarowicz	HineI (HinfIII)	CGAAT[f]	Type III[e]		0	5	1	155
Haemophilus influenzae R_b	C.A. Hutchison	HinbIII (HindIII)	AAGCTT	6	11	6	0	1	141,149
Haemophilus influenzae R_c	A. Landy, G. Leidy	HincII (HindII)	GTPyPuAC	34	>20	7	13	2	112
Haemophilus influenzae R_d	S.H. Goodgal (exo⁻ mutant)	HindI	CA*C	specific methylase					172;173
		HindII	GTPy↑PuA*C	34	>20	7	13	2	194;101; 172;173
		HindIII	A*↑AGCTT	6	11	6	0	1	152;152; 172;173
		HindIV	GA*C	specific methylase					172;173
Haemophilus influenzae R_f	C.A. Hutchison	HinfI	G↑ANTC	>50	>50	10	21	10	141;92,147
		HinfII (HindIII)	AAGCTT	6	11	6	0	1	133
		HinfIII	CGAAT[f]	Type III[e]		0	5	1	100,155

Microorganism	Source	Enzyme[a]	Sequence[b]	Number of cleavage sites[c]					References
				λ	Ad2	SV40	φX	pBR	
Haemophilus influenzae H-1	M. Takanami	HinHI (HaeII)	PuGCGCPy	>30	>30	1	8	11	210
Haemophilus influenzae P_1	S. Shen	HinP_1I (HhaI)	G↑CGC	>50	>50	2	18	31	182
Haemophilus influenzae S_1	S. Shen	HinS_1 (HhaI)	GCGC	>50	>50	2	18	31	182
Haemophilus influenzae S_2	S. Shen	HinS_2 (HhaI)	GCGC	>50	>50	2	18	31	182r
Haemophilus influenzae JC9	A. Piekarowicz	HinJCI (HindII)	GTPy↑PuAC	34	>20	7	13	2	156
		HinJCII (HindIII)	AAGCTT	6	11	6	0	1	156
Haemophilus para-haemolyticus	C.A. Hutchison	HphI	GGTGA(8/7)	>50	>50	4	9	12	141;105
Haemophilus parainfluenzae	J. Setlow	HpaI	GTT↑AA*C	13	6	4	3	0	181;56,1
		HpaII	C↑C*GG	>50	>50	1	5	26	181;56,132; 132
Haemophilus suis	ATCC 19417	HsuI (HindIII)	A↑AGCTT	6	11	6	0	1	149
Herpetosiphon giganteus HP1023	J.H. Parish	HgiAI	G$\binom{A}{T}$GC$\binom{A}{T}$↑C	24	>20	0	3	8	24
Herpetosiphon giganteus Hpg 5	H. Reichenbach	HgiBI (AvaII)	G↑$\binom{A}{T}$CC	>17	>30	6	1	8	84
Herpetosiphon giganteus Hpg 9	H. Reichenbach	HgiCI	G↑GPyPuCC	13	>25	1	3	9	84,108
		HgiCII (AvaII)	G↑G$\binom{A}{T}$CC	>17	>30	6	1	8	84
		HgiCIII (SalI)	G↑TCGAC	2	3	0	0	1	84
Herpetosiphon giganteus Hpa2	H. Reichenbach	HgiDI (AcyI)	GPu↑CGPyC	>14	>14	0	7	6	84,108
		HgiDII (SalI)	G↑TCGAC	2	3	0	0	1	84
Herpetosiphon giganteus Hpg 24	H. Reichenbach	HgiEI (AvaII)	G↑G$\binom{A}{T}$CC	>17	>30	6	1	8	84
		HgiEII	ACC$(N)_6$GGT	?	?	1	1	2	84

Microorganism	Source	Enzyme[a]	Sequence[b]	Number of cleavage sites[c]					References
				λ	Ad2	SV40	φX	pBR	
Herpetosiphon giganteus Hpg 14	H. Reichenbach	HgiFI	?	?	15	?	?	?	136
Herpetosiphon giganteus Hpa 1	H. Reichenbach	HgiGI (AcyI)	GPu↑CGPyC	>14	>14	0	7	6	108
Herpetosiphon giganteus HP1049	J.H. Parish	HgiHI (HgiCI)	G↑GPyPuCC	>13	>25	1	3	9	23
		HgiHII (AcyI)	GPu↑CGPyC	>14	>14	0	7	6	23
		HgiHIII (AvaII)	G↑G($^{A}_{T}$)CC	>17	>30	6	1	8	23
Herpetosiphon giganteus HFS101	H. Foster	HgiJI	?	?	?	?	?	?	23
		HgiJII	GPuGCPy↑C	7	>35	2	0	2	23
Herpetosiphon giganteus Hpg 32	H. Reichenbach	HgiKI	?	>18	>20	?	?	?	136
Klebsiella pneumoniae OK8	J. Davies	KpnI	GGTAC↑C	2	8	1	0	0	192;212
Mastigocladus laminosus	CCAP 1447/1	MlaI (AsuII)	TT↑CGAA	7	1	0	0	0	44
Microbacterium thermosphactum	ATCC 11509	MthI (MboI)	GATC	>50[d]	>50	8	0	22	113
Micrococcus luteus	IFO 12992	MluI	A↑CGCGT	?	?	0	2	0	202
Micrococcus radiodurans	ATCC 13939	MraI (SacII)	CCGCGG	4	>25	0	1	0	227
Microcoleus species	D. Comb	MstI	TGC↑GCA	>10	>15	0	1	4	33;62
		MstII (SauI)	CC↑TNAGG	2	7	0	0	0	177
Moraxella bovis	ATCC 10900	MboI	↑GATC	>50[d]	>50	8	0	22	58
		MboII	GAAGA(8/7)	>50	>50	16	11	11	58;20,47
Moraxella bovis	ATCC 17947	MbvI	?	?	?	?	?	?	98
Moraxella glueidi LG1	J. Davies	MglI	?	?	?	?	?	?	192
Moraxella glueidi LG2	J. Davies	MglII	?	?	?	?	?	?	192

Microorganism	Source	Enzyme[a]	Sequence[b]	Number of cleavage sites[c]					References
				λ	Ad2	SV40	φX	pBR	
Moraxella kingae	ATCC 23331	MkiI (HindIII)	AAGCTT	6	11	6	0	1	98
Moraxella nonliquefaciens	ATCC 19975	MnoI (HpaII)	C↑CGG	>50	>50	1	5	26	149;5
		MnoII (MnnIII)	?	>10	>6	3	?	?	149
		MnoIII (MboI)	GATC	>50[d]	>50	8	0	22	149
Moraxella nonliquefaciens	ATCC 17953	MnlI	CCTC(7/7)	>50	>50	51	34	26	237;177
Moraxella nonliquefaciens	ATCC 17954	MnnI (HindII)	GTPyPuAC	34	>20	7	13	2	77
		MnnII (HaeIII)	GGCC	>50	>50	19	11	22	77
		MnnIII	?	>10	>6	3	?	?	77
		MnnIV (HhaI)	GCGC	>50	>50	2	18	31	77
Moraxella nonliquefaciens	ATCC 19996	MniI (HaeIII)	GGCC	>50	>50	19	11	22	98
		MniII (HpaII)	CCGG	>50	>50	1	5	26	98
Moraxella osloensis	ATCC 19976	MosI (MboI)	GATC	>50[d]	>50	8	0	22	58
Moraxella phenylpyruvica	ATCC 17955	MphI (EcoRII)	CC($^{A}_{T}$)GG	>35[d]	>35	16	2	6	98
Moraxella species	R. J. Roberts	MspI (HpaII)	C↑CGG	>50	>50	1	5	26	219,177
Myxococcus stipitatus Mxs2H	H. Reichenbach	MsiI (XhoI)	CTCGAG	1	6	0	1	0	141,149
		MsiII	?	?	?	?	?	?	136
Myxococcus virescens V-2	H. Reichenbach	MviI	?	1	?	?	?	?	143
		MviII	?	?	?	?	?	?	143
Neisseria caviae	NRCC 31003	NcaI (HinfI)	GANTC	>50	>50	10	21	10	224
Neisseria cinerea	NRCC 31006	NciI (CauII)	CC($^{C}_{G}$)GG[g]	>15	>15	0	1	10	228;88

Microorganism	Source	Enzyme[a]	Sequence[b]	Number of cleavage sites[c]					References
				λ	Ad2	SV40	ϕX	pBR	
Neisseria denitrificans	NRCC 31009	NdeI	CA↑TATG	?	?	2	0	1	224;177
		NdeII (MboI)	GATC	>50[d]	>50	8	0	22	224
Neisseria flavescens	NRCC 31011	NflI (MboI)	GATC	>50d	>50	8	0	22	224
		NflII	?	?	?	?	?	?	224
		NflIII	?	?	?	?	?	?	224
Neisseria gonorrhoea	G. Wilson	NgoI (HaeII)	PuGCGCPy	>30	>30	1	8	11	231
Neisseria gonorrhoea	CDC 66	NgoII (HaeIII)	GGCC	>50	>50	19	11	22	36,37
Neisseria gonorrhoea KH 7764-45	L. Mayer	NgiIII (SacII)	CCGCGG	4	>25	0	1	0	151
Neisseria mucosa	NRCC 31013	NmuI (NaeI)	GCCGGC	2	>13	1	0	4	224
Neisseria ovis	NRCC 31020	NovI	?	?	?	?	?	?	224
		NovII (HinfI)	GANTC	>50	>50	10	21	10	224
Nocardia aerocolonigenes	ATCC 23870	NaeI	GCC↑GGC	2	>13	1	0	4	35
Nocardia argentinensis	ATCC 31306	NarI	GG↑CGCC	2	10	0	2	4	34
Nocardia blackwellii	ATCC 6846	NblI (PvuI)	CGAT↑CG	3	7	0	0	1	177
Nocardia corallina	ATCC 19070	NcoI	C↑CATGG	6	14	3	0	0	113
Nocardia dassonvillei	ATCC 21944	NdaI (NarI)	GG↑CGCC	2	10	0	2	4	31
Nocardia opaca	ATCC 21507	NopI (SalI)	G↑TCGAC	2	3	0	0	1	177
		NopII (SalII)	?	?	?	?	?	?	113
Nocardia otitidis-caviarum	ATCC 14630	NotI	?	0	6	0	0	0	13
Nocardia rubra	ATCC 15906	NruI	TCG↑CGA	7	6	0	2	1	32

Microorganism	Source	Enzyme[a]	Sequence[b]	Number of cleavage sites[c]					References
				λ	Ad2	SV40	ΦX	pBR	
Nocardia uniformis	ATCC 21806	NunI	?	?	?	?	?	?	113
		NunII (NarI)	GG↑CGCC	2	14	0	2	4	113
Nostoc species	PCC 6705	NspBI	?	?	?	?	?	?	46
		NspBII	GC($^{G}_{C}$)G↑C	?	?	2	17	21	46
Nostoc species	PCC 7524	NspCI	PuCATG↑Py	>15	>15	2	0	4	46
Nostoc species	PCC 7413	NspHI (NspCI)	PuCATG↑Py	>15	>15	2	0	4	46
Oerskovia xanthineolytica	R. Shekman	OxaI (AluI)	AGCT	>50	>50	35	24	16	199
		OxaII	?	?	?	?	?	?	199
Proteus vulgaris	ATCC 13315	PvuI	CGAT↑CG	3	7	0	0	1	65
		PvuII	CAG↑CTG	15	22	3	0	1	65
Providencia alcalifaciens	ATCC 9886	PalI (HaeIII)	GGCC	>50	>50	19	11	22	59
Providencia stuartii 164	J. Davies	PstI	CTGCA↑G	18	25	2	1	1	192;22
Pseudoanabaena species	ATCC 27263	PspI (AsuI)	GGNCC	>30	>30	11	2	15	144
Pseudomonas aeruginosa	G. A. Jacoby	PaeR7	?	1	5	0	0	0	83
Pseudomonas facilis	M. VanMontagu	PfaI (MboI)	GATC	>50[d]	>50	8	0	22	219
Pseudomonas maltophila	D. Comb	PmaI (PstI)	CTGCAG	18	25	2	1	1	177
Rhizobium leguminosarum 300	J. Beringer	RleI	?	6	>10	?	?	?	233
Rhizobium lupini #1	W. Heumann	RluI	?	1	8	?	?	?	232,82
Rhizobium meliloti	J. L. Denarie	RmeI	?	8	>10	?	?	?	82
Rhodospirillum rubrum	J. Chirikjian	RrbI	?	?	4	5	1	?	120
Rhodopseudomonas sphaeroides	R. Lascelles	RspI (PvuI)	CGATCG	3	7	0	0	1	9

Microorganism	Source	Enzyme[a]	Sequence[b]	Number of cleavage sites[c]					References
				λ	Ad2	SV40	φX	pBR	
Rhodopseudomonas sphaeroides	S. Kaplan	RshI (PvuI)	CGAT↑CG	3	7	0	0	1	127
Rhodopseudomonas sphaeroides	S. Kaplan	RsaI	GT↑AC	>50	>50	11	11	3	128
Rhodopseudomonas sphaeroides	S. Kaplan	RsrI (EcoRI)	GAATTC	5	5	1	0	1	55
Salmonella infantis	A. deWaard	SinI (AvaII)	GG(A/T)CC	>17	>30	6	1	8	126
Serratia marcescens S_b	C. Mulder	SmaI	CCC↑GGG	3	12	0	0	0	70;49
Serratia species SAI	B. Torheim	SspI	?	?	?	?	?	?	213
Sphaerotilus natans C	A. Pope	SnaI	GTATAC	2	?	0	0	1	158
Spiroplasma citri ASP2	M.A. Stephens	SciNI (HhaI)	G↑CGC	>50	>50	2	18	31	195
Staphylococcus aureus 3A	E.E. Stobberingh	Sau3A (MboI)	↑GATC	>50[d]	>50	8	0	22	204
Staphylococcus aureus PS96	E.E. Stobberingh	Sau96I (AsuI)	G↑GNCC	>30	>30	11	2	15	205
Staphylococcus saprophyticus	ATCC 13518	SsaI	?	>10	?	?	?	?	113
Streptococcus cremoris F	C. Daly	ScrFI	CCNGG	>50	>50	16	3	16	52
Streptococcus durans	A. Janulaitis	SduI	G(G/A/T)GC(C/A/T)C	?	?	4	3	10	95
Streptococcus faecalis var. zymogenes	R. Wu	SfaI (HaeIII)	GG↑CC	>50	>50	19	11	22	234
Streptococcus faecalis GU	J. Chirikjian	SfaGU I (HpaII)	CCGG	>50	>50	1	5	26	30
Streptococcus faecalis ND547	D. Clewell	SfaNI	GCATC(5/9)	>50	>30	6	12	22	179

Microorganism	Source	Enzyme[a]	Sequence[b]	Number of cleavage sites[c]					References
				λ	Ad2	SV40	ϕX	pBR	
Streptomyces achromogenes	ATCC 12767	SacI	GAGCT↑G	2	7	0	0	0	2
		SacII	CCGC↑GG	4	>25	0	1	0	2
		SacIII	?	>100	>100	?	?	?	2
Streptomyces albus	CM1 52766	SalPI (PstI)	CTGCA↑G	18	25	2	1	1	27;25
Streptomyces albus subspecies pathocidicus	KCC S0166	SpaI (XhoI)	CTCGAG	1	6	0	1	0	188
Streptomyces albus G	J.M. Ghuysen	SalI	G↑TCGAC	2	3	0	0	1	3
		SalII	?	>20	?	?	?	?	3
Streptomyces aureofaciens IKA 18/4	J. Timko	SauI	CC↑TNAGG	2	7	0	0	0	211
Streptomyces bobili	ATCC 3310	SboI (SacII)	CCGCGG	4	>25	0	1	0	188;209
Streptomyces caespitosus	H. Takahashi	ScaI	AGTACT	6	5	0	0	1	106
Streptomyces cupidosporus	KCC S0316	ScuI (XhoI)	CTCGAG	1	6	0	1	0	188
Streptomyces exfoliatus	KCC S0030	SexI (XhoI)	CTCGAG	1	6	0	1	0	188
		SexII	?	2	?	?	?	?	188
Streptomyces fradiae	ATCC 3355	SfrI (SacII)	CCGCGG	4	>25	0	1	0	188;209
Streptomyces ganmycicus	KCC S0759	SgaI (XhoI)	CTCGAG	1	6	0	1	0	188
Streptomyces goshikiensis	KCC S0294	SgoI (XhoI)	CTCGAG	1	6	0	1	0	188
Streptomyces griseus	ATCC 23345	SgrI	?	0	7	0	?	?	2
Streptomyces hygroscopicus	T. Yamaguchi	ShyTI	?	2	?	?	?	?	188
Streptomyces hygroscopicus	F. Walter	ShyI (SacII)	CCGCGG	4	>25	0	1	0	225
Streptomyces lavendulae	ATCC 8664	SlaI (XhoI)	C↑TCGAG	1	6	0	1	0	208

Microorganism	Source	Enzyme[a]	Sequence[b]	Number of cleavage sites[c]					References
				λ	Ad2	SV40	ϕX	pBR	
Streptomyces luteoreticuli	KCC S0788	SluI (XhoI)	CTCGAG	1	6	0	1	0	209
Streptomyces oderifer	ATCC 6246	SodI	?	?	?	?	?	?	113
		SodII	?	?	?	?	?	?	113
Streptomyces phaeo-chromogenes	F. Bolivar	SphI	GCATG↑C	4	9	2	0	1	53
Streptomyces stanford	S. Goff, A. Rambach	SstI (SacI)	GAGCT↑C	2	7	0	0	0	66;142
		SstII (SacII)	CCGC↑GG	4	>25	0	1	0	66
		SstIII (SacIII)	?	>100	>100	?	?	?	66
		SstIV (BclI)	TGATCA	7	5	1	0	0	87
Streptomyces tubercidicus	H. Takahashi	StuI	AGG↑CCT	5	12	7	1	0	187
Streptoverticillium flavopersicum	Upjohn UC 5066	SflI (PstI)	CTGCA↑G	18	25	2	1	1	98
Thermoplasma acidophilum	D. Searcy	ThaI (FnuDII)	CG↑CG	>50	>50	0	14	23	137
Thermopolyspora glauca	ATCC 15345	TglI (SacII)	CCGCGG	4	>25	0	1	0	63
Thermus aquaticus YTI	J.I. Harris	TaqI	T↑CGA*	>50	>50	1	10	7	174;174;175
		TaqII	?	>30	>30	4	6	?	149
Thermus aquaticus	S.A. Grachev	TaqXI (EcoRII)	CC*↑AGG	>35[d]	>35	16	2	6	67
Thermus flavus AT62	T. Oshima	TflI (TaqI)	TCGA	>50	>50	1	10	7	175
Thermus thermophilus HB8	T. Oshima	TthHB8 I (TaqI)	TCGA*	>50	>50	1	10	7	176,222;176, 222;175;175
Thermus thermophilus strain 23	T. Oshima	TtrI (Tth111 I)	GACNNNGTC	2	>12	0	0	1	191
Thermus thermophilus strain 110	T. Oshima	TteI (Tth111 I)	GACNNNGTC	2	>12	0	0	1	191

Microorganism	Source	Enzyme[a]	Sequence[b]	Number of cleavage sites[c] λ	Ad2	SV40	φX	pBR	References
Thermus thermophilus strain 111	T. Oshima	Tth111 I	GAGN↑NNGTC	2	>12	0	0	1	191,191
		Tth111 II	CAAPuCA	>30	>30	12	11	5	190,190
		Tth111 III	?	?	?	?	?	?	189
Tolypothrix tenuis	W. Siegelman	TtnI (HaeIII)	GGCC	>50	>50	19	11	22	200
Xanthomonas amaranthicola	ATCC 11645	XamI (SalI)	GTCGAC	2	3	0	0	1	3
Xanthomonas badrii	ATCC 11672	XbaI	T↑CTAGA	1[d]	4	0	0	0	238
Xanthomonas holcicola	ATCC 13461	XhoI	C↑TCGAG	1	6	0	1	0	64
		XhoII	Pu↑GATCPy	>20	>20	?	0	8	153;63
Xanthomonas malvacearum	ATCC 9924	XmaI (SmaI)	C↑CCGGG	3	12	0	0	0	49
		XmaII (PstI)	CTGCAG	18	25	2	1	1	49
		XmaIII	C↑GGCCG	2	10	0	0	1	109
Xanthomonas manihotis strain 7AS1	B-C. Lin	XmnI	GAANN↑NNTTC	>11	?	0	3	2	123,177
Xanthomonas nigromaculans	ATCC 23390	XniI (PvuI)	CGATCG	3	7	0	0	1	77
Xanthomonas oryzae	M. Ehrlich	XorI (PstI)	CTGCAG	18	25	2	1	1	226
		XorII (PvuI)	CGAT↑CG	3	7	0	0	1	226;65
Xanthomonas papavericola	ATCC 14180	XpaI (XhoI)	C↑TCGAG	1	6	0	1	0	64

[a] When two enzymes recognize the same sequence, i.e., are isoschizomers, the prototype (i.e., the first example isolated) is indicated in parentheses.

[b] Recognition sequences are written from 5' -> 3', only one strand being given, and the point of cleavage is indicated by an arrow (↑). When no arrow appears, the precise cleavage site has not been determined. For example, G↑GATCC is an abbreviation for

```
5' G↑G A T C C 3'
3' C C T A G↑G 5'.
```

For enzymes such as HgaI, MboII, etc., which cleave away from their recognition sequence the sites of cleavage are indicated in parentheses. For example HgaI GACGC(5/10) indicates cleavage as shown below

```
5'GACGCNNNNN↑     3'
3'CTGCGNNNNNNNNNN↑5'.
```

In all cases the recognition sequences are oriented so that the cleavage sites lie on their 3' side.

Bases appearing in parentheses signify that either base may occupy that position in the recognition sequence. Thus, AccI cleaves the sequences GTAGAC, GTATAC, GTACGAC, and GTCTAC. Where known, the base modified by the corresponding specific methylase is indicated by an asterisk. $\overset{*}{A}$ is N^6-methyladenosine. $\overset{*}{C}$ is 5-methylcytosine.

[c] These columns indicate the frequency of cleavage by the various specific endonucleases on bacteriophage lambda DNA (λ), adenovirus-2 DNA (Ad2), simian virus 40 DNA (SV40), ϕX174 Rf DNA, and pBR322 DNA (pBR). In the latter three cases, the sites were checked by computer search of the published sequences.

[d] In most _E. coli_ strains, bacteriophage λ DNA is partially modified against the action of AtuBI, AtuII, AtuCI, BclI, BssGII, BstGI, BstGII, BstXI, BstEIII, CpaI, CpeI, DpnII, EcaII, EclII, EcoRII, FnuAII, FnuCI, MboI, MnoIII, MosI, MphI, and NdeII. XbaI also rarely gives complete digestion presumably because the XbaI recognition sequence overlaps a dam recognition site. It should be noted that MthI, FnuEI, PfaI, and Sau3A are not inhibited by _dam_ methylation; BstNI, TaqXI, and ApyI are not inhibited by mec methylation.

[e] EcoPI, EcoP15, HineI, and HinfIII have characteristics intermediate between those of the type=I and type=II restriction endonucleases. They are designated type III in accordance with the suggestion of Kauc and Piekarowicz (100).

[f] Both HinfIII and HineI cleave about 25 bases 3' of the recognition sequence.

[g] NciI leaves termini carrying a 3'-phosphate group (88).

```
5' G↑G A T C C 3'
3' C C T A G↑G 5'.
```

For enzymes such as HgaI, MboII, etc., which cleave away from their recognition sequence the sites of cleavage are indicated in parentheses. For example HgaI GACGC(5/10) indicates cleavage as shown below

```
5'GACGCNNNNN↑     3'
3'CTGCGNNNNNNNNNN↑5'.
```

In all cases the recognition sequences are oriented so that the cleavage sites lie on their 3' side.

Bases appearing in parentheses signify that either base may occupy that position in the recognition sequence. Thus, AccI cleaves the sequences GTAGAC, GTATAC, GTACGAC, and GTCTAC. Where known, the base modified by the corresponding specific methylase is indicated by an asterisk. $\overset{*}{A}$ is N^6-methyladenosine. $\overset{*}{C}$ is 5-methylcytosine.

[c] These columns indicate the frequency of cleavage by the various specific endonucleases on bacteriophage lambda DNA (λ), adenovirus-2 DNA (Ad2), simian virus 40 DNA (SV40), ϕX174 Rf DNA, and pBR322 DNA (pBR). In the latter three cases, the sites were checked by computer search of the published sequences.

[d] In most *E. coli* strains, bacteriophage DNA is partially modified against the action of AtuBI, AtuII, AtuCI, BclI, BssGII, BstGI, BstGII, BstXI, BstEIII, CpaI, CpeI, DpnII, EcaII, EclII, EcoRII, FnuAII, FnuCI, MboI, MnoIII, MosI, MphI, and NdeII. XbaI also rarely gives complete digestion presumably because the XbaI recognition sequence overlaps a dam recognition site. It should be noted that MthI, FnuEI, PfaI, and Sau3A are not inhibited by *dam* methylation; BstNI, TaqXI, and ApyI are not inhibited by mec methylation.

[e] EcoPI, EcoP15, HineI, and HinfIII have characteristics intermediate between those of the type=I and type=II restriction endonucleases. They are designated type III in accordance with the suggestion of Kauc and Piekarowicz (100).

[f] Both HinfIII and HineI cleave about 25 bases 3' of the recognition sequence.

[g] NciI leaves termini carrying a 3'-phosphate group (88).

REFERENCES

1. Agarwal, K. (unpubl.).

2. Arrand, J.R., P.A. Myers, and R.J. Roberts (unpubl.).

3. Arrand, J.R., P.A. Myers, and R.J. Roberts. 1978. J. Mol. Biol. 118:127.

4. Bachi, B., J. Reiser, and V. Pirrotta. 1979. J. Mol. Biol. 128:143.

5. Baumstark, B.R., R.J. Roberts, and U.L. RajBhandary. 1979. J. Biol. Chem. 254:8943.

6. Bickle, T.A. and K. Ineichen. 1980. Gene 9:205.

7. Bickle, T., R. Yuan, V. Pirrotta, and K. Ineichen (unpubl.).

8. Bigger, C.H., K. Murray, and N.E. Murray. 1973. Nature New Biol. 244:7.

9. Bingham, A.H.A., A. Atkinson, and J. Darbyshire (unpubl.).

10. Bingham, A.H.A., T. Atkinson, D. Sciaky, and R.J. Roberts. Nucleic Acids Res. 5:3457.

11. Bingham, A.H.A. and J. Darbyshire (unpubl.).

12. Bingham, A.H.A., R.J. Sharp, and T. Atkinson (unpubl.).

13. Borsetti, R., D. Wise, and I. Schildkraut (unpubl.).

14. Boyer, H.W., L.T. Chow, A. Dugaiczyk, J. Hedgpeth, and H.M. Goodman. 1973. Nature New Biol. 244:40.

15. Brockes, J.P. 1973. Biochem. J. 133:629.

16. Brockes, J.P., P.R. Brown, and K. Murray. 1972. Biochem. J. 127:1.

17. Bron, S., K. Murray, and T.A. Trautner. 1975. Mol. Gen. Genet. 143:13.

18. Bron, S. and K. Murray. 1975. Mol. Gen. Genet. 143:25.

19. Brown, N.L. (unpubl.).

20. Brown, N.L., C.A. Hutchison, and M. Smith. 1980. J. Mol. Biol. 140:143.

21. Brown, N.L. and M. Smith. 1977. Proc. Natl. Acad. Sci. 74:3213.

22. Brown, N.L. and M. Smith. 1976. FEBS Lett. 65:284.

23. Brown, N.L. and P.R. Whitehead (unpubl.).

24. Brown, N.L., M. McClelland, and P.R. Whitehead. 1980. Gene 9:49.

25. Carter, J.A., K.F. Chater, C.J. Bruton, and N.L. Brown. 1980. Nucleic Acids Res. 8:4943.

26. Catterall, J.F. and N.E. Welker. 1977. J. Bacteriol. 129:1110.

27. Chater, K.F. 1977. Nucleic Acids Res. 4:1989.

28. Chirikjian, J.G., A. George, and L.A. Smith. 1978. Fed. Proc. 37:1415.

29. Clarke, C.M. and B.S. Hartley. 1979. Biochem. J. 177:49.

30. Coll, E. and J. Chirikjian (unpubl.).

31. Comb, D.G., E.J. Hess, and G. Wilson (unpubl.).

32. Comb, D.G. and I. Schildkraut (unpubl.).

33. Comb, D.G., I. Schildkraut, and R.J. Roberts (unpubl.).

34. Comb, D.G., I. Schildkraut, G. Wilson, and L. Greenough (unpubl.).

35. Comb, D.G. and G. Wilson (unpubl.).

36. Clanton, D.J., J.M. Woodward, and R.V. Miller. 1978. J. Bacteriol. 135:270.

37. Clanton, D.J., W.S. Riggsby, and R.V. Miller. 1979. J. Bacteriol. 137:1299.

38. DeWaard, A. and M. Duyvesteyn. 1980. Arch. Microbiol. 128:242.

39. DeWaard, A., J. Korsuize, C.P. van Beveren, and J. Maat. 1978. FEBS Lett. 96:106.

40. DeWaard, A., C.P. van Beveren, M. Duyvesteyn, and H. van Ormondt. 1979. FEBS Lett. 101:71.

41. DiLauro, R. (unpubl.).

42. Dugaiczyk, A., J. Hedgpeth, H.W. Boyer, and H.M. Goodman. 1974. Biochemistry 13:503.

43. Duncan, C.H., G.A. Wilson, and F.E. Young. 1978. J. Bacteriol. 134:338.

44. Duyvesteyn, M.G.C. and A. deWaard. 1980. FEBS Lett. 111:423.

45. Duyvesteyn, M.G.C., J. Korsuize, and A. deWaard. 1981. Plant Mol. Biol. 1:75.

46. Duyvesteyn, M.G.C., J. Reaston, and A. deWaard (unpubl.).

47. Endow, S.A. 1977. J. Mol. Biol. 114:441.

48. Endow, S.A. and R.J. Roberts (unpubl.).

49. Endow, S.A. and R.J. Roberts. 1977. J. Mol. Biol. 112:521.

50. Eskin, B. and S. Linn. 1972. J. Biol. Chem. 247:6183.

51. Fisherman, J., T.R. Gingeras, and R.J. Roberts, (unpubl.).

52. Fitzgerald, G. and C. Daly (unpubl.).

53. Fuchs, L.Y., L. Covarrubias, L. Escalante, S. Sanchez, and F. Bolivar. 1980. Gene 10:39.

54. Fuchs, C., E.C. Rosenvold, A. Honigman, and W. Szybalski. 1978. Gene 4:1.

55. Gardner, J.F., L.K. Cohen, S.P. Lynn, and S. Kaplan (unpubl.).

56. Garfin, D.E. and H.M. Goodman. 1974. Biochem. Biophys. Res. Comm. 59:108.

57. Geier, G.E. and P. Modrich. 1979. J. Biol. Chem. 254:1408.

58. Gelinas, R.E., P.A. Myers, and R.J. Roberts. 1977. J. Mol. Biol. 114:169.

59. Gelinas, R.E., P.A. Myers, and R.J. Roberts (unpubl.).

60. Gelinas, R.E., P.A. Myers, G.A. Weiss, K. Murray, and R.J. Roberts. 1977. J. Mol. Biol. 114:433.

61. Gelinas, R.E. and R.J. Roberts (unpubl.).

62. Gingeras, T.R., J.P. Milazzo, and R.J. Roberts. 1979. Nucleic Acids Res. 5:4105.

63. Gingeras, T.R. and R.J. Roberts (unpubl.).

64. Gingeras, T.R., P.A. Myers, J.A. Olson, F.A. Hanberg, and R.J. Roberts. 1978. J. Mol. Biol. 118:113.

65. Gingeras, T.R., L. Greenough, I. Schildkraut, and R.J. Roberts. 1981. Nucleic Acids Res. 9:4525.

66. Goff, S.P. and A. Rambach. 1978. Gene 3:347 (and unpubl.).

67. Grachev, S.A., S.V. Mamaev, A.I. Gurevich, A.V. Igoshin, M.N. Kolosov, and A.G. Slyusarenko. 1981. Bioorg. Khim. 7:628.

68. Grandioni, R.P. and D. Comb (unpubl.).

69. Greenaway, P.J. 1980. Biochem. Biophys. Res. Commun. 95:1282.

70. Greene, R. and C. Mulder (unpubl.).

71. Greene, P.J., M.C. Betlach, H.W. Boyer, and H.M. Goodman. 1974. Methods Mol. Biol. 7:87.

72. Grosveld, G.C. 1982. Gene (in press).

73. Gunthert, U., K. Storm, and R. Bald. 1978. Eur. J. Biochem. 90:581.

74. Haberman, A. 1974. J. Mol. Biol. 89:545.

75. Haberman, A., J. Heywood, and M. Meselson. 1972. Proc. Natl. Acad. Sci 69:3138.

76. Hadi, S.M., B. Bachi, J.C.W. Shepherd, R. Yuan, K. Ineichen, and T.A. Bickle. 1979. J. Mol. Biol. 134:655.

77. Hanberg, F., P.A. Myers, and R.J. Roberts (unpubl.).

78. Hartmann, H. and W. Goebel. 1977. FEBS Lett. 80:285.

79. Hattman, S., J.E. Brooks, and M. Masurekar. 1978. J. Mol. Biol. 126:367.

80. Hattman, S., T. Keisler, and A. Gottehrer. 1978. J. Mol. Biol. 124:701.

81. Hedgpeth, J., H.M. Goodman, and H.W. Boyer. 1972. Proc. Natl. Acad. Sci. 69:3448.

82. Heumann, W. 1979. Curr. Top. Microbiol. Immunol. 88:1.

83. Hinkle, N.F. and R.V. Miller. 1979. Plasmid 2:387.

84. Hobom, G., H. Mayer, and H. Schutte (unpubl.).

85. Hobom, G., E. Schwarz, M. Melzer, and H. Mayer. 1981. Nucleic Acids Res. 9:4823.

86. Hoshino, T., T. Uozumi, S. Horinouchi, A. Ozaki, T. Beppu, and K. Arima. 1977. Biochim. Biophys. Acta 479:367.

87. Hu, A.W., D. Kuebbing, and R.J. Blakesley. 1978. Fed. Proc. 38:780 and (unpubl.).

88. Hu, A.W. and A.H. Marschel (unpubl.).

89. Hughes, S.G., T. Bruce, and K. Murray (unpubl.).

90. Hughes, S.G., T. Bruce, and K. Murray. 1980. Biochem. J. 185:59.

91. Hughes, S.G. and K. Murray. 1980. Biochem. J. 185:65.

92. Hutchison, C.A. and B.G. Barrell (unpubl.).

93. Ikawa, S., T. Shibata, and T. Ando. 1976. J. Biochem. (Tokyo) 80:1457.

94. Ikawa, S., T. Shibata, and T. Ando. 1979. Agric. Biol. Chem. 43:873.

95. Janulaitis, A., L. Marcinkeviciene, M. Petrusyte, and A. Mironov. 1981. FEBS Lett. 134:172.

96. Janulaitis, A.A., M.P. Pyatrushite, B.P. Jaskeleviciene, A.C. Kraev, and K.G. Scriabin. 1981. Dokl. Akad. Nauk. SSSR 577:749.

97. Janulaitis, A.A., P.J. Stakenas, B.P. Jaskeleviciene, E.N. Lebedenko, and Y.A. Berlin. 1980. Bioorg. Khim. 6:1746.

98. Jiang, B.D. and P. Myers (unpubl.).

99. Kan, N.C., J.A. Lautenberger, M.H. Edgell, and C.A. Hutchison, III. 1979. J. Mol. Biol. 130:191.

100. Kauc, L. and A. Piekarowicz. 1978. Eur. J. Biochem. 92:417.

101. Kelly, T. J., Jr. and H.O. Smith. 1970. J. Mol. Biol. 51:393.

102. Kholmina, G.V., B.A. Rebentish, Y.S. Skoblov, A.A. Mironov, N.K. Yankovsky, Y.I. Kozlov, L.I. Glatman, A.F. Moroz, and V.G. Debabov. 1980. Dokl. Akad. Nauk. SSSR 253:495.

103. Khosaka, T., T. Sakurai, H. Takahashi, and H. Saito. Gene (in press).

104. Kiss, A., B. Sain, E. Csordas-Toth, and P. Venetianer. 1977. Gene 1:323.

105. Kleid, D., Z. Humayun, A. Jeffrey, and M. Ptashne. 1976. Proc. Natl. Acad. Sci. 73:293.

106. Kojima, H., H. Takahashi, and H. Saito (unpubl.).

107. Koncz, C., A. Kiss, and P. Venetianer. 1978. Eur. J. Biochem. 89:523.

108. Kroger, M., H. Mayer, H. Schutte, and G. Hobom (unpubl.).

109. Kunkel, L.M., M. Silberklang, and B.J. McCarthy. 1979. J. Mol. Biol. 132:133.

110. Lacks, S. and B. Greenberg. 1975. J. Biol. Chem. 250:4060.

111. Lacks, S. and B. Greenberg. 1977. J. Mol. Biol. 114:153.

112. Landy, A., E. Ruedisueli, L. Robinson, C. Foeller, and W.W. Ross. 1974. Biochemistry 13:2134.

113. Langdale, J.A., P.A. Myers, and R.J. Roberts (unpubl.).

114. Lau, R.H. and W.F. Doolittle. 1980. FEBS Lett. 121:200.

115. Lautenberger, J.A., M.H. Edgell, and C.A. Hutchison, III. 1980. Gene 12:171.

116. Lautenberger, J.A., N.C. Kan, D. Lackey, S, Linn, M.H. Edgell, and C.A. Hutchison, III. 1978. Proc. Natl. Acad. Sci. 75:2271.

117. Lautenberger, J.A. and S. Linn. 1972. J. Biol. Chem. 247:6176.

118. Lautenberger, J.A., C.T. White, N.L. Haigwood, M.H. Edgell, and C.A. Hutchinson, III. 1980. Gene 9:213.

119. LeBon, J.M., C. Kado, L.J. Rosenthal, and J. Chirikjian. 1978. Proc. Natl Acad. Sci. 75:4097.

120. LeBon, J., T. LeBon, R. Blakesley, and J. Chirikjian (unpubl.).

121. Leung, D.W., A.C.P. Lui, H. Merilees, B.C. McBride, and M. Smith. 1979. Nucleic Acids Res. 6:17.

122. Levi, C. and T. Bickle (unpubl.).

123. Lin, B-C., M-C. Chien, and S-Y. Lou. 1980. Nucleic Acids Res. 8:6189.

124. Lui, A., B.C. McBride, and M. Smith (unpubl.).

125. Lui, A.C.P., B.C. McBride, G.F. Vovis, and M. Smith. 1979. Nucleic Acids Res. 6:1.

126. Lupker, H.S.C. and B.M.M. Dekker. 1981. Biochim. Biophys. Acta 654:297.

127. Lynn, S.P., L.K. Cohen, J.F. Gardner, and S. Kaplan. 1979. J. Bacteriol. 138:505.

128. Lynn, S.P., L.K. Cohen, S. Kaplan, and J.F. Gardner. 1980. J. Bacteriol. 142:380.

129. Makula, R.A. (unpubl.).

130. Makula, R.A. and R.B. Meagher. 1980. Nucleic Acids Res. 8:3125.

131. Mann, M.B., R.N. Rao, and H.O. Smith. 1978. Gene 3:97.

132. Mann, M.B. and H.O. Smith. 1977. Nucleic Acids Res. 4:4211.

133. Mann, M.B. and H.O. Smith (unpubl.).

134. Mayer, H., R. Grosschedl, H. Schutte, and G. Hobom. 1981. Nucleic Acids Res. 9:4833.

135. Mayer, H. and J. Klaar (unpubl.).

136. Mayer, H. and H. Schutte (unpubl.).

137. McConnell, D.J., D.G. Searcy, and J.G. Sutcliffe. 1978. Nucleic Acids Res. 5:1729.

138. Meagher, R.B. (unpubl.).

139. Meselson, M. and R. Yuan. 1968. Nature 217:1110.

140. Middleton, J.H., M.H. Edgell, and C.A. Hutchison, III. 1972. J. Virol. 10:42.

141. Middleton, J.H., P.V. Stankus, M.H. Edgell, and C.A. Hutchison, III (unpubl.).

142. Muller, F., S. Stoffel, and S.G. Clarkson (unpubl.).

143. Morris, D.W. and H.H. Parish. 1976. Arch. Microbiol. 108:227.

144. Mulligan, B.J. and M. Szekeres (unpubl.).

145. Murray, K., J.S. Brown, and S.A. Bruce (unpubl.).

146. Murray, K., S.G. Hughes, J.S. Brown, and S.A. Bruce. 1976. Biochem. J. 159:317.

147. Murray, K. and A. Morrison (unpubl.).

148. Murray, K., A. Morrison, H.W. Cooke, and R.J. Roberts (unpubl.).

149. Myers, P.A. and R.J. Roberts (unpubl.).

150. Nardone, G. and R. Blakesley. 1981. Fed. Proc. 40:1848.

151. Norlander, L., J.K. Davies, P. Hagblom, and S. Normark. 1981. J. Bacteriol. 145:788.

152. Old, R., K. Murray, and G. Roizes. 1975. J. Mol. Biol. 92:331.

153. Olson, J. A., P.A. Myers, and R.J. Roberts (unpubl.).

154. Piekarowicz, A. (unpubl.).

155. Piekarowicz, A., T.A. Bickle, J.C.W. Shepherd, and K. Ineichen (in press).

156. Piekarowicz, A., A. Stasiak, and J. Stanczak. 1980. Acta Microbiol. Pol. 29:151.

157. Pirrotta, V. 1976. Nucleic Acids Res. 3:1747.

158. Pope, A., S.P. Lynn, and J.F. Gardner (unpubl.).

159. Pugatsch, T. and H. Weber (unpubl.).

160. Ravetch, J. V., K. Horiuchi, and N.D. Zinder. 1978. Proc. Natl. Acad. Sci. 75:2266.

161. Reiser, J. and R. Yuan. 1977. J. Biol. Chem. 252:451.

162. Roberts, R.J. 1980. Nucleic Acids Res. 8:r63.

163. Roberts, R.J. 1980. Gene 8:329.

164. Roberts, R.J., J.B. Breitmeyer, N.F. Tabachnik, and P.A. Myers. 1975. J. Mol. Biol. 91:121.

165. Roberts, R.J., P.A. Myers, A. Morrison, and K. Murray. 1976. J. Mol. Biol. 102:157.

166. Roberts, R.J., P.A. Myers, A. Morrison, and K. Murray. 1976. J. Mol. Biol. 103:199.

167. Roberts, R.J., G.A. Wilson, and F.E. Young. 1977. Nature 265:82.

168. Roizes, G., P-C. Nardeux, and R. Monier. 1979. FEBS Lett. 104:39.

169. Roizes, G., M. Pages, C. Lecou, M. Patillon, and A. Kovoor. 1979. Gene 6:43.

170. Roizes, G., M. Patillon, and A. Kovoor. 1977. FEBS Lett. 82:69.

171. Rosenvold, E. C. and W. Szybalski. 1979. (unpubl.) cited in Gene 7:217.

172. Roy, P.H. and H.O. Smith. 1973. J. Mol. Biol. 81:427.

173. Roy, P.H. and H.O. Smith. 1973. J. Mol. Biol. 81:445.

174. Sato, S., C.A. Hutchison, and J.I. Harris. 1977. Proc. Natl. Acad. Sci. 74:542.

175. Sato, S., K. Nakazawa, and T. Shinomiya. 1980. J. Biochem. 88:737.

176. Sato, S. and T. Shinomiya. 1978. J. Biochem. 84:1319.

177. Schildkraut, I. (unpubl.).

178. Schildkraut, I. and D. Comb (unpubl.).

179. Sciaky, D. and R.J. Roberts (unpubl.).

180. Seurinck, J. and M. Van Montagu (unpubl.).

181. Sharp, P.A., B. Sugden, and J. Sambrook. 1973. Biochemistry 12:3055.

182. Shen, S., Q. Li, P. Yan, B. Zhou, S. Ye, Y. Lu, and D. Wang. 1980. Sci. Sin. 23:1435.

183. Shibata, T. and T. Ando. 1975. Mol. Gen. Genet. 138:269.

184. Shibata, T. and T. Ando. 1976. Biochim. Biophys. Acta 442:184.

185. Shibata, T., S. Ikawa, C. Kim, and T. Ando. 1976. J. Bacteriol. 128:473.

186. Shimatake, H. and M. Rosenberg (unpubl.).

187. Shimotsu, H., H. Takahashi, and H. Saito. 1980. Gene 11:219.

188. Shimotsu, H., H. Takahashi, and H. Saito. 1980. Agric. Biol. Chem. 44:1665.

189. Shinomiya, T. (unpubl.).

190. Shinomiya, T., M. Kobayashi, and S. Sato. 1980. Nucleic Acids Res. 8:3275.

191. Shinomiya, T. and S. Sato. 1980. Nucleic Acids Res. 8:43.

192. Smith, D.I., F.R. Blattner, and J. Davies. 1976. Nucleic Acids Res. 3:343.

193. Smith, H.O. and D. Nathans. 1973. J. Mol. Biol. 81:419.

194. Smith, H.O. and K.W. Wilcox. 1970. J. Mol. Biol. 51:379.

195. Stephens, M.A. (unpubl.).

196. Smith, J. and D. Comb (unpubl.).

197. Smith, L., R. Blakesley, and J. Chirikjian (unpubl.).

198. Stobberingh, E.E., R. Schiphof, and J.S. Sussenbach. 1977. J. Bacteriol. 131:645.

199. Stotz, A. and P. Philippson (unpubl.).

200. Streips, U. and B. Golemboski (unpubl.).

201. Sugisaki, H. 1978. Gene 3:17.

202. Sugisaki, H. and S. Kanazawa. Gene (in press).

203. Sugisaki, H. and M. Takanami. 1973. Nature New Biol. 246:138.

204. Sussenbach, J.S., C.H. Monfoort, R. Schiphof, and E.E. Stobberingh. 1976. Nucleic Acids Res. 3:3193.

205. Sussenbach, J.S., P.H. Steenbergh, J.A. Rost, W.J. van Leeuwen, and J.D.A. van Embden. 1978. Nucleic Acids Res. 5:1153.

206. Sutcliffe, J.G. and G.M. Church. 1978. Nucleic Acids Res. 5:2313.

207. Syddall, R.J., K. McGrath, J. Byron, and C. Stachow (unpubl.).

208. Takahashi, H., M. Shimizu, H. Saito, Y. Ikeda, and H.A. Sugisaki. 1979. Gene 5:9.

209. Takahashi, H. (unpubl.).

210. Takanami, M. 1974. Methods in Mol. Biol. 7:113.

211. Timko, J., A.H. Horwitz, J. Zelinka, and G. Wilcox. 1981. J. Bacteriol. 145:873.

212. Tomassini, J., R. Roychoudhury, R. Wu, and R.J. Roberts. 1978. Nucleic Acids Res. 5:4055.

213. Torheim, B. (unpubl.).

214. Trautner, T. A. (unpubl.).

215. Tu, C-P. D., R. Roychoudhury, and R. Wu. 1976. Biochem. Biophys. Res. Commun. 72:355.

216. Van den Hondel, J.J., G.A. Van Arkel, M.G.C. Duyvesteyn, and A. deWaard (unpubl.).

217. Van Heuverswyn, H. and W. Fiers. 1980. Gene 9:195.

218. Van Montagu, M. (unpubl.).

219. Van Montagu, M., D. Sciaky, P.A. Myers, and R.J. Roberts (unpubl.).

220. Van Ormondt, H., J.A. Lautenberger, S. Linn, and A. deWaard. 1973. FEBS Lett. 33:177.

221. Vanyushin, B.F. and A.P. Dobritsa. 1975. Biochim. Biophys. Acta 407:61.

222. Venegas, A., R. Vicuna, A. Alonso, F. Valdes, and A. Yudelevich. 1980. FEBS Lett. 109:156.

223. Venetianer, P. (unpubl.).

224. Visentin, L.P., R.J. Watson, S. Martin, and M. Zuker, M. (unpubl.).

225. Walter, F., M. Hartmann, and M. Roth. 1978. Abstracts of 12th FEBS Symposium, Dresden.

226. Wang, R.Y.-H., J.G. Shedlarski, M.B. Farber, D. Kuebbing, and M. Ehrlich. 1980. Biochim. Biophys. Acta 606:371.

227. Wani, A.A., R.E. Stephens, S.M. D'Ambrosio, and R.W. Hart. ASM Abstracts 1981. Abstract # 1778.

228. Watson, R., M. Zuker, S.M. Martin, and L.P. Visentin. 1980. FEBS Lett. 118:47.

229. Wilson, G.A. and F.E. Young. 1975. J. Mol. Biol. 97:123.

230. Wilson, G.A. and F.E. Young. 1976. In Microbiology (ed. D. Schlessinger), p. 350. American Society Microbiology, Washington, D.C.

231. Wilson, G.A. and F.E. Young (unpubl.).

232. Winkler, K. 1979. Diploma Dissertation.

233. Winkler, K. and A. Rosch (unpubl.).

234. Wu, R., C.T. King, and E. Jay. 1978. Gene 4:329.

235. Yoshimori, R.N. 1971. Ph.D. Thesis.

236. Zabeau, M. and R.J. Roberts (unpubl.).

237. Zabeau, M., R. Greene, P.A. Myers, and R.J. Roberts (unpubl.).

238. Zain, B.S. and R.J. Roberts. 1977. J. Mol. Biol. 115:249.

APPENDIX B
Tabulation of Some Well-characterized Enzymes with Deoxyribonuclease Activity

Stuart Linn
Department of Biochemistry
University of California
Berkeley, California 94720

In 1976 D. Kowalski and M. Laskowski compiled a comprehensive list of nucleases (57). To reproduce a complete survey of this magnitude would be beyond the scope of this book in view of the extensive productivity in the study of nucleases during the intervening years. Instead, the following list of deoxyribonuclease activities is an update, subject to several constraints.

The restriction endonucleases (Chapters 5 and 6 and Appendix A) and the DNases of *Escherichia coli* (Chapter 12) are not included, as they are tabulated elsewhere. Moreover, similar enzymes from analogous cell types or viruses are only exemplified. In the extreme case, the ATP-dependent DNases of the exonuclease V (or *rec*BC) type are omitted, as they are discussed in Chapter 2 and are exemplified by the *E. coli rec*BC enzyme. Similar omissions apply to several other repair and replication nucleases that are exemplified by *E. coli* enzymes and are discussed in the relevant chapters of this book.

Only enzymes that have been characterized with regard to the phosphomonoester group produced are tabulated. Similarly, only exonucleases of known polarity are included, whereas viral nucleases are included only if it is known that they are virally coded.

In some instances, entries in this list duplicate entries in Appendix C, because they have both DNase and RNase activity. Repair endonucleases have been limited to those enzymes whose substrate has been chemically defined as to the exact alteration recognized and whose precise cleavage location relative to that damage has been determined. Such criteria limit the tabulation to the apurinic/apyrimidinic endonucleases and point out a needed avenue of future research (see Chapter 4.)

A conclusion easily drawn from the tabulation is the need for a more systematic nomenclature for the mammalian enzymes. There are two "DNase V" enzymes, for example, and comparison between organisms and tissues is difficult, though it has been attempted by Sierakowska and Shugar (115).

Another interesting observation from the tabulation is that, barring ignorance of the literature, to date no sugar-specific exodeoxyribonuclease has been reported that forms 3′-phosphomonoesters.

Lastly, should any reader feel that her/his important study or favorite enzyme was omitted, please allow me to plead limitation of time, space, and knowledge as defense.

Enzyme	Characteristics	References
I. EXODEOXYRIBONUCLEASES PRODUCING 5'-PHOSPHOMONOESTERS		
Lambda exonuclease	Acts in recombination. Hydrolysis in the 5'-3' direction to yield mononucleotides. Preference for double-stranded (DS) DNA or oligonucleotides. Does not attack single-strand (SS) breaks. Often complexed to ß-protein.	77,117,54
Pneumococcus exonuclease	DS specific; acts in a 3'-5' direction to produce mononucleotides. Has DNA 3'-phosphomonoesterase activity. Acts from nicks.	64
Phage SP3 exonuclease	Hydrolysis in the 5'-3' direction to yield dinucleotides. Prefers SS DNA.	122
Phage T2 and T4 induced exonuclease	Degrades oligonucleotides to mononucleotides. Requires a free 5' terminus.	91,112,43, 113,55
Phage T4-induced exonuclease(s)	Hydrolyzes SS or DS DNA in the 5'-3' direction to yield oligonucleotides. Excises pyrimidine dimers from UV-irradiated DNA that has been incised by T4 UV endonuclease.	110
Phage T5-induced exonuclease	Attacks SS or DS DNA. Hydrolysis from the 5' terminus produces oligonucleotides of an average length of 3. Does not attack circular DNA. Attacks UV-irradiated DNA. Can initiate hydrolysis at SS breaks.	20,21
Phage T7-induced exonuclease	Hydrolyzes DS DNA producing 50% acid-soluble material, composed almost entirely of mononucleotides. Attack from the 5' terminus. The rate of hydrolysis is not affected by the presence of 5'-P. From a 5'-hydroxyl terminus, a dinucleoside monophosphate is formed. Marked preference for DS DNA. Attacks at nicks as well as external termini. Gene 6 product.	50,51,52
Phage T7 gene 5 exonuclease	Stepwise hydrolysis of SS DNA in 3'-5' direction to yield mononucleotides. When bound to host thioredoxin, takes on DNA polymerase activity. Exonuclease acts in 3'-5' direction. Inhibited by dNTPs.	42,1
M. luteus ''UV exonuclease''	SS specific. Hydrolysis in 5'-3' or 3'-5' directions. Excises pyrimidine	46,29

Enzyme	Characteristics	References
	dimers as a trinucleotide; otherwise forms mononucleotides. Mutants lacking the enzyme are recombination deficient.	
Mammalian DNase III	Hydrolyzes SS and DS DNA in a 3'-5' direction to yield mono- and dinucleotides.	72
Mammalian DNase IV	Hydrolyzes DNA in a 5'-3' direction to yield mononucleotides. Removes pyrimidine dimers. Can act at nicks.	73,70,71
Mammalian DNase V	Hyrolyzes DS DNA in both 5'-3' and 3'-5' directions to yield mononucleotides. Binds ß-polymerase. Acts from nicks.	82
Mammalian DNase VII	Hydrolyzes SS and nicked DS DNA in a 3'-5' direction to yield mononucleotides.	40
Mammalian DNase VIII	Hydrolyzes 5'-SS tails and nicks of duplex DNA in a 5'-3' direction to form oligonucleotides.	28,93
Mammalian ''correxonuclease''	Hydrolyzes nicked DS DNA and SS DNA in the 3'-5' and 5'-3' directions to form oligonucleotides.	17
Rabbit bone marrow DNA polymerase δ exonuclease	Hydrolyzes ''activated DNA'' or poly(dA-dT) in a 3'-5' direction to yield mononucleotides.	9,31
Herpes simplex virus DNase	Hydrolyzes DS DNA in the 3'-5' and 5'-3' directions to yield mononucleotides. Has endonuclease activity at nicks and gaps.	39
II. ENDODEOXYRIBONUCLEASES PRODUCING 5'-PHOSPHOMONOESTERS		
Phage T4-induced endonuclease II	Attacks DS DNA by predominantly making SS breaks. Fragments produced are about 1000 nucleotides long. Tenfold greater activity on DS DNA than on SS DNA. T4 DNA (glucosylated or nonglucosylated) is resistant. Involved in host DNA breakdown.	108
Phage T4-induced endonuclease III	Preference for SS DNA. Produces oligonucleotides. 5'-Terminal nucleotides are not specific. Does attack denatured T4 DNA. No acid-soluble material produced from poly(dC).	107

Enzyme	Characteristics	References
Phage T4-induced endonuclease IV	Products average 50 nucleotides long and contain exclusively 5'-pC; cleaves fd and SS B DNA but has no effect on SS T4 DNA (glucosylated or nonglucosylated). About 200-fold greater activity against SS DNA than against DS DNA. Involved in host DNA breakdown.	170,109
Phage T7-induced endonuclease I	Hydrolyzes SS DNA 150-fold faster than DS DNA; acid-soluble material is produced only from SS substrates. 5'-end groups are predominantly pyrimidine residues. Extensive hydrolysis of SS T7, DS T7, and DS E. coli DNAs yield products of average single-strand mw of 6×10^3, 1×10^5, and 6×10^3, respectively. Inhibited by high concentration of RNA. Missing in gene 3 mutants that fail to break down host DNA.	11,12,106
Phage T7-induced endonuclease II	Exhaustive digestion of native T7 DNA produces 13 SS cleavages. Has no activity toward SS DNA. 5'-termini produced contain all four bases.	10
Diplococcus pneumoniae major endonuclease	Produces oligonucleotides. Acts on SS and DS DNA. Membrane localized. Involved in entry of transforming DNA, possibly by ''translocating'' the DNA while degrading one strand.	64,66,65
Streptococcal DNase	Products are predominantly tri- to hexanucleotides. Dinucleotides are resistant. Attacks native DNA much faster than denatured.	28
Bacillus subtilis endonuclease	50-Fold more activity on SS than DS DNA. Forms oligonucleotides. Periplasmic; absent in spores. Reduced activity in a class of rec^- mutants.	13,14
Chlamydomonas non-random endonuclease	Makes nonrandom SS scissions in duplex DNA to give discrete products. Gaps are formed. May be in chloroplasts. T and G at 5' termini.	8
Yeast DNase	750-Fold preference for SS over DS DNA. Predominant products are oligonucleotides of an average length of 4 residues.	96
Mammalian DNase I (pancreatic DNase)	Preference for DS DNA. In the early stages, DS DNA is nicked preferentially at dA and dT. Products have an average length of 4. With poly(dA-dT) only even-numbered products d(pTpA) are formed.	68

Enzyme	Characteristics	References
	Divalent cation affects the preferentially hydrolyzed linkage. Actin specifically inhibits and forms a 1:1 complex with the enzyme.	
Bovine small intestinal mucosa endonuclease	Hydrolyzes SS DNA 20 times faster than DS DNA. Nicks duplex DNA to a limit of several kb. No base specificity at 5' terminus. Bound to chromatin. Ca^{2+} inhibited. May be similar to the KB cell endonuclease described in reference 22. The latter reference also summarizes all known KB-cell DNases.	86
Calf thymus ''DNase V'' (SS-specific KB cell DNase)	Prefers SS DNA. Nicks DS DNA. Products of SS DNA are oligonucleotides. Probably same KB cell enzyme as reported in reference 123.	128
Calf thymus Ca^{2+}-Mg^{2+}-dependent endonuclease	Requires Ca^{2+} and Mg^{2+}. Chromatin associated. Affected by ADP ribosylation. Similar enzyme in rat liver nuclei. Prefers DS DNA. Oligonucleotide products.	85,38
Nonrandom mammalian endonuclease	Cleaves DS DNA by DS breaks rapidly to discrete products of 1.5×10^5 - 1.5×10^6 daltons, then more slowly to products near 120 bp. Hydrolyzes SS DNA without apparent specificity.	78
Vaccinia virus acid DNase	Mixed endo- and exonuclease. Highly specific for SS DNA. Produces mono- and oligonucleotides. Associated with viral core. Properties resemble that of S1 nuclease except that it is DNA specific. Optimal pH of 4.5. Two subunits, each of approximately 50,000.	103,102
Vaccinia virus alkaline DNase	Highly specific for SS DNA. Endonuclease forming oligonucleotides. Associated with the viral core. Interferes with host DNA replication by degrading nascent SS DNA.	97,98
Retrovirus-associated endonuclease	Associated with proteolytic cleavage products of RNA-directed DNA polymerases. Puts nicks into duplex DNA to form large oligonucleotides. ATP activated.	89
III. ENDODEOXYRIBONUCLEASES PRODUCING 3'-PHOSPHOMONOESTERS		
Mammalian DNase II (spleen endonuclease)	Usually obtained from spleen or thymus, but similar activities in other tissues. Optimal pH near 5. Attacks DS DNA faster	5

Enzyme	Characteristics	References
	than SS. Cuts two strands simultaneously. In early stages of reaction, preference for dGpdC. Localized in the lysosomes. Trinucleotides bearing 5'-monophosphate are resistant.	
Crab testes DNase	Attacks DS DNA by making predominantly DS breaks. Preference for dNpdT. Produces di- and trinucleotides.	27,105
Agate snail DNase	Hydrolyzes poly(dA) and poly(dT) or DS DNA at runs of dA or of dT. Products are small oligonucleotides.	131
Aspergillus DNase K1	Preference for Pu-Pu and SS DNA. Predominant product are mononucleotides, some oligonucleotides.	48,111
Salmon testes DNase	DS DNA is a better substrate than SS DNA. Products formed are oligonucleotides. G predominates at the 3' terminus.	114
IV. SUGAR-NONSPECIFIC EXONUCLEASES PRODUCING 5'-PHOSPHOMONOESTERS		
Venom phosphodiesterase	Present in many snake venoms. Attacks ribose, deoxyribose, and arabinose derivatives from the 3' terminus. Attacks poly(adenosine diphosphate ribose) endonucleolytically. Chains terminated in 5'-phosphate are best substrates. Removal of phosphate decreases susceptibility tenfold; introduction of 3'-phosphate decreases another 50-fold. Has NTPase activity. Optimal pH near 9. With chains terminated in 3'-phosphate, 5' and 3' termini can be quantitatively determined as N and pNp, respectively. Nicks superhelical DNA.	67
Hog kidney phosphodiesterase	Attacks from the 3' terminus producing mononucleotides. Monophosphatase activity is present.	101
Carrot exonuclease	Attacks from the 3' terminus producing mononucleotides. DS DNA is resistant. Requires divalent cation.	34,35
Avena leaf phosphodiesterase	Attacks from the 3' terminus producing mononucleotides. Resembles venom phosphodiesterase. DS DNA is resistant.	126
N. crassa SS-specific exonuclease	Attacks from both 5' and 3' ends to yield mainly mononucleotides and some	99,80

Enzyme	Characteristics	References
	small oligonucleotides. Has a trace of endonuclease activity. Requires Mg^{2+}. mw = 72 kilodaltons.	
V. SUGAR NONSPECIFIC EXONUCLEASES PRODUCING 3'-PHOSPHOMONOESTERS		
Spleen exonuclease (spleen phosphodiesterase)	Preference for SS- or partially degraded substrates. Attacks from the 5'-hydroxyl terminus producing mononucleotides. Chains bearing 5'-phosphate are resistant at pH 4.8, but slowly hydrolyzed at pH 6.2. Widely used for the identification of the 3'-hydroxyl terminus after treatment with monophosphatase.	6
B. subtilis extra cellular exonuclease	Hydrolyzes high-molecular-weight RNA. Attacks SS DNA from the 5'-terminus, whereas DS DNA is initially attacked from 3' terminus. Produces 3'-mononucleotides. Hydrolyzes denatured DNA faster than native. Good 5'-end group reagent. Requires Ca^{2+}. Slight endonuclease contamination.	53,45,44
Salmon testes exonuclease	Attacks from the 5' terminus, producing mononucleotides.	79
L. acidophilus phosphodiesterase	Produces mononucleotides from the 5'-hydroxyl terminus.	18
Ustilago maydis nuclease ß	Exo-endonuclease. Hydrolyzes from the 5' terminus to form mono- and small oligonucleotides. Distributive. Contains 5'-nucleotidase activity.	104
VI. SUGAR-NONSPECIFIC ENDONUCLEASES PRODUCING 5'-PHOSPHOMONOESTERS		
Sl nuclease (Aspergillus)	Endo- and exonuclease. Produces predominantly mononucleotides. Attacks superhelical DNA to produce linear molecules. Cleaves nicked duplex DNA opposite nicks. Inflicts a limited number of cleavages in native viral DNA. Optimal pH 4.5. Requires Zn^{2+}.	2,120,127, 30,129
N. crassa endonuclease	Very high specificity for SS substrates. Produces oligo- and mononucleotides. Of the mononucleotides, pG predominates. Attacks superhelical DNA to produce nicked circles. Optimal pH 7.5-8.5. Active in absence of divalent cation.	75,74,49,4

Enzyme	Characteristics	References
Mung bean nuclease	Endonucleolytically attacks RNA and DNA. Also has exonucleolytic activity on oligonucleotides. Preference for SS DNA over DS DNA. Produces mono- and oligonucleotides. Inflicts a limited number of cleavages in native viral DNA. Splits B DNA into two halves. Optimal pH 4.8-5.5. A similar enzyme exists in wheat seedlings. Both have inseparable 3'-nucleotidase activity.	58,61,56,33, 60,59
BAL 31 nuclease (Alteromonas)	Single-strand-specific endonuclease that also cleaves duplex supercoiled circles and duplex DNA at sites of damage. Linear duplex DNAs are shortened from both termini so as to form shortened duplexes that are susceptible to blunt ligation. pH optimum near 8.5.	32,69
Lamb brain nuclease	Specific for SS substrates. Produces hexanucleotides and larger.	37
Ustilago maydis nuclease α	Very similar to N. crassa endonuclease in properties. Cleaves damaged duplex DNA and supercoiled DNA. Progressively shortens duplexes from both termini (unlike N. crassa enzyme). Ultimate products are mono- and small oligonucleotides. Low levels in certain recombination-deficient mutants.	41
N. crassa mitochondrial nuclease	Produces oligo- and mononucleotides. Of the mononucleotides, pG and pC predominate. Requires divalent cation. Optimal pH 6.5-7.5.	76
N. crassa endoexonuclease	With DNA, a SS endonuclease and a DS exonuclease. An endonuclease with RNA. Requires divalent cation, inhibited by ATP. Inactive 88K precursor forms active 50-60K enzyme.	19,63
Silkworm nuclease	ApN preferentially cleaved; GpN discriminated against. Produces oligo- and a small number of mononucleotides. Requires Mg^{2+}. Optimal pH 10.	83
Potato nuclease I	5' Terminus 75% G; 3' terminus 55% A, 40% T. Produces oligonucleotides. Requires Mg^{2+}. Optimal pH 6-7. Inseparable from a 3'-nucleotidase activity.	7,90
Sheep kidney nuclease	Specific for SS substrates. Produces	47

Enzyme	Characteristics	References
	mononucleotides. Native DNA and poly(A)-poly(U) are resistant.	
Azotobacter nuclease	Prefers ApA linkages; discriminates against GpG. Produces oligonucleotides of predominantly 2-4 nucleotides.	118,109
Penicillium citrinum P1 nuclease	Mixed, endo- and exonuclease. Initially produces oligonucleotides that are further degraded to mononucleotides. Requires Zn^{2+}. Optimal temperature 70°C. Optimal pH 4.5-6.0. The same enzyme hydrolyzes 2'- and 3'-mononucleotides to nucleosides and P_i. Ribo 3'-mononucleotides are hydrolyzed 30-fold faster than deoxyribo.	24,25,26
Avena leaf nuclease	Preference for SS DNA. Produces mono- and oligonucleotides. Approximately 33K. Zn^{2+} selectively inhibits RNase activity. EDTA inhibits DNase and RNase activities.	130
Tobacco culture nuclease (extracellular)	Produces mono- and oligonucleotides (predominantly of 5 residues or less). Highly preferential for SS DNA.	92
Serratia marcescens nuclease (extracellular)	No preference for bases or for DS versus SS DNA. Produces mainly di- to tetranucleotides. DNA and RNA are equally susceptible.	87,88
VII. SUGAR-NONSPECIFIC ENDONUCLEASES PRODUCING 3'-PHOSPHOMONOESTERS		
Micrococcal (staphylococcal) nuclease (Staphylococcus aureus)	Original attack endonucleolytic; soon thereafter, exonucleolytic action starts. Initial products have Ap and Tp termini. Attacks SS faster than DS DNA. Shows autoacceleration. Proximal 5'-phosphomonoester inhibits, whereas proximal 3'-phosphomonoester enhances rate. Absolute requirement for Ca^{2+}; as fragments become shorter, higher Ca^{2+} concentration allows regulation of the size of products. Strongly inhibited by pAp and pTp, which limits usefulness of this enzyme for identification of termini of oligonucleotides of four and shorter. Optimal pH = 9. Probably best-studied nuclease chemically and physically.	3,15,124,125
Chlamydomonas endonuclease	Denatured DNA is hydrolyzed 200-300-fold faster than native. Optimum pH = 8.8; activated by Ca^{2+}, inhibited by NaCl, but	116

Enzyme	Characteristics	References
	not by KCl.	
VIII. DAMAGE-SPECIFIC ENDODEOXYRIBONUCLEASES		
M. luteus ''UV-endonuclease''	Cleaves baseless sites in duplex DNA to form 3'-baseless sugar and 5'-nucleoside phosphate termini. Has associated pyrimidine dimer glycosylase.	36
T4 ''UV-endonuclease'' (endonuclease V)	T4 v gene product. Cleaves baseless sites in duplex DNA to form 3'-baseless sugar and 5'-nucleoside phosphate termini. Has associated pyrimidine dimer glycosylase.	16,100,84
M. luteus AP endonuclease	Two forms that cleave AP DNA to generate 3'-hydroxynucleotide and 5'-sugar phosphate termini. Active without divalent cations present, but inhibited by EDTA. Other similar bacterial enzymes are tabulated in reference 23.	94,95
Human fibroblast AP endonuclease	Generates 3'-baseless sugar and 5'-nucleoside phosphate termini. Absent in xeroderma pigmentosum group-D cultured fibroblasts.	62,81
Human fibroblast AP endonuclease II	Generates 3'-hydroxynucleotide and 5'-sugar phosphate termini. Other similar eukaryotic enzymes are tabulated in reference 23 and studies of localization of the rat liver activities are noted in reference 121.	62,81

REFERENCES

1. Adler, S. and P. Modrich. 1979. J. Biol. Chem. 64:11605.
2. Ando, T. 1966. Biochim. Biophys. Acta 114:158.
3. Anfinsen, C., P. Cuatrecasas, and H. Taniuchi. 1971. In The enzymes (ed. P.D. Boyer), 3rd edition, vol. 4, p. 177. Academic Press, New York.
4. Bartok, K. and D. Denhardt. 1976. J. Biol. Chem. 251:530.
5. Bernardi, G. 1971. In The Enzymes (ed. P.D. Boyer), 3rd edition, vol. 4, p. 271. Academic Press, New York.
6. Bernardi, A. and G. Bernardi. 1971. In The Enzymes (ed. P.D. Boyer), 3rd edition, vol. 4, p. 329. Academic Press, New York.
7. Bjork, W. 1965. Biochim. Biophys. Acta 95:652.
8. Burton, W.G., R.J. Roberts, P.A. Myers, and R. Sager. 1977. Proc. Natl. Acad. Sci. 74:2687.
9. Byrnes, J.J., K.M. Downey, U.L. Black, and A.G. So. 1976. Biochemistry 15:2817.
10. Center, M.S. 1972. J. Biol. Chem. 247:146.
11. Center, M.S. 1972. J. Biol. Chem. 245:6285.
12. Center, M.S. and C.C. Richardson. 1970. J. Biol. Chem. 245:6292.
13. Cobianchi, F., G. Ciarrocchi, C. Attolini, and A. Falaschi. 1978. Eur. J. Biochem. 84:533.
14. Cobianchi, F., C. Attolini, A. Falaschi, and G. Ciarrocchi. 1980. J. Bacteriol. 141:968.
15. Cotton, F.A. and E.E. Hazen, Jr. 1971. In The enzymes (ed. P.D. Boyer), 3rd edition, vol. 4, p. 153. Academic Press, New York.
16. Demple, B. and S. Linn. 1980. Nature 287:203.
17. Doniger, J. and L. Grossman. 1976. J. Biol. Chem. 251:4579.
18. Fiers, W. and H.G. Khorana. 1963. J. Biol. Chem. 238:2798.
19. Fraser, M.J., R. Tjeerde, and K. Matsumoto. 1976. Can J. Biochem. 54:971.
20. Frenkel, G.D. and C.C. Richardson. 1971. J. Biol. Chem. 246:4839.
21. Frenkel, G.D. and C.C. Richardson. 1971. J. Biol. Chem. 246:4848.
22. Frenkel, G.D., K. Randles, and N. Burns. 1981. Nucleic Acids

Res. 9:6635.

23. Friedberg, E.C., T. Bonura, E.H. Radany, and J.D. Love. 1981. In The enzymes (ed. P.D. Boyer), vol. 14, p. 251. Academic Press, New York.
24. Fujimoto, M., A. Kuninaka, and H. Yoshino. 1974. Agric. Biol. Chem. 38:777.
25. Fujimoto, M., A. Kuninaka, and H. Yoshino. 1974. Agric. Biol. Chem. 38:785.
26. Fujimoto, M., A. Kuninaka, and H. Yoshino. 1974. Agric. Biol. Chem. 38:1555.
27. Georgatsos, J.G. 1965. Biochim. Biophys. Acta 95:544.
28. Georgatsos, J.G., W.C. Unterholzner, and M. Laskowski, Sr. 1962. J. Biol. Chem. 237:2626.
29. Ghanges, G.S. and R. Wu. 1974. J. Biol. Chem. 249:7550.
30. Ghanges, G.S. and R. Wu. 1975. J. Biol. Chem. 250:4601.
31. Goscin, L.P. and J.J. Byrnes. 1982. Biochemistry 21:2513.
32. Gray, H.B., Jr., D.A. Ostrander, J.L. Hodnett, R.J. Legerski, and D.L. Robberson. 1975. Nucleic Acids Res. 2:1459.
33. Hanson, D.M. and J.L. Fairley. 1969. J. Biol. Chem. 244:2440.
34. Harvey, C., L. Malsman, and A.L. Nussbaum. 1967. Biochemistry 6:3689.
35. Harvey, C.L., K.C. Olson, and R. Wright. 1970. Biochemistry 9:921.
36. Haseltine, W.A., L.K. Gordon, C.P. Lindan, R.H. Graftstrom, N.L. Shaper, and L. Grossman. 1980. Nature 285:634.
37. Healy, J.W., D. Stollar, and L. Levine. 1966. In Procedures in nucleic acid research (eds. Cantoni and Davies), p. 188. Harper and Row, New York.
38. Hewish, D.R. and L.A. Burgoyne. 1973. Biochem. Biophys. Res. Commun. 52:475.
39. Hoffman, P.J. 1981. J. Virol. 38:1005.
40. Hollis, G.F. and L. Grossman. 1981. J. Biol. Chem. 256:8074.
41. Holloman, W.K., T.C. Rowe, and J.R. Rusche. 1981. J. Biol. Chem. 256:5087.
42. Hori, K., D.F. Mark, and C.C. Richardson. 1979. J. Biol. Chem. 254:11591.

43. Jorgensen, S.E. and J.F. Koerner. 1966. J. Biol. Chem. 241:3090.
44. Kanamori, N., N.R. Cozzarelli, and R. Okazaki. 1974. Biochim. Biophys. Acta 335:173.
45. Kanamori, N., K. Sakabe, and R. Okazaki. 1974. Biochim. Biophys. Acta 335:155.
46. Kaplan, J.C., S.R. Kushner, and L. Grossman. 1971. Biochem. 10:3315.
47. Kasai, K. and M. Grunberg-Manago. 1967. Eur. J. Biochem. 1:152.
48. Kato, M. and Y. Ikeda. 1968. J. Biochem. 64:321.
49. Kato, A.C., K. Bartok, M.J. Fraser, and D. Denhardt. 1973. Biochim. Biophys. Acta 308:68.
50. Kerr, C. and P.D. Sadowski. 1972. J. Biol. Chem. 247:305.
51. Kerr, C. and P.D. Sadowski. 1972. J. Biol. Chem. 247:311.
52. Kerr, C. and P.D. Sadowski. 1975. Virology 65:1281.
53. Kerr, I.M., J.R. Chien, and I.R. Lehman. 1967. J. Biol. Chem. 242:2700.
54. Kmiec, E. and W.K. Holloman. 1981. J. Biol. Chem. 256:12636.
55. Koerner, J.F. 1970. Annu. Rev. Biochem. 39:291.
56. Kole, R., H. Sierakowska, H. Szemplinska, and D. Sugar. 1974. Nucleic Acids Res. 1:699.
57. Kowalski, D. and M. Laskowski, Sr. 1976. In Handbook of biochemistry and molecular biology, 3rd Edition (ed. G.D. Fasman), Nucleic acids volume II, p. 491. CRC Press, Cleveland.
58. Kowalski, D., W.D. Kroeker, and M. Laskowski, Sr. 1976. Biochem. 15:4457.
59. Kroeker, W.D. and J.L. Fairley. 1975. J. Biol. Chem. 250:3773.
60. Kroeker, W.D., D.M. Hanson, and J.L. Fairley. 1975. J. Biol. Chem. 250:3767.
61. Kroeker, W.D., P. Kowalski, and M. Laskowski, Sr. 1976. Biochemistry 15:4463.
62. Kuhnlein, U., B. Lee, E.E. Penhoet, and S. Linn. 1978. Nucleic Acids Res. 5:951.
63. Kwong, S. and M.J. Fraser. 1978. Can. J. Biochem. 56:370.
64. Lacks, S. and B. Greenberg. 1967. J. Biol. Chem. 242:3108.
65. Lacks, S. and M. Neuberger. 1975. J. Bacteriol. 124:1321.
66. Lacks, S., B. Greenberg, and M. Neuberger. 1975. J. Bacteriol.

123:222.

67. Laskowski, M., Sr. 1971. In The enzymes (ed. P.D. Boyer), 3rd edition, vol. 4, p. 313. Academic Press, New York.
68. Laskowski, M., Sr. 1971. In The enzymes (ed. P.D. Boyer), 3rd edition, vol. 4, p. 289. Academic Press, New York.
69. Legerski, R.J., J.L. Hodnett, and H.B. Gray, Jr. 1978. Nucleic Acids Res. 5:1445.
70. Lindahl, T. 1971. Eur. J. Biochem. 18:407.
71. Lindahl, T. 1971. Eur. J. Biochem. 18:415.
72. Lindahl, T., J.A. Gally, and G.M. Edelman. 1969. J. Biol. Chem. 244:5014.
73. Lindahl, T., J.A. Gally, and G.M. Edelman. 1969. Proc. Natl. Acad. Sci. 62:597.
74. Linn, S. 1967. Methods Enzymol. 12:247.
75. Linn, S. and I.R. Lehman. 1965. J. Biol. Chem. 240:1287.
76. Linn, S. and I.R. Lehman. 1966. J. Biol. Chem. 241:2694.
77. Little, J.W. 1967. J. Biol. Chem. 242:679.
78. McKenna, W.G., J.J. Maio, and F.L. Brown. 1981. J. Biol. Chem. 256:6435.
79. Menon, K.M.J. and M. Smith. 1970. Biochemistry 9:1584.
80. Mills, C. and M.J. Fraser. 1973. Can. J. Biochem. 51:888.
81. Mosbaugh, D.W. and S. Linn. 1980. J. Biol. Chem. 255:11743.
82. Mosbaugh, D.W. and R.R. Meyer. 1980. J. Biol. Chem. 255:10239.
83. Mukai, J.-I. 1965. Biochem. Biophys. Res. Commun. 21:562.
84. Nakabeppu, Y. and M. Sekiguchi. 1981. Proc. Natl. Acad. Sci. 78:2742.
85. Nakamura, M., Y. Sakaki, N. Watanabe, and Y. Takagi. 1981. J. Biochem. 89:143.
86. Nakayama, J., T. Fujiyoshi, M. Nakamura, and M. Anai. 1981. J. Biol. Chem. 256:1636.
87. Nestle, M. and W.K. Roberts. 1969. J. Biol. Chem. 244:5213.
88. Nestle, M. and W.K. Roberts. 1969. J. Biol. Chem. 244:5219.
89. Nissen-Meyer, J., A.J. Raae, and I.F. Nes. 1981. J. Biol. Chem. 256:7985.
90. Nomura, A., M. Suno, and Y. Mizuno. 1971. J. Biochem. 70:993.
91. Oleson, A.E. and J.F. Koerner. 1964. J. Biol. Chem. 239:2935.

92. Oleson, A.E., A.M. Janski, and E.T. Clark. 1974. _Biochim. Biophys. Acta_ _366_:89.
93. Pedrini, A.M. and L. Grossman (unpubl.).
94. Pierre, J. and J. Laval. 1980. _Biochemistry_ _19_:5018.
95. Pierre, J. and J. Laval. 1980. _Biochemistry_ _19_:5024.
96. Pinon, R. 1970. _Biochemistry_ _9_:2839.
97. Pogo, B.G.T. and S. Dales. 1969. _Proc. Natl. Acad. Sci._ _63_:820.
98. Pogo, B.G.T. and S. Dales. 1973. _Proc. Natl. Acad. Sci._ _70_:1726.
99. Rabin, E.Z., H. Tenenhouse, and M.J. Fraser. 1972. _Biochim. Biophys. Acta_ _259_:50.
100. Radany, E.H. and E.C. Friedberg. 1980. _Nature_ _286_:182.
101. Razzell, W.E. 1961. _J. Biol. Chem._ _236_:3031.
102. Rosemond-Hornbeak, H. and B. Moss. 1974. _J. Biol. Chem._ _249_:3292.
103. Rosemond-Hornbeak, H., E. Paoletti, and B. Moss. 1974. _J. Biol. Chem._ _249_:3287.
104. Rusche, J.R., T.C. Rowe, and W.K. Holloman. 1980. _J. Biol. Chem._ _255_:9117.
105. Sabeur, G., P.J. Sicard, and G. Aubel-Sadron. 1974. _Biochemistry_ _13_:3203.
106. Sadowski, P.D. 1971. _J. Biol. Chem._ _246_:209.
107. Sadowski, P.D. and I. Bakyta. 1972. _J. Biol. Chem._ _247_:405.
108. Sadowski, P.D. and J. Hurwitz. 1969. _J. Biol. Chem._ _244_:6182.
109. Sadowski, P.D. and J. Hurwitz. 1969. _J. Biol. Chem._ _244_:6192.
110. Shimizu, K. and M. Sekiguchi. 1976. _J. Biol. Chem._ _251_:2613.
111. Shishido, K., M. Kato, and Y. Ideda. 1968. _J. Biochem._ _65_:479.
112. Short, E.C., Jr. and J.F. Koerner. 1965. _Proc. Natl. Acad. Sci._ _54_:595.
113. Short, E.C., Jr. and J.F. Koerner. 1969. _J. Biol. Chem._ _244_:1487.
114. Sieliwanowicz, B., M. Yamamoto, L. Stasiuk, and M. Laskowski, Sr. 1975. _Biochemistry_ _14_:39.
115. Sierakowska, H. and D. Shugar. 1977. _Prog. Nucleic Acid Res. Mol. Biol._ _20_:59.
116. Small, G.D. and R.B. Sparks, Jr. 1972. _Arch. Biochem. Biophys._ _153_:171.

117. Sriprakash, K.S., N. Lundh, M.M.-O. Huh, and C.M. Radding. 1975. J. Biol. Chem. 250:5438.

118. Stevens, A. and R.J. Hilmoe. 1960. J. Biol. Chem. 235:3016.

119. Stevens, A. and R.J. Hilmoe. 1960. J. Biol. Chem. 235:3023.

120. Sutton, W.D. 1971. Biochim. Biophys. Acta 240:522.

121. Thibodeau, L. and W.G. Verly. 1980. Eur. J. Biochem. 107:555.

122. Trilling, D.M. and H.V. Aposhian. 1968. Proc. Natl. Acad. Sci. 60:214.

123. Tsuro, T., M. Arens, R. Padmanabhan, and M. Green. 1978. J. Biol. Chem. 253:3400.

124. Tucker, P.W., E.E. Hazen, and F.A. Cotton. 1979. Mol. Cell Biochem. 23:67.

125. Tucker, P.W., E.E. Hazen, and F.A. Cotton. 1979. Mol. Cell Biochem. 23:131.

126. Udvardy, J., E. Marre, and G.L. Farkas. 1970. Biochim. Biophys. Acta 206:392.

127. Vogt, V.M. 1973. Eur. J. Biochem. 33:192.

128. Wang, E.-C. and J.J. Furth. 1977. J. Biol. Chem. 252:116.

129. Weigand, R.C., N.G. Godson, and C.M. Radding. 1975. J. Biol. Chem. 250:8848.

130. Wyen, N.V., S. Erdei, and G.L. Farkas. 1971. Biochim. Biophys. Acta 232:472.

131. Yanagawa, H., Y. Ogawa, and F. Egami. 1981. J. Biochem. 90:1479.

APPENDIX C
Ribonucleases

Howard M. Frankfort
Hugh D. Robertson
Rockefeller University
New York, New York 10021

Name	Reaction	Reports of similar activities/other name	Reference
I. EXORIBONUCLEASES			
Ribonuclease II	Exonucleolytic cleavage in 3'-5' direction to yield 5'-phosphomononucleotides. Preference for single-stranded (SS) RNA.	Lactobacillus planterum RNase, mouse nuclear RNase, oligoribonuclease of Escherichia coli.	20, 29, 41
Exoribonuclease H	Exonucleolytic cleavage to 5'-phosphomonoester oligonucleotides in both 5'-3' and 3'-5' direction. Attacks RNA in duplex with DNA strand. Found in certain oncornaviruses and animal cells.		50
Yeast ribonuclease	Exonucleolytic cleavage to 3' phosphomononucleotides.	RNase U4	30
Venom exonuclease	Exonucleolytic in the 3'-5' direction to yield 5' phosphomononucleotides. Preference for single-stranded substrate. Also active on DNA. The venom enzyme is also active with superhelical turns.	Hog kidney phosphodiesterase; Lactobacillus exonuclease.	24
Spleen exonuclease	Exonucleolytic cleavage in the 5'-3' direction to yield 3'-phosphomononucleotides. Also active on DNA. Preference for single-stranded substrate.	Lactobacillus acidophilus nuclease, Bacillus subtilis nuclease, salmon testis nuclease.	6
Ribonuclease D	3' Trimming of E. coli tRNA precursors.		12
II. ENDORIBONUCLEASES			
Ribonuclease (Physarum polycephalum)	Endonucleolytic cleavage to 5'-phosphomonoester.	Pig liver nuclease, HeLa cell RNase; E. coli RNase, bovine adrenal cortex RNase.	18

Name	Reaction	Reports of similar activities/other name	Reference
Ribonuclease alpha	Endonucleolytic cleavage to 5'-phosphomonoester.		28
Bacillus cereus ribonuclease	Endonucleolytic cleavage at U and C residues specifically producing 3'-phosphomono- and oligonucleotides.		25
Phy M	Endonucleolytic cleavage. Specific for A + U in 7 M urea at 50°C. Prefers A, U, G at 37°C.		9
Ribonuclease L	Endonucleolytic cleavage. Specific for U residues producing 3'-phosphomono-and oligonucleotides. Activated by 2'-5' A_n n>2.	Ribonuclease F	11, 39
Endoribonuclease H (calf thymus)	Endonucleolytic cleavage to 5'-phosphomonoester. Acts on RNA of RNA:DNA hybrids.	Similar enzymes from: E. coli, chicken embryo, human KB cells, rat liver, Ustilago maydis, human leukemic cells, Saccharomyces cerevisiae (H2), and Tetrahymena pyriformis.	43, 15
Ribonuclease Tl	Two-stage endonucleolytic cleavage to 3'-phosphomono- and oligonucleotides ending in Gp with 2', 3'-cyclic phosphate intermediates.	Guanyloribonuclease, Aspergillus oryzae RNase, RNase Nl and N2; similar enzymes: Neurospora crassa RNase Nl and N2, Ustilago sphaerogena RNase, B. subtilis RNase, microbial RNase I.	19, 47
Ribonuclease (B. subtilis)	Endonucleolytic cleavage to 2', 3'-cyclic nucleotides.	Similar enzymes: Azotobacter agilis RNase, Proteus mirabilis RNase.	27, 53, 54
Ribonuclease T2	Two-stage endonucleolytic cleavage to 3'-phosphomononucleotides and oligonucleotides with 2', 3'-cyclic phosphate intermediates.	Ribonuclease II. Similar enzymes: plant RNase, E. coli RNase I, RNase N_2, microbial RNase II.	17, 36, 49

Name	Reaction	Reports of similar activities/other name	Reference
Ribonuclease U2	Two-stage endonucleolytic cleavage to 3'-phosphomono- and oligonucleotides ending in Ap or Gp with 2', 3'-cyclic phosphate intermediates.	Similar enzymes: RNase U3, Pleospora RNase, Trichoderma koning, RNase III.	13, 14, 48
Pancreatic ribonuclease	Endonucleolytic cleavage to 3'-phosphomono- and oligonucleotides ending with Cp or Up with 2', 3'-cyclic phosphate intermediates.	RNase, RNase I. Similar enzymes: venom RNase, Thiobacillus thioparus RNase, Xenopus laevis RNase, Rhizopus oligosporus RNase.	3, 5
Endonuclease Sl (Aspergillus)	Endonucleolytic cleavage to 5'-phosphomono- and oligonucleotide end products. Also active on DNA. Preference for single-stranded substrate.	Single-stranded nucleate endonuclease. Deoxyribonuclease Sl. Similar enzymes: N. crassa nuclease 1423, mung bean nuclease, Penicillium citrinum nuclease Pl.	1, 46, 51
Endonuclease (Serratia marcescens)	Endonucleolytic cleavage to 3'-phosphomono- and oligonucleotide end products. Hydrolysis of double- or single-stranded substrate.	Similar enzymes: silkworm nuclease, potato nuclease, Azotobacter nuclease.	26, 44, 45 52
Micrococcal endonucleases	Endonucleolytic cleavage to 3'-phosphomono- and oligonucleotide end products. Also active on DNA. Hydrolyses of double- or single-stranded substrate.	Similar enzymes: Chlamydomonas nuclease, spleen phosphodiesterase, spleen endonuclease.	2
Ribonuclease III	Endonucleolytic cleavage to 5'-phosphomonoester. Specific for DS RNA. In vivo processing of E. coli tRNA, rRNA, and mRNA precursors.	Similar enzymes from: calf thymus, chicken, HeLa cells, human KB cells.	7, 8, 16, 31, 35, 37
Ribonuclease P	Endonucleolytic cleavage to 5'-phosphomonoester. Processing in vivo of E. coli tRNA precursors.	Similar enzymes from: human KB, HeLa cells, yeast.	10, 21, 22

Name	Reaction	Reports of similar activities/other name	Reference
RNase M5	Endonucleolytic cleavage to 5'-phosphomono- and oligonucleotides. Processing in vivo of B. subtilis 5S RNA.	E. coli	33, 42
RNase E	Endonucleolytic cleavage of 30S E. coli rRNA precursors. Processing in vivo of E. coli 30S rRNA.		4
Yeast tRNA splicing endo-nuclease	Endonucleolytic cleavage to 3'-phosphomono-oligonucleotides. Processing of in vivo yeast tRNA precursors.	Xenopus laevis RNase XLaI	32, 34
Ribonuclease P2	Cleavage of multimeric intermediates in E. coli tRNA precursors.		38
Ribonuclease O	Cleavage of multimeric intermediates in E. coli tRNA precursors.		40
Ribonuclease F	Cleavage of multimeric intermediates in E. coli tRNA processing.		4

REFERENCES

1. Ando, T. 1966. *Biochem. Biophys. Acta* *114*:158.

2. Anfinsen, C., P. Cuatrecasas, and H. Taniuchi. 1971. In *The enzymes*, (ed. P.D. Boyer), 3rd edition, vol. 4, p. 177. Academic Press, New York.

3. Anfinsen, C.B. and F.H. White, Jr. 1961. In *The enzymes* (ed. P.D. Boyer, H. Lardy, and K. Myrback), 2nd edition, vol. 3, p. 95. Academic Press, New York.

4. Apiron, D., B.K. Ghora, G. Plautz, T.K. Misra, and P. Gegenheimer. 1980. In *Transfer RNA: Biological aspects* (ed. D. Soll, J.N. Abelson, and P.R. Schimmel), vol. 9, pt. B, p. 139. Cold Spring Harbor Laboratory, Cold Spring Harbor, New York.

5. Beard, J.R. and W.K. Razzel. 1964. *J. Biol. Chem.* *239*:4186.

6. Bernardi, A. and G. Bernardi. 1971. In *The enzymes* (ed. P.D. Boyer), 3rd edition, vol. 4, p. 329. Academic Press, New York.

7. Busen, W. and P. Hausen. 1975. *Eur. J. Biochem.* *52*:179.

8. Crouch, R.J. 1974. *J. Biol. Chem.* *249*:1314.

9. Donis-Keller, H. 1980. *Nucleic Acids Res.* *8*:3133.

10. Ferrari, S., C.O. Yehle, H.D. Robertson, and E. Dickson. 1980. *Proc. Natl. Acad. Sci.* *77*:2395.

11. Floyd-Smith, G., E. Slattery, and P. Lengyel. 1981. *Science* *212*:1030.

12. Ghosh, R.K. and M.P. Deutscher. 1980. In *Transfer RNA: Biological aspects*, (ed. D. Soll, J.N. Abelson, and P.R. Schimmel), p. 59. Cold Spring Harbor Laboratory, Cold Spring Harbor, New York.

13. Glitz, D.G. and C.A. Decker. 1964. *Biochemistry* *3*:1391.

14. Glitz, D.G. and C.A. Decker. 1964. *Biochemistry* *3*:1399.

15. Haberkern, R.C. and G.L. Cantoni. 1973. *Biochemistry* *12*:2389.

16. Hall, S.H. and R.J. Crouch. 1977. *J. Biol. Chem.* *252*:4092.

17. Heppel, L.A. 1966. In *Procedures in nucleic acid research* (ed. G.L. Cantoni, and D.R. Davies), p. 31. Harper & Row, New York.

18. Hiramara, M., T. Uchida, and F. Egami. 1969. *J. Biochem.* (Tokyo). *65*:701.

19. Kasai, K., T. Uchida, F. Egami, K. Yoshida, and M. Nomoto. 1969. *J. Biochem.* (Tokyo). *60*:389.

20. Keir, H.M., R.H. Mathog, and C.E. Carter. 1964. *Biochemistry* *3*:1188.

21. Kline, L., S. Nishikawa, and D. Soll, D. 1981. *J. Biol. Chem.* *256*:5058.

22. Kile, R. and S. Altman. 1981. *Biochemistry* *20*:1902.

23. Koski, R.A., A.L.M. Bothwell, and S. Altman. 1976. *Cell* *9*:101.

24. Laskowski Sr., M. 1966. In *Procedures in nucleic acid research* (ed. G.L. Cantoni, and D.R. Davis), p. 85. Harper & Row, New York.

25. Lockard, R.E., B. Alzner-Deweerd, J.E. Heckman, J. MacGee, M.W. Tabor, and U.L. RajBhandary. 1978. *Nucleic Acids Res.* *5*:37.

26. Mikulski, A.J. and M. Laskowski, Sr. 1970. *J. Biol. Chem.* *245*:5026.

27. Nishimura, H. and B. Maruo. 1960. *Biochem. Biophys. Acta* *40*:355.

28. Norton, J. and J.S. Roth. 1967. *J. Biol. Chem.* *242*:2029.

29. Nossal, N.G. and M. Singer. 1968. *J. Biol. Chem.* *243*:913.

30. Ohtaka, Y., T. Uchida, and T. Sakai. 1963. *J. Biochem.* (Tokyo) *54*:322.

31. Ohtsuki, K., Y. Groner, and J. Hurwitz. 1977. *J. Biol. Chem.* *252*:483.

32. Otsuka, A., A. DePaolis, and G.P. Tocchini-Valentini. 1981. *Mol. Cell. Biol.* *1*:269.

33. Pace, N.R., B. Meyhack, B. Pace, and M.L. Sogin. 1980. In *Transfer RNA: Biological aspects*, (ed. D. Soll, J.N. Abelson, and P.R. Schimmel), p. 155. Cold Spring Harbor Laboratory, Cold Spring Harbor, New York.

34. Peebles, C. and J. Abelson. 1981. (pers. comm.).

35. Rech, J., G. Cathala, and P. Jeanteur. 1976. *Nucleic Acids Res.* *3*:2055.

36. Reddi, K.K. and L.J. Mauser. 1965. *Proc. Natl. Acad. Sci.* *53*:607.

37. Robertson, H.D., R.E. Webster, and N.D. Zinder. 1968. *J. Biol. Chem.* *243*:82.

38. Schedl, P., P. Primakoff, and J. Roberts. 1974. In *Processing of RNA* (ed. J.J. Dunn), Brookhaven Symposia in Biology, No. 26, p. 53. Brookhaven National Laboratory, Upton, New York.

39. Schmidt, A., A. Zilberstein, L. Shulman, P. Federman, H. Berissi, and M. Revel. 1978. _FEBS Lett_. _95_:257.

40. Shimura, Y. and H. Sakano. 1977. In _Nucleic acid-protein recognition_ (ed. H.J. Vogel), p. 293. Academic Press, New York.

41. Spahr, P.F. 1964. _J. Biol. Chem._ _239_:3716.

42. Stahl, D. and N.R. Pace. 1981. (pers. comm.).

43. Stavrianopoulos, J.G. and E. Chargaff. 1973. _Proc. Natl. Acad. Sci._ _70_:1959.

44. Stevens, A. and R.J. Hilmoe. 1960. _J. Biol. Chem._ _235_:3016.

45. Stevens, A. and R.J. Hilmoe. 1960. _J. Biol. Chem._ _235_:3023.

46. Sutton, W.D. 1971. _Biochem. Biophys. Acta_ _240_:522.

47. Takahashi, K. 1961. _J. Biochem._ (Tokyo) _49_:1.

48. Uchida, T. and F. Egami. 1961. In _The enzymes_ (ed. P.D. Boyer), 3rd edition, vol. 4, p. 205. Academic Press, New York.

49. Uchida, T. and F. Egami. 1967. _J. Biochem._ (Tokyo) _61_:44.

50. Verma, I.M. 1975. _J. Virol._ _15_:843.

51. Vogt, V.M. 1973. _Eur. J. Biochem._ _33_:192.

52. Wechter, W.J., A.J. Mikulski, and M. Laskowski, Sr. 1968. _Biochem. Biophys. Res. Commun._ _30_:318.

53. Yamasaki, M. and K. Arima. 1967. _Biochem. Biophys. Acta_ _139_:202.

54. Yamasaki, M. and K. Arima. 1969. _Biochem. Biophys. Res. Commun._ _37_:430.

Index